Biodiversity in Environmental Assessment: Enhancing Ecosystem Services for Human Well-Being

Human-induced development activities are introduced with insufficient attention to their consequences for our living environment, even in cases where environmental assessments have been carried out. This apparent lack of attention to biodiversity in environmental assessment is rooted in the difficulties we have in adequately addressing biodiversity within the scope, time frame, and budget allocated for assessments. This book provides a conceptual background and practical approaches to overcome these difficulties. It integrates the objectives of the Convention on Biological Diversity, its ecosystem approach, and the conceptual framework of the Millennium Ecosystem Assessment into a comprehensive approach to biodiversity in environmental assessment. It highlights the need to consider the value of biodiversity based on its use by each stakeholder, addresses the importance of both social and economic development to reach the Millennium Development Goals, and provides insights into ways to balance present and future needs.

ROEL SLOOTWEG is an ecologist with a PhD in Environmental Science. His work focuses on natural resources management, with particular interest in the linkages between nature conservation, natural resources management, and social and economic development. His experience in environmental assessment includes assessment studies with strong emphasis on stakeholder participation, scoping and review activities, capacity development, and the development of new approaches for better assessment. Dr Slootweg has worked on four continents. He presently owns a consultancy company: SevS natural and human environment consultants, Oegstgeest, the Netherlands.

ASHA RAJVANSHI is Professor and Head of the Environmental Impact Assessment Cell at the Wildlife Institute of India, Dehradun, India. She holds a doctoral degree in Environmental Science. As a teacher, trainer, and EA practitioner she has made significant contributions in professionalizing EIA through development of EIA instruments and best practice guidance tools and conducting training programmes in different continents, specially targeted at mainstreaming biodiversity in the EA framework. She also provides advisory support for environmental decision making through her membership at apex bodies of the Government of India. She is currently the Chair of the biodiversity section of the International Association for Impact Assessment (IAIA).

VINOD B. MATHUR is Professor and Dean at the Wildlife Institute of India, Dehradun, India. He holds professional qualifications in forestry and a doctoral degree in wildlife ecology from Oxford University. He has vast professional experience in conducting environmental assessments of developmental projects and in reviewing EIA reports and Environmental Management Plans. His area of special interest is wildlife ecology and management. He is a member of the International Association for Impact Assessment (IAIA), the IUCN World Commission on Protected Areas (WCPA), and the IUCN Commission on Ecosystem Management (CEM). As a trainer, he has organized several IAIA capacity building programmes on mainstreaming biodiversity in impact assessment.

AREND KOLHOFF is senior technical secretary at the international department of the Netherlands Commission for Environmental Assessment. He has a degree in human geography from the Utrecht University and is currently working on PhD research that aims to identify guiding principles for the development of better performing EIA systems in developing countries. He has fifteen years working experience with the Commission in about twenty countries as a trainer and advisor on EIA and SEA capacity development activities.

ECOLOGY, BIODIVERSITY AND CONSERVATION

The world's biological diversity faces unprecedented threats. The urgent challenge facing the concerned biologist is to understand ecological processes well enough to maintain their functioning in the face of the pressures resulting from human population growth. Those concerned with the conservation of biodiversity and with restoration also need to be acquainted with the political, social, historical, economic, and legal frameworks within which ecological and conservation practice must be developed. The new *Ecology, Biodiversity, and Conservation* series will present balanced, comprehensive, up-to-date, and critical reviews of selected topics within the sciences of ecology and conservation biology, both botanical and zoological, and both 'pure' and 'applied'. It is aimed at advanced final-year undergraduates, graduate students, researchers, and university teachers, as well as ecologists and conservationists in industry, government, and the voluntary sectors. The series encompasses a wide range of approaches and scales (spatial, temporal, and taxonomic), including quantitative, theoretical, population, community, ecosystem, landscape, historical, experimental, behavioural, and evolutionary studies. The emphasis is on science related to the real world of plants and animals rather than on purely theoretical abstractions and mathematical models. Books in this series will, wherever possible, consider issues from a broad perspective. Some books will challenge existing paradigms and present new ecological concepts, empirical or theoretical models, and testable hypotheses. Other books will explore new approaches and present syntheses on topics of ecological importance.

The Ecology of Phytoplankton
C. S. Reynolds

Invertebrate Conservation and Agricultural Ecosystems
T. R. New

Risks and Decisions for Conservation and Environmental Management
Mark Burgman

Nonequilibrium Ecology
Klaus Rohde

Ecology of Populations
Esa Ranta, Veijo Kaitala and Per Lundberg

Ecology and Control of Introduced Plants
Judith H. Myers, and Dawn Bazely

Systematic Conservation Planning
Chris Margules, and Sahotra Sarkar

Large Scale Landscape Experiments
David Lindenmayer

Assessing the Conservation Value of Fresh Waters
Philip J. Boon, and Catherine M. Pringle

Bird Conservation and Agriculture
Jeremy Wilson, Andrew Evans, and Phillip Grice

Insect Species Conservation
Tim New

Cave Biology
Aldemaro Romero

Biodiversity in Environmental Assessment

Enhancing Ecosystem Services for Human Well-Being

ROEL SLOOTWEG

SevS natural and human environment consultants, Oegstgeest, The Netherlands

ASHA RAJVANSHI

Wildlife Institute of India, Dehradun, India

VINOD B. MATHUR

Wildlife Institute of India, Dehradun, India

AREND KOLHOFF

Netherlands Commission for Environmental Assessment, Utrecht, The Netherlands

CAMBRIDGE
UNIVERSITY PRESS

CAMBRIDGE
UNIVERSITY PRESS

University Printing House, Cambridge CB2 8BS, United Kingdom

One Liberty Plaza, 20th Floor, New York, NY 10006, USA

477 Williamstown Road, Port Melbourne, VIC 3207, Australia

4843/24, 2nd Floor, Ansari Road, Daryaganj, Delhi - 110002, India

79 Anson Road, #06-04/06, Singapore 079906

Cambridge University Press is part of the University of Cambridge.

It furthers the University's mission by disseminating knowledge in the pursuit of education, learning and research at the highest international levels of excellence.

www.cambridge.org
Information on this title: www.cambridge.org/9780521716550

© Cambridge University Press 2010

First published 2010

A catalogue record for this publication is available from the British Library

Library of Congress Cataloging in Publication data
Biodiversity in environmental assessment : enhancing ecosystem services for human well-being / Roel Slootweg . . . [et al.].
    p.   cm. – (Ecology, biodiversity and conservation)
Includes bibliographical references and index.
ISBN 978-0-521-88841-7 (hardback)
1. Biodiversity conservation.   2. Environmental monitoring.   3. Environmental indicators.   4. Human ecology.   I. Slootweg, Roel, 1958–   II. Title.   III. Series.
QH75.B53245   2009
363.7′063 – dc22   2009036657

ISBN 978-0-521-88841-7 Hardback
ISBN 978-0-521-71655-0 Paperback

Contents

List of contributors   *page* ix

Foreword by the Executive Secretary of the Convention on
     Biological Diversity – Ahmed Djoghlaf   xi

Preface   xiii

List of abbreviations   xvi

Part I. Setting the stage

1   **Introduction**   3
    *Roel Slootweg, Asha Rajvanshi, Vinod B. Mathur, and
Arend Kolhoff*

2   **Interpretation of biodiversity**   14
    Roel Slootweg

3   **Biodiversity conservation and development:
challenges for impact assessment**   59
    Asha Rajvanshi and Vinod B. Mathur

Part II. Assessment tools

4   **The impact assessment framework**   87
    Roel Slootweg and Peter P. Mollinga

5   **Environmental assessment**   125
    Arend Kolhoff, Bobbi Schijf, Rob Verheem, and Roel Slootweg

6   **Biodiversity in Environmental Impact
Assessment**   154
    Asha Rajvanshi, Vinod B. Mathur, and Roel Slootweg

7   **Biodiversity-inclusive Strategic Environmental
Assessment**   205
    Roel Slootweg

Part III. Emerging issues

8 Reconciling conservation and development:
   the role of biodiversity offsets   255
   Asha Rajvanshi and Vinod B. Mathur

9 Valuation of ecosystem services: lessons from
   influential cases   287
   Pieter van Beukering and Roel Slootweg

Epilogue – Topics in need of further elaboration   328
   *Roel Slootweg, Asha Rajvanshi, Vinod B. Mathur, and
   Arend Kolhoff*
Annex: valuation of ecosystem services: influential cases   334
   Pieter van Beukering, Roel Slootweg, and Desirée Immerzeel
References   398
Index   434

Contributors

Pieter van Beukering is an environmental economist at the Institute for Environmental Studies (IVM) of the VU University, Amsterdam. His main research interests are economic valuation, natural resource management, solid waste management, renewable energy, and poverty and environment. Most of his work takes place in developing countries. Van Beukering is also a lecturer in environmental economics at the Vrije Universiteit, predominantly teaching for the Master's course 'Environmental Resource Management (ERM)'. He runs a consultancy firm, Van Beukering consulting, founded in 1999, which specialises in the economics of coral reefs and tropical forests.

Desirée Immerzeel is a consultant at AidEnvironment, Amsterdam. She has a Master's Degree in Natural Resources Management from Utrecht University. She conducted research on the use of nontimber forest products (NTFPs) in Guatemala and on biodiversity values at the Caribbean island of Montserrat. Her practical experience lies in the field of international nature conservation and environmental economics, with expertise in the use of decision support tools (cost–benefit analysis, GIS, and multicriteria analysis). Her ambitions lie in the combination of sustainable development and nature conservation.

Peter P. Mollinga is Senior Researcher at the Center for Development Research (ZEF) at Bonn University, Germany. He is also Convenor of SaciWATERs (South Asia Consortium for Interdisciplinary Water Resources Studies), based in Hyderabad, India, and one of the three founding editors of *Water Alternatives*, an interdisciplinary journal on water, politics, and development. His research fields are water governance and water politics, and interdisiplinary and transdisciplinary approaches to natural resources management. He consults on education and capacity building for integrated water resources management in India and Indonesia.

Bobbi Schijf is a technical secretary at the Netherlands Commission for Environmental Assessment, Utrecht, the Netherlands, where she is involved in reviewing Dutch assessment reports, as well as in capacity development for EIA and SEA in the context of development cooperation. Before joining the Commission, her work focussed on EU accession support in eastern and central Europe. As a trainer and policy advisor, she took part in projects for the introduction or strengthening of SEA in Turkey, Lithuania, Georgia, and Romania. Mrs Schijf has a PhD from the University of Otago (New Zealand), where she researched the influence of impact assessment information on decision making.

Rob Verheem is deputy director of the Netherlands Commission for Environmental Assessment, Utrecht, the Netherlands. He is in charge of the international work of this Commission, mainly focusing on strengthening of EIA and SEA practice in developing countries and countries in transition. Rob has more than 20 years of experience in both EIA and SEA, with a focus on application of environmental assessment for strategic planning and decision making. Before, he worked at the Netherlands Ministry of Environment and the European Commission. In 2004 he received IAIA's individual award for his contribution to the development of SEA.

Foreword

Clean water and air, food, and shelter are the cornerstones of human well-being. Thanks to the intricate interactions between the millions of species that make up Earth's biodiversity and the many ecosystem services they provide, we can meet our basic, and not so basic, needs. Yet, the ecological footprint of humanity already exceeds Earth's capacity to regenerate and continue delivering these goods and services. Large-scale human interventions that seriously impact both biodiversity and ecosystem functions continue to increase. New policies, plans, and programmes are introduced with insufficient attention to the consequences for our living environment, even in cases where environmental assessments have been carried out. This apparent lack of attention to biodiversity in environmental assessment is not so much a deliberate decision to ignore natural processes but is rooted in the difficulties we face in adequately addressing biodiversity within the time frame and budget allocated for assessments.

This book provides a conceptual background and practical approaches to overcome these apparent difficulties. It fully integrates the objectives of the Convention on Biological Diversity, its ecosystem approach, and the conceptual framework of the Millennium Ecosystem Assessment into a comprehensive approach to biodiversity in environmental assessment. It highlights the need to consider the value of biodiversity based on its use by each stakeholder, addresses the importance of both social and economic development to reach the Millennium Development Goals, and provides insights into ways to balance present and future needs. The authors have drawn together helpful case studies that demonstrate how biodiversity can and should be integrated into environmental and strategic assessments. They provide a powerful argument that biodiversity can and must be considered to ensure that projects, programmes, and policies are culturally, environmentally, and socially acceptable. In this respect the book provides a valuable source of information for academics, and practitioners as well as decision makers.

Ahmed Djoghlaf
Executive Secretary, Convention on Biological Diversity

Preface

The incorporation of biodiversity-related issues in impact assessment has received considerable attention from the global conservation community comprising practitioners, academics, planners, and decision makers. The secretariat of the Convention on Biological Diversity (CBD) and the Ecology and Biodiversity section of the International Association for Impact Assessment (IAIA) have been active in developing guidelines for the better integration of biodiversity in both Environmental Impact Assessment (EIA) and Strategic Environmental Assessment (SEA). Other bodies like the Convention on Wetlands of International Importance (Ramsar, 1971), Convention on the conservation of European wildlife and nature habitats (Bern, 1979), and the Convention on Migratory Species (Bonn, 1979) have also been influential on the topic. Special biodiversity issues of impact assessment-related scientific journals (*Impact Assessment and Policy Appraisal* and *Journal of Environmental Assessment Policy and Management*) have contributed in highlighting the need for integrating biodiversity concerns in impact assessment. With these specific efforts, a desirable convergence of interests has now come about between international biodiversity related initiatives and the world of impact assessment. Despite this, literature search and the recent reviews of impact assessment studies suggest that biodiversity considerations on the whole are not as well served by impact assessment practice as they should be. There consequently is an urgent need to bring together the varied experience gathered in the areas of ecology/biodiversity conservation and impact assessment.

The only comprehensive book available on biodiversity in impact assessment so far is *Ecological Impact Assessment* by Jo Treweek (1999). As the focus of this book was limited to Environmental Impact Assessment and its internationally accepted procedural steps, an update is genuinely needed after rapid and recent developments in Strategic Environmental Assessment. Furthermore, the earlier book is written from a European nature conservation background and consequently lacks the perspective of human development needs, emphasised by the Biodiversity Convention, Millennium Ecosystem Assessment, and Millennium Development Goals. In a globalising and increasingly interrelated world, this book is the logical and necessary follow up to Treweek's important pioneering work. The group of four author/editors, who have been actively involved in promoting the science and practice of impact assessment through their varied roles as teachers, trainers, and practitioners, have taken a lead in this direction.

The targeted audience of this book can be found in the industrialised world (often referred to as western countries, even although some lie very far south or east) as well as in the developing world and includes scientists from biological, economic, and social sciences as well as practitioners involved in the design of (conceptual and regulatory) frameworks for environmental assessment. The book also provides a useful insight for business and corporate groups who are keen to acknowledge the relevance of encouraging development that also benefits biodiversity. Of course, the book provides relevant material to students who are the ones that have to internalise the new concepts presented in this book and (hopefully) act as the real agents of change throughout their professional careers.

The book is academic in nature, but firmly rooted in everyday practice. It is not a practical how-to-do guideline, but it provides a view on how to address questions related to biodiversity and human development needs in an environmental assessment context. Where possible, case material is used to illustrate and support arguments and logic. Text boxes appear throughout the book to provide supporting case evidence. The case material has been included in various sections to enrich the book with practical examples; this material is often hidden in inaccessible 'grey' literature, such as planning documents, project proposals and appraisals, environmental impact statements, or evaluation reports, and is seldom referred to in the formal scientific literature. Of course, the authors have extensive practical experience in conducting studies in all major sectors on four continents and draw extensively from their experience throughout the book.

Many of the ideas expressed in this book evolved from spirited discussions with friends and environmental professionals during the annual meetings of the International Association for Impact Assessment (IAIA), the world's largest network of impact assessment experts. These discussions repeatedly highlighted the need for a book on this subject that has been long overdue and actually lured four of us to make a book become a reality. Friends and colleagues with whom we had more frequent intellectual discussions will see their influence throughout this book.

We drew extensively from the invaluable and rich source of information of many professionals represented on the different sections of IAIA, who shared their scientific writings, experiences, and case studies with us. We are especially most grateful to all those who have contributed to the work of the biodiversity section of IAIA since its first meeting in 1997. Many will find their work referred to throughout the book.

Our sincere words of appreciation are for Professor Michael Usher, Series Editor for Cambridge University Press who not only invited us to write this book but has been an enthusiastic source of energy and encouragement. He helped in shaping the initial idea for the book and subsequently provided several rounds of comments on each chapter, on matters of style, organization, and coverage, as well as detailed comments on the subject matter of the book. Alan Crowden of Cambridge University Press instilled a sense of urgency for hastening our writing. His remark 'that most books that take long are the ones that never get finished at all' helped us regain our speed when the going was slow for one or another reason.

For two of us (Asha and Vinod), the Wildlife Institute of India (WII) provided the stimulating academic environment for writing of this book. Mr P. R. Sinha, Director, Wildlife Institute of India, reposed immense trust in our abilities to handle both the writing of the book and our professional responsibilities.

We acknowledge the Netherlands Commission for Environmental Assessment for supporting this endeavour. We thank Mr Ahmed Djoghlaf, Executive Secretary of the Convention on Biological Diversity (CBD), for considering this book worthy of his appreciative Foreword.

Our families were caring and understanding and provided the tranquility and peace of mind for writing the book especially when the authors were located in different parts of the world with different time zones. Our computers never showed sign of fatigue and the e-communication technology never let us down during several back and forth consultations that were inevitable in every stage of the book production.

Publishing with Cambridge University Press, the oldest and one of the largest academic publishers in the world, is an honour.

And finally, we thank you, the reader, for picking up this book and using it to expanding the horizon of impact assessment by popularising the innovative thoughts and perspectives that we have tried sharing through this book. I hope you find something in here useful!

Abbreviations

BAP	Biodiversity Action Plan
CBA	Cost–Benefit Analysis
CBD	Convention on Biological Diversity
CEAA	Canadian Environmental Assessment Agency
CEQ	Council on Environmental Quality
CIDA	Canadian International Development Agency
CoP	Conference of Parties (governing body of an international convention)
CSIR	Council for Scientific and Industrial Research (South Africa)
CVM	Contingent Valuation Method
DFID	Department of International Development (UK)
DSS	Decision Support System
E&P Forum	Oil Industry International Exploration and Production Forum
EBI	Energy and Biodiversity Initiative
EC	European Commission (implementing body of the European Union)
EC	European Council (main decision-making body of the European Union)
EcIA	Ecological Impact Assessment
EIA	Environmental Impact Assessment
EIS	Environmental Impact Statement
EMP	Environmental Management Plan
EPA	Environmental Protection Agency
ESIA	Environmental and Social Impact Assessment
EU	European Union
EVRI	Environmental Valuation Reference Inventory
GIS	Geographic Information System
HIA	Health Impact Assessment
IA	Impact Assessment

IAIA	International Association of Impact Assessment
ICMM	International Council on Mining and Minerals
IEEM	Institute of Ecology and Environmental Management
IFC	International Finance Corporation
IIED	International Institute for Environment and Development
IPCC	Intergovernmental Panel on Climate Change
IPIECA	International Petroleum Industry Environmental Conservation Association
IPTRID	International Programme for Technology and Research in Irrigation and Drainage (at the UN Food and Agricultural Organisation)
IUCN	International Union for the Conservation of Nature/ The World Conservation Union
NBSAP	National Biodiversity Strategy and Action Plan
MA	Millennium Ecosystem Assessment
MCA	Multi-Criteria Analysis
MDGs	Millennium Development Goals
MMSD	Mining, Minerals and Sustainable Development project
NEN	National Ecological Network
NEPA	USA National Environmental Policy Act
NGOs	Non-Governmental Organisations
NSW	New South Wales (Australia)
OECD	Organisation for Economic Cooperation and Development
OECD-DAC	OECD Development Assistance Committee
PES	Payment for Ecosystem Services
RSPB	Royal Society for the Protection of Birds (Birdlife UK)
SAIEA	Southern African Institute for Environmental Assessment
SAP	Species Action Plan
SEA	Strategic Environmental Assessment
SEAN	Strategic Environmental Analysis
SIA	Social Impact Assessment
SIDA	Swedish International Development Cooperation Agency
PPP	People, Planet, Profit (triple bottom line of sustainability, used in the corporate sector)

	Plans, Polices, and Programmes (decision-making levels where SEA applies)
TEV	Total Economic Value (of an ecosystem)
UN	United Nations
UNCED	United Nations Conference on Environment and Development
UNDP	United Nations Development Programme
UNEP	United Nations Environmental Programme
VROM	Netherlands Ministry of Housing, Spatial Planning and Environment
WCED	World Commission on Environment and Development
WII	Wildlife Institute of India
WTP	Willingness to Pay

Part I
Setting the stage

1 · Introduction

Roel Slootweg, Asha Rajvanshi, Vinod B. Mathur, and Arend Kolhoff

> Biodiversity matters to everyone. Its loss impoverishes the environment and reduces its capacity to support people now and in the future. Impact assessment can help to ensure that development is compatible with the conservation and sustainable use of biodiversity.

These are the opening words of the biodiversity in impact assessment principles formulated by the International Association for Impact Assessment (IAIA, 2005) and perfectly set the stage for a book that stresses the dependency of humanity on benefits from biodiversity, explores how present and future environmental securities are linked with biodiversity, and stimulates the need to balance the need for conservation with that for human development through sustainable use of biodiversity.

Through his simple quote in *Closing Circle* – 'Everything is connected to everything else', Barry Commoner (1971) conveyed the importance of interconnectedness between the different components of the living world. Human activities do not occur in a vacuum but are an inherent part of complex biological systems, such as food chains, and large-scale abiotic processes, such as the water cycle or climate change. This interconnectedness helps us to understand that most ecological systems are complex, making it difficult to come to a consensus on cause-and-effect relationships. If we are to develop truly sustainable economies and ensure the perpetuity of the ecosystem benefits that drive economies and human well-being, we must have a better grasp of the intricate relationship between the environment and the factors that bring about changes. Moreover, we must make sure that available knowledge is used in the best possible way to support day-to-day decision making on large human interventions. There is far too much at stake, financially, socially, and environmentally, if we ignore the connectedness between development and conservation objectives.

Increasing evidence that biodiversity and ecosystem services linked to biodiversity are in rapid decline has put biodiversity on the agenda of decision makers. Consequently biodiversity has emerged as a major priority for environmental assessors. This book positions impact assessment as a decision support tool and promotes it as an objective-oriented tool to enhance sustainability in decision making on proposed human activities. In this respect, it makes one more modest addition to the existing volumes of the books on the subject. Yet at the same time it is novel in approaching environmental assessment from the perspectives of both conservation community and developers, trying to balance between the 'harvesting' and 'harnessing' of biodiversity resources. To guide decision making, the valuation of biodiversity and its related ecosystem services (in ecological, social, or economic terms) is strongly recommended in the overall assessment of biodiversity-related impacts of developments.

The primary objective of environmental assessment is to aid decision making. To ensure its effectiveness, information gathered by any assessment has to be tailored to the needs of decision makers. Yet, the concepts and language used by the impact assessment community are often not suited to the needs of the decision-making process. Decision makers, more often than not, have no particular interest in biodiversity; they are facing multiple demands from sections in society, all trying to influence decision making. Biodiversity, as the environment in general, does not have a voice to express itself. Environmental assessment was conceived to give the environment a voice. Yet, biodiversity still is badly represented in environmental assessment and decision making. It is either considered to be a 'trifle' or a 'difficult' subject to deal with. Logically, the latter holds to be more correct as is also supported by the analysis of the state of knowledge about biodiversity by Metrick and Weitzman (1998) – 'As a society, we have not even come close to defining what the objective is. What is biodiversity? In what units is it to be measured? . . . We have to make up our minds here what it is we are optimising. This is the essential problem confounding the preservation of biodiversity today'. On the other hand, the Convention on Biological Diversity has provided us with a widely endorsed set of objectives with respect to biodiversity management, including approaches to attain these objectives. So in our view, the problem moreover is to translate these objectives and approaches for environmental assessment to work in practice.

This book is the first attempt to fully integrate the objectives of the Biodiversity Convention, its ecosystem approach (oriented towards the management of biodiversity), and the outputs of Millennium

Ecosystem Assessment (oriented towards the proper valuation of biodiversity) into one comprehensive approach for biodiversity-inclusive environmental assessment. We expand from ecology and traditional conservation language, venturing into the need for social and economic development and the act of balancing between present and future.

A fundamental point of departure of this book is that it does not stress the need for an altogether new assessment procedure, but it advocates a more pragmatic and rational treatment of biodiversity within the existing impact assessment framework for better development decisions. Environmental Impact Assessment (EIA) for the assessment of concrete projects and Strategic Environmental Assessment (SEA) for the assessment of impacts resulting from programmes, plans, and policies are widely recognised and used instruments. Such instruments provide important power; yet, with respect to biodiversity there is an urgent need to do a better job and to improve procedures. In this respect we try to keep a middle road between two extremes, eloquently described by Nooteboom (2007):

where no procedures at all may lead to the "boiled frog syndrome" (the frog doesn't jump out of the water pan as it is slowly boiled – the frog has no procedure that gives warning signals), an overdose of procedures leads to the "frozen deer syndrome' (the deer stays in the car's headlights – all options for action are rejected by an overdose of checks). Both animals are not sustainable.

The world is struggling to comprehend the implications of the broad concept of biodiversity as agreed by the international Convention on Biological Diversity (CBD). Different perspectives hamper a uniform interpretation of biodiversity, thus confusing decision makers. Added to this, the scientific language of the biodiversity community is unappealing to the outside world. As Pritchard (2005) states, 'despite the clear role for impact assessment being spelled out in several convention texts . . . for the greater part of the history of both the conventions and EIA, there has been a striking separateness between these two worlds in terms of their processes and the people involved'. There is an obvious need for an unambiguous interpretation of biodiversity, and there is a need for an approach which translates biodiversity into decision maker's language. This book is an attempt to do so. International conventions, such as the CBD, and agreements, such as the Millennium Development Goals, provide the framework and give direction. Recent developments in ecology and environmental assessment provide the necessary scientific background and the tools to influence decision making for the better. Last but not least, the book aims to provide direction to the scientific agenda; gaps in knowledge are identified and suggestions for further scientific work are provided.

Organisation of the book

The book can be divided into three main parts. Part I provides the appropriate background for appreciating the meaning and importance of biodiversity. This part sets the stage with Chapter 2, introducing in detail the concepts underlying biodiversity and Chapter 3, positioning biodiversity in the context of international agreements on poverty reduction and sustainable development. In Part II, which contains four chapters, the book introduces the range of environmental assessment tools. Chapter 4 is conceptual in nature and introduces the impact assessment framework, which relates biodiversity to human well-being and provides insight in the complex cause–effect chains that may lead to desired or undesired effects of human interventions. Chapter 5 provides a general introduction to Environmental Impact Assessment (EIA) of projects and Strategic Environmental Assessment (SEA) of policies, plans, and programmes, with a special emphasis on recent insights in the performance of these instruments. Chapter 6 extensively explains how biodiversity can be addressed in EIA, following the internationally accepted procedural steps. Chapter 7 addresses biodiversity in SEA. As SEA is linked to changeable planning processes, the approach of the chapter is more conceptual and less procedural. Part III of the book, consisting of two chapters, goes deeper into the practical world of environmental assessment. Two emerging issues are introduced which seem to be particularly powerful in putting biodiversity on the agenda of decision makers. Chapter 8 explains biodiversity offsets, a mechanism for securing conservation in the face of growing development pressure, increasingly being used by the private sector. Chapter 9 provides lessons from ten influential cases in which the valuation of ecosystem services has had a tangible influence on strategic decision making. The book ends with an annex containing the ten case studies referred to in Chapter 9. These cases provide supporting evidence of how many of the concepts introduced throughout the book can be effectively used in practice, and indeed result in better representation of biodiversity in decision making. Ten additional short cases provide additional information for those interested in reviewing a range of practical experiences from around the world.

Summary of chapters

Chapter 2 introduces biological diversity (or biodiversity). It aims to provide an unequivocal interpretation of biodiversity, based on internationally accepted definitions. The chapter starts with an overview of historically developed differences in perspectives: biodiversity conservation versus biodiversity as provider of livelihoods. These perspectives are merged by the Convention on Biological Diversity (CBD) in its three objectives, taking into account present needs, but maintaining future options, and introducing the principle of equity. This provides an important message: biodiversity is about people and how people manage it for their own well-being (here, there, and in the future). The CBD provides an approach to put this into more concrete terms, the ecosystem approach. The ecosystem approach is transparent and participatory, putting emphasis on the role of stakeholders. These principles are shared with environmental assessment; the ecosystem approach thus provides an obvious link between biodiversity and environmental assessment.

If biodiversity is about people, biodiversity has to be linked to people. The Millennium Ecosystem Assessment provides us with the vocabulary and the concepts to do this. Ecosystem services translate biodiversity into concepts understandable to people. *Ecosystem services* (goods and services provided by biodiversity) can be linked to stakeholders. For impact assessment, this provides an important mechanism to translate biodiversity into decision makers' language. Valuing of biodiversity is in this respect an important mechanism; the role of stakeholders in expressing values is highlighted. The valuing itself is subject of a separate chapter.

Chapter 2 provides the approaches and the language to describe biodiversity in relevant terms from an environmental assessment and decision-making perspective. The chapter also highlights recent developments in ecology on three aspects of biodiversity: (i) composition, (ii) structure, and (iii) key processes, with a (nonexhaustive) review of relevant literature. These aspects provide the mechanism to assess how human activities interfere with biodiversity, at genetic, species, or ecosystem level. Fundamental principles, such as the 'no net loss principle' and the 'precautionary principle', are explained with examples of how these principles can be practically dealt with.

Chapter 3 addresses the dominant feeling that biodiversity and development are two opposing themes. Nonindustrialised countries face the dilemma of addressing the present needs of poor sections of society while maintaining the potential of biodiversity to meet the needs and aspirations of future generations. The chapter gives an overview of the history of the 'triple bottom line' concept of sustainable development that encompasses ecological, social, and economic sustainability, and then ventures into newer models of conservation through development that can help bridge the long-standing, conservation-development divide. In this chapter the Millennium Development Goals (MDGs) are taken as the point of departure, providing evidence that biodiversity underpins all MDGs and could 'be the basis for ensuring freedom and equity for all' (see Chapter 3). A major portion of the chapter is spent on the exploration of linkages between biodiversity and each of the MDGs.

Chapter 4 introduces the 'impact assessment framework'. The desire to integrate environmental, social, and economic aspects in assessments of projects, plans, programmes, and policies provides the stimulus for such an integrating framework. In practice, the worlds of environmental impact assessment (in its strict meaning of assessing biophysical impacts only), social impact assessment, and economic cost–benefit analysis largely continue to operate in their separate realms and experience great difficulty in working in a multidisciplinary environment. The framework aims to provide insight into the relations between human society and the biophysical environment and the way in which both biophysical environment and human society are being influenced and managed. The core element of the conceptual approach is the characterisation and classification of ecosystem services provided by the biophysical environment and the assessment of their value for sustaining human livelihoods. Values by definition are assigned by stakeholders; in other words, ecosystem services can be linked to stakeholders.

The impact assessment framework is a framework of thinking. It is not intended to be a 'standardised procedure' or a predictive analytical model. It is a device for the facilitation and systematising of the interaction between the different disciplines involved in an assessment process. It mediates between different types of knowledge: natural and social science knowledge, lay and expert knowledge, knowledge about facts, and knowledge about values. It thus does not produce or predict 'solutions' by itself, but its active use by those involved in a certain problem situation can help to find sensible and feasible ways forward. In situations where interdisciplinarity or transdisciplinarity are required for effective analysis and decision making, as is the case in most environmental assessment situations, the problem of boundary crossing presents itself. The chapter ends with an overview of how boundaries can effectively be crossed through the use of a boundary concept (ecosystem services), a boundary object (the impact assessment framework), and boundary settings (institutional arrangements).

Chapter 5 provides essential knowledge on both EIA and SEA for those not being fully informed on the instruments of environmental assessment. It is not a handbook text, but highlights recent developments and state of the art thoughts. Environmental assessment has been around for more than 40 years and is practised in most countries around the world. The principle behind environmental assessment is deceptively simple: it directs decision-makers to 'look before they leap'. When there is a clear insight into the environmental consequences, decision makers are in a better position to direct development into a more sustainable course. At its best, environmental assessment does not merely provide information, but it brings parties together. The chapter explains the internationally accepted procedure, with a series of well-defined steps. Crucial elements in the process are highlighted (alternatives, how to deal with gaps in knowledge, public review, participation, etc.). Special attention is paid to the effectiveness of EIA and the conditions that can guarantee good practice.

The practice of SEA is less easily demarcated than that of EIA. There are a large number of assessment tools in planning that do not necessarily carry the label SEA, but have strong similarities. However, the fundamental differences between approaches are fewer than might be assumed from existing publications. There is no generally agreed SEA procedure as such and no 'one-size-fits-all' approach. As planning processes vary greatly from context to context SEA needs to be applied flexibly. However, there is general agreement about the activities that make up an SEA process. These are discussed in some detail with special emphasis on the state of the art: what is needed for effective SEA? The chapter ends highlighting three current trends in environmental assessment thinking and practice: (i) increased attention to the assessment context, (ii) integration of effects for sustainability assessment, and (iii) tailoring the assessment to the decision process.

Chapter 6 provides extensive first-hand background documentation on the EIA guidelines adopted by the Convention on Biological Diversity. The chapter is structured according to a generalised and internationally accepted sequence of steps. The chapter is a practical application of the concepts introduced in Chapter 2 (biodiversity) and in Chapter 4 (impact assessment framework); case examples are used to illustrate both concepts and practice. Special emphasis is given to the screening and scoping stages of EIA, for two reasons. First, the need for an impact assessment study has to be defined by good screening criteria and procedures; second, the impact assessment study has to be carried out in such a manner that all relevant issues are properly dealt with. Because scoping determines the contents and quality of the terms of reference of the impact study, good scoping procedures and guidance on the scoping process are of fundamental importance. The chapter also contains an extensive overview of recent initiatives in different sectors to enhance biodiversity in project planning, impact assessment, and operations.

Chapter 7 gives, similar to Chapter 6, extensive background information to the CBD guidance on biodiversity in SEA. It is not structured according to a procedure (as with Chapter 6 on the EIA) because good practice SEA should ideally be fully integrated into a planning (or policy development) process. Since planning processes differ widely, there is, by definition, no one-size-fits-all sequence of procedural steps in SEA. The chapter answers three basic questions. First, WHY is special attention to biodiversity in SEA and decision making needed? This to convince decision makers that biodiversity is a relevant issue. A second question is WHAT biodiversity issues are relevant to SEA? Not all biodiversity can be studied in SEA; on the contrary, the problem usually is how to limit an assessment in such a way that it is done in a timely way, and costs and efforts involved are reasonable. The third and last question is HOW to address biodiversity in SEA? This section is based on the conceptual approaches described in Chapters 2 and 4. To be able to make a judgement whether a policy, plan, or programme has potential biodiversity impact, three conditions are defined that 'trigger' the need for special attention to biodiversity. When any one or a combination of these conditions applies, special attention to biodiversity is required. The approach is based on the analysis of a significant number of cases which are referred to throughout the text.

Chapter 8 explores rapid recent developments predominantly taking place in the private sector. Finding innovative ways to link biodiversity conservation with development becomes a challenge and urgency for conservation organisations, businesses as well as voluntary bodies, governments, and civil society. The mitigation step in EIA frameworks, targeted for integrating biodiversity, provides options for preventing and minimising the impacts of development projects on biodiversity by utilising an array of strategies, policy instruments, economic incentives, and market solutions for compensating the residual impacts. The concept of 'biodiversity offset' as a compensation measure is relatively new and therefore lacks a universally acceptable definition. In simple wording biodiversity offsets are creatively designed mechanisms to achieve either 'no net environmental loss' or a 'net environmental benefit'.

The chapter provides an overview of the various global directives and country specific regulatory mechanisms in place for promoting legal and voluntary approaches for applying biodiversity offsets. Different forms of biodiversity offset are presented and supported with appropriate examples, including onsite, offsite, and third party offsets and a range of options including conservation-oriented actions to widely applicable market-based approaches such as conservation banking, development of tradable rights and biodiversity credits, direct payments for resources/services, or creation of trust funds and monetary bonds for financing impact mitigation. Practical experience of the design, implementation, and evaluation of biodiversity offsets shared through several case examples provide a useful input to the chapter. These examples illustrate that the various mechanisms of mainstreaming biodiversity conservation in business plans are largely aimed at creating mutually beneficial opportunities for both business and biodiversity. Business groups are beginning to appreciate the benefits from applying offsets and are taking leads to demonstrate that responsible biodiversity stewardship is a fundamental business issue for managing risks, capitalising on opportunities, and improving the corporate performance in environmentally and socially responsible manners. The chapter also presents the many risks and constraints that pose methodological challenges and practical difficulties in the design and implementation of offsets. Despite this, the conclusion is drawn that the objective of offset is ideologically sound, and there is a clear need to overcome these barriers for more and better conservation outcomes for biodiversity to occur by identifying possible routes to achieving better levels of success.

Chapter 9 contains a re-edited text from a recent publication with the same title. In order to put biodiversity into decision-makers language, we have already emphasised the need to translate biodiversity in terms of ecosystem services and to link these services to (present and future) stakeholders. Ecosystem services are the benefits people obtain from ecosystems. Ecosystem services have received significant attention since the appearance of the Millennium Ecosystem Assessment. A growing body of knowledge is developing on ecosystem services and on the valuation of these services. Yet, cases where valuation of ecosystem services has actually made a difference in real-life policies or plans still remain scarce, or in any case hidden. So far, the SEA community has hardly used the opportunities provided by ecosystem services to translate the environment into societal benefits.

Therefore, a number of influential cases were documented, where the recognition, quantification, and valuation of ecosystem services have significantly contributed to strategic decision making. In all cases, the use of the ecosystem services concept supported decision making by providing better information on the consequences of new policies or planned developments. In several cases SEA or a process similar to SEA was followed. Yet, in all cases valuation of ecosystem services, in one form or another, resulted in major policy changes or decision making on strategic plans or investment programmes. The analysis of cases reveals that the role of ecosystem services in decision making can range from simple recognition of services, via semiquantified valuation techniques to full-fledged monetisation of ecosystem services. The presented evidence suggests that at higher strategic level there is less need for fully quantified and monetised information. Seven main messages are derived from the analysis of cases, aimed at decision makers, SEA practitioners, environmental economists, and ecologists.

The Annex to the book presents the ten case study documents underlying Chapter 9. Even although the reason to collect these cases was to evaluate the role of valuation of ecosystem services in decision making, the case studies provide a detailed view on how many of the concepts and approaches introduced in this book can be used in practice. The ten cases presented are:

(1) Water Conservation and Irrigation Rehabilitation, Egypt (voluntary SEA).
(2) Wetland Restoration Strategy, Aral Sea region (SEA-like process).
(3) Strategic Catchment Assessment, South Africa (part of SEA process).
(4) Making Space for Water, United Kingdom (experimental SEA).
(5) Climate policies and the *Stern Review* (study to inform policy making).
(6) Natural gas extraction in the Wadden Sea, Netherlands (study to inform EIA and SEA processes).
(7) Management of marine parks, Netherlands Antilles (sustainable financing).
(8) Watershed rehabilitation and services provision, Costa Rica (payments for ecosystem services).

(9) Water transfer, Spain (advocacy study to influence decision making).
(10) Exxon Valdez oil spill, Alaska USA (damage assessment).

Apart from these cases, ten additional cases are presented in boxes to provide supporting evidence.

The References contain the citations of all book chapters. A significant number of references come from nonscientific literature, and therefore are not accessible through regular scientific channels. As much as possible we have tried to provide access to these sources of information, in most cases by providing relevant websites.

2 · Interpretation of biodiversity

Roel Slootweg

Introduction

A multitude of tools and techniques are currently used by companies, governments, certifying agencies, and the like, to predict, measure, or report on the human impacts on biodiversity. Environmental impact assessment (EIA) and strategic environmental assessment (SEA) are the focus of attention in this book, but other instruments exist, such as environmental audits, sustainability reporting, and certification schemes. These instruments are not always based on an unambiguous interpretation of biodiversity (Slootweg, 2005). A rapid comparison of the objectives of the Convention on Biological Diversity (CBD) with a number of biodiversity-related methods for certification and assessment shows that no one instrument addresses all aspects of biodiversity as defined by the CBD. General omissions include: (i) a partial focus on only one level of diversity, mostly the species level and often ecosystem level, but hardly ever the genetic level of diversity; (ii) a focus on either conservation of biodiversity, and sometimes on sustainable use, but largely overlooking the third objective of the CBD on equitable sharing of revenues obtained from biodiversity, and never including all three objectives simultaneously; (iii) a general lack of identification and involvement of stakeholders; and finally, (iv) a lack of attention to the potential positive effects of human activities on biodiversity (opportunities for enhancement) (Slootweg *et al.*, 2003).

As the CBD has provided clear definitions and supporting documents on what biodiversity is, it should be possible to define a more consistent approach to the incorporation of biodiversity in assessment instruments. The CBD was negotiated under the auspices of United Nations Environmental Programme (UNEP) and signed at the Rio Summit by more than 150 nations. It came into force on 29 December 1993 after the requisite number of 30 ratifications. It is a framework treaty in two senses: first, in that its provisions are generally expressed as overall goals and policies rather than precise obligations, and second, more specifically it adopts a holistic approach by not setting targets or including lists of species or areas to be protected, but by setting out general rights and obligations (Glazewski and Paterson, 2005). This chapter provides an overview of the ways in which biodiversity is understood – it describes the knowledge minimally required to address biodiversity within the broader context of environmental assessment, and it addresses major areas of uncertainty. The definitions and principles provided by the CBD are followed as closely as possible. Convention texts describe how parties to the convention have defined biodiversity, the objectives of biodiversity management, and the approaches to biodiversity management. From a scientific point of view, a significant contribution to the unambiguous understanding of biodiversity has been made by the Millennium Ecosystem Assessment (2003), which will be extensively cited throughout this book.

The biodiversity convention: a broad view

The title of this section, 'a broad view', refers to the fact that many nonbiodiversity experts in the environmental assessment field may view the presented description of biodiversity as an all-encompassing concept, which includes many aspects of environmental assessment that already are common practice without necessarily being described as biodiversity. This chapter will show that biodiversity indeed is a broad concept. The preamble of the CBD states that biodiversity has an economic, a social, and an ecological dimension. As Glazewski and Paterson (2005) point out,

the CBD spans a broad area – from the conservation of endangered species, to protecting indigenous knowledge, to dealing with the safety ramifications of genetic modification. As such, the CBD is a landmark in the environment and development field, as it takes for the first time a holistic and integrated, rather than a species-based approach, to the conservation and sustainable utilisation of natural resources.

And indeed, present-day impact assessment already effectively deals with many aspects of biodiversity, but as stated above, often in a fragmented manner without a clear explanation (or even understanding) of how the internationally accepted and adopted definitions and objectives of the CBD are being reflected (Kolhoff and Slootweg, 2005; Treweek, 2001).

Box 2.1. Use of some terms in the Convention on Biological Diversity (Article 2)

- *Biological diversity* means the variability among living organisms from all sources including inter alia, terrestrial, marine and other aquatic ecosystems and the ecological complexes of which they are part; this includes diversity within species, between species and of ecosystems.
- *Biological resources* includes genetic resources, organisms or parts thereof, populations, or any other biotic component of ecosystems with actual or potential use or value for humanity.
- *Domesticated or cultivated species* means species in which the evolutionary process has been influenced by humans to meet their needs.
- *Ecosystem* means a dynamic complex of plant, animal and microorganism communities and their non-living environment interacting as a functional unit.
- *Ex-situ conservation* means the conservation of components of biological diversity outside their natural habitats.
- *Genetic resources* means genetic material of actual or potential value.
- *Habitat* means the place or type of site where an organism or population naturally occurs.
- *In-situ conservation* means the conservation of ecosystems and natural habitats and the maintenance and recovery of viable populations of species in their natural surroundings and, in the case of domesticated or cultivated species, in the surroundings where they have developed their distinctive properties.
- *Sustainable use* means the use of components of biological diversity in a way and at a rate that does not lead to the long-term decline of biological diversity, thereby maintaining its potential to meet the needs and aspirations of present and future generations.

The definition of biodiversity provided by the convention (see Box 2.1) describes biodiversity as the variety of life on Earth at all levels (CBD, 1992). In the definition three different levels of biodiversity are described: genetic, species, and ecosystem diversity. However, in reality, biological diversity is a continuum of layers from the genetic diversity of individual specimens to the Earth as one functioning ecosystem. Among ecologists the hierarchy of biological organisation is often described in terms of *gene – organism – population – species – community – ecosystem – biome – biosphere*. Boundaries between levels are not as clear-cut as the simplified three-tier system of the CBD, or any other system, suggests. Yet, the CBD *genes – species – ecosystems* division provides a clear distinction among the most fundamental levels in biodiversity. Genetic diversity is the driver behind evolutionary processes. Species and species diversity are the visible result of this evolution, while ecosystems are expressions of the nonrandom manner in which species coexist, interact with and among each other, and interact with their physical environment.

Genetic diversity starts at the description of the genome of a single individual. Genetic diversity among individual specimens from the same species has been described by Williams and Humphries (1996) as the 'fundamental currency of diversity', just as genetic variability is a trait often associated with the ability of populations of species to adapt to changing environmental conditions. Among separate populations of a species, the range in genetic variability can differ, leading to the recognition of different strains, races, varieties, and subspecies within a species. This genetic variability has been used for the creation of different races or cultivars in agriculture and horticulture and for breeds of animals in animal husbandry. Reproductive isolation of such populations is the first step towards the creation of new species (speciation), creating a fluid boundary between genetic and species diversity. Genetic variability is the field of interest of *bioprospecting*, described as the search for wild species, their genes, and their products with actual or potential use to humans with a view to their exploitation (Glazewski and Paterson, 2005).

Species diversity is the most commonly understood expression of biological diversity. Estimations of the number of species on Earth range from five million to many millions. However, fewer than two million species have actually been scientifically described. Hence, the scientifically described species represent only a fraction of the total number of species on Earth. Species diversity can be described in terms of the number of species present in certain area, the (un)evenness of their distribution (a measure for the abundance of a species), or in terms of their evolutionary relatedness (phylogenetic diversity). With respect to the latter, one can discuss whether an area with a large number of closely related species is more biodiverse and thus represents a larger range in genetic diversity compared to an area with fewer species that represents a larger number of families or orders. Whittaker (1972) provided yet another manner of describing species diversity by distinguishing three measures of biodiversity over spatial scales: alpha diversity, beta diversity, and gamma diversity. *Alpha diversity* refers to the diversity within a particular area (i.e. species richness). The difference in species diversity between two or more areas is called *beta diversity* (i.e. the number of species unique to each area). *Gamma diversity* is a measure of the overall diversity of various ecosystems within a region. (See Magurran (2003) for further in-depth reading on measurement of species diversity.)

Worldwide biodiversity is declining on two scales, in beta diversity (difference between regions) and gamma diversity (global diversity), whereas alpha diversity (local diversity) is increasing in many locations due to the introduction of exotic or invading species. Sax *et al.* (2002) have demonstrated that the number of species naturalisations exceeds extinctions on islands worldwide. So apart from just counting species, the importance of native species over nonnative species has to be emphasised in assessing species richness in a specific location. Summarising, the world is facing a loss of rare species and the spread of invasive species (Thompson and Starzomski, 2006). A further variety of mathematical diversity indices has been developed, but elaboration would go beyond the scope of this chapter. All of the above provides evidence of the strong focus of the scientific community on species diversity, acting almost as a synonym to biodiversity. Unfortunately, most existing biological measures, especially those reflecting species richness or various aspects of species diversity, do not reflect many important aspects of biodiversity, especially those that are significant for the delivery of ecosystem services (Hassan *et al.*, 2005: 111).

A term closely linked to species is *habitat* – the place or type of site where an organism or a population of this organism naturally occurs (CBD, 1992: Article 2 – Use of terms). According to Oindo *et al.*, (2003) spatial heterogeneity is one of the most popular hypotheses used to explain patterns of species richness. However, their study revealed that the association between species richness and habitat diversity indices is strongly dependent on the scale of observation. Using a combination of remote sensing techniques and field observations, they found the strongest relationship between habitat and species diversity for African herbivores at a relatively intermediate scale (20 km × 20 km). Most probably this will be completely different for other animal or plant taxa. In other words, from a biological perspective it is very complex to exactly define relevant boundaries of a study area. Therefore, the boundaries in an environmental assessment study need to be defined from a completely different perspective. The CBD ecosystem approach provides conceptual and procedural leads on how to do this, which is discussed in the next section. The idea will be further conceptualised in Chapter 4.

Species are not randomly distributed over the Earth but occur in communities of species. When considered in combination with its physical environment, such a community is generally referred to as an *ecosystem*. Following the CBD definition (see Box 2.1), ecosystems can be characterised by a set of physical components (soil, minerals, water, etc.) and a combination of populations of different species. Energy, water, and minerals flow through and between ecosystems. Ecosystems can change over time as a result of external factors (allogenic succession due to, for example, climate change, landslides, or volcanic activity) or internal factors (autogenic succession). Delineation of ecosystems is difficult, because boundaries in nature are characterised by fluid transitions or gradients. The CBD ecosystem approach (CBD, 2000a and 2004a) further states that

the CBD definition of an ecosystem does not specify any particular spatial unit or scale. Thus, the term "ecosystem" does not, necessarily, correspond to the terms '"biome" or "ecological zone", but can refer to any functioning unit at any scale. Indeed, the *scale of analysis and action should be determined by the problem being addressed* [emphasis by the author]. It could, for example, be a grain of soil, a pond, a forest, a biome or the entire biosphere.

In this sense the ecosystem definition of the convention is very flexible and ideal for environmental assessment, because ecosystems can be defined according to the scope and required spatial and temporal ranges of the assessment study. In this context an ecosystem is sometimes described as 'problemshed', an area which encompasses all the elements of the issue at stake, no matter where they may be located (H. Prins, personal communication, 2002). The CBD definition differs from the conventional use of the terminology. For example, landscape ecologists consider landscapes as being composed of ecosystems suggesting a hierarchy of recognisable scales. Complex hierarchies of ecological land classifications have been developed, for example, to describe ecosystems at various levels of detail ranging from ecozones or ecoregions at a mapping scale of >1:50,000,000 to ecoelements at a mapping scale of <1:5,000. From the perspective of environmental assessment, the description provided by the ecosystem approach provides the best clue, because it is the scale and nature of the issue at stake that determines the required level of study. Further on in this chapter, we will elaborate on the ecosystem approach and its consequences for environmental assessment.

Countries that have signed the CBD (called 'parties' in convention terminology) are required to implement policies to protect biodiversity at the three distinguished levels, and monitor the components of biological diversity listed below, paying particular attention to those requiring urgent conservation measures and those which offer the greatest potential for sustainable use (CBD, 1992: annex 1):

(1) *Ecosystems and habitats*: containing high diversity, large numbers of endemic or threatened species, or wilderness; required by migratory species; of social, economic, cultural, or scientific importance; or, which are representative, unique, or associated with key evolutionary or other biological processes;

(2) *Species and communities* which are: threatened; wild relatives of domesticated or cultivated species; of medicinal, agricultural, or other economic value; or social, scientific, or cultural importance; or importance for research into the conservation and sustainable use of biological diversity, such as indicator species; and

(3) Described *genomes and genes* of social, scientific, or economic importance.

Apart from the three levels of diversity the CBD has put forward, three main objectives, which can easily be linked to the social, ecological, and economic pillars of sustainability, are often dubbed 'people', 'planet', and 'profit' in the business community when referring to activities within the framework of corporate social responsibility. The first objective is the *conservation of biological diversity* in order to maintain Earth's life-support systems and to maintain future options for human development. This is the ecological or 'planet' pillar of sustainability. The second objective is the *sustainable use of components of biodiversity* in order to provide livelihoods to people, without jeopardising the opportunities for development of future generations. This can be considered the economic or 'profit' pillar of sustainability. The third objective is the *fair and equitable sharing* of the benefits arising from the use of genetic resources. This last objective was triggered by a discussion on the use of traditional knowledge of local (often tribal) communities. Large multinational pharmaceutical companies have tapped the knowledge of traditional healers on the use of medicinal herbs, often leading to significant profits for these companies, while the providers of the knowledge, often living in great poverty, did not receive anything for their contribution. Since the establishment of the biodiversity convention, many examples of benefit sharing have become available, which recognise the rights of local communities to share in benefits obtained using their knowledge, but also recognise the rights of countries to share in benefits arising from the commercial and other utilization of genetic resources from these countries (CBD, 2000b and 2004b). This third objective can be linked to the social, or 'people' pillar of sustainability. Figure 2.1 summarises the convention definition of biodiversity and the objectives of the convention. Box 2.2 provides Guiding Principles drafted by the International Association for Impact Assessment on how to address the three objectives of the convention in impact assessment.

As with the three pillars of sustainability, the three objectives of the convention are intimately linked. Conservation of biodiversity now, will, of course, maintain the possibility for new, yet unknown uses in future; equitable sharing of benefits will contribute to sustainability of biodiversity use; sustainable use contributes to intergeneration equity, and so forth. In this respect a word of caution is needed on the present popularity of the 'three pillars' notion of sustainability. This representation is often misused by decision makers as a mechanism to create trade-offs between the pillars: conservation, it is said, is either sacrificed at the cost of economic development or hinders social development. In reality, the problem lies in the factor time and in the distribution of power and benefits. Making trade-offs between the pillars is similar to taking the benefits 'here' and 'now', while transferring the costs to 'later' and 'elsewhere' (we want the benefits now at the cost of future generations or of remote stakeholders elsewhere).

According to Article 6 each Contracting Party of the convention shall, in accordance with its particular conditions and capabilities, develop national strategies, plans, or programmes for the conservation and sustainable use of biological diversity. Similarly, the conservation and sustainable use of biological diversity should be integrated into relevant sectoral or cross-sectoral plans, programmes, and policies. These national plans are the so-called NBSAPs, or National Biodiversity Strategy and Action Plans. Each party is expected to produce an NBSAP or similar document and submit this to the convention. NBSAPs can be of significant value to environmental assessment studies, because these documents are valuable sources of information on biodiversity status and policies in a country. The integration of biodiversity objectives into relevant sectoral or cross-sectoral plans, programmes, and policies is the terrain of Strategic Environmental Assessment. Article 14(b) specifically addresses this issue, even although at the moment of the writing of the convention text the term 'strategic environmental assessment' was not commonly used and consequently does not show in the text (see Box 2.3).

Figure 2.1 Combining three objectives of the biodiversity convention with three levels of biodiversity results in a diagram with nine boxes. The top row shows conservation issues at the three levels of diversity: (from left to right) the protected ecosystem of the Iguaçu falls at the border of Brazil and Argentina; species diversity in the largely unprotected Pantanal wetlands in Brazil, the worlds largest freshwater wetland; closely related threatened butterflies of tropical South America. The middle row on sustainable use shows a man-made wetland ecosystem in the Netherlands (drained peatlands), highly valued for its landscape features and multiple recreational services; a monoculture production forest in France showing to be very vulnerable – a severe storm has virtually destroyed the entire forest; a vegetable market on the isle of Madeira showing the rich variety of local horticulture products. The Bottom row on equitable sharing shows beaches near Nazaré (Portugal) where traditional fishermen and the tourism industry compete for the same ecosystem (sun drying versus sunbathing); fishing day in the floodplains of the Benue river (Cameroon) – a hydropower dam has destroyed yearly floods resulting in loss of income for traditional women groups; market in Kungrad near the Aral sea in Uzbekistan, a region considered to be the cradle of many of our agricultural cultivars – destruction of the Aral Sea and its surrounding lands may have destroyed 'wild' ancestors of our cultivars. Six of the boxes above have a direct link to people, while all nine have a link with future generations. Conclusion: biodiversity is about people!

Box 2.2. IAIA Principles on Biodiversity in Impact Assessment

Guiding principles on the objectives of the CBD by the International Association for Impact Assessment (IAIA, 2005):

(1) Aim for *Conservation* and *"No Net Loss"* of Biodiversity. The biodiversity-related Conventions are based on the premise that further loss of biodiversity is unacceptable. Biodiversity must be conserved to ensure it survives, continuing to provide services, values and benefits for current and future generations. Take the following approach to help achieve no net loss of biodiversity: 1. Avoid irreversible losses of biodiversity; 2. Seek alternative solutions that minimize biodiversity losses; 3. Use mitigation to restore biodiversity resources; 4. Compensate for unavoidable loss by providing substitutes of at least similar biodiversity value; 5. Seek opportunities for enhancement. This approach can be called "positive planning for biodiversity." It helps achieve no net loss by ensuring priorities and targets for biodiversity at international, national, regional and local levels are respected, and a positive contribution to achieving them is made. Damage is avoided to unique, endemic, threatened or declining species, habitats and ecosystems; to species of high cultural value to society, and to ecosystems providing important services.

(2) Seek *Sustainable Use* of Biodiversity Resources. Use impact assessment to identify, protect and promote sustainable use of biodiversity so that yields/harvests can be maintained over time. Recognise the benefits of biodiversity in providing essential life support systems and ecosystem services such as water yield, water purification, breakdown of wastes, flood control, storm and coastal protection, soil formation and conservation, sedimentation processes, nutrient cycling, carbon storage, and climatic regulation as well as the costs of replacing these services. In a developing country context, this principle is likely to be a key priority – i.e. for biodiversity to be conserved and protected in this context, it is essential that it is linked to the issue of securing sustainable livelihoods for local people based on biodiversity resources.

(3) Ensure *Equitable Sharing*. Ensure traditional rights and uses of biodiversity are recognized in Impact Assessment and the benefits from commercial use of biodiversity are shared fairly. Consider the

needs of future as well as current generations (inter-generational needs): seek alternatives that do not trade in biodiversity "capital" to meet short-term needs, where this could jeopardize the ability of future generations to meet their needs.

Box 2.3. CBD (1992) Article 14(b) on impact assessment

Each contracting party is requested, as far as possible and as appropriate, to:

(a) Introduce appropriate procedures requiring **environmental impact assessment** of its proposed projects that are likely to have significant adverse effects on biological diversity with a view to avoiding or minimizing such effects and, where appropriate, allow for public participation in such procedures;

(b) Introduce appropriate arrangements to ensure that the environmental consequences of its **programmes and policies** that are likely to have significant adverse impacts on biological diversity are duly taken into account;

(c) Promote, on the basis of reciprocity, notification, exchange of information and consultation on activities under their jurisdiction or control which are likely to significantly affect adversely the biological diversity of other States or areas **beyond the limits of national jurisdiction**, by encouraging the conclusion of bilateral, regional, or multilateral arrangements, as appropriate;

(d) In the case of imminent or grave danger or damage, originating under its jurisdiction or control, to biological diversity within the area under jurisdiction of other States or in areas beyond the limits of national jurisdiction, **notify immediately the potentially affected States** of such danger or damage, as well as initiate action to prevent or minimize such danger or damage; and

(e) Promote national arrangements for **emergency responses** to activities or events, whether caused naturally or otherwise, which present a grave and imminent danger to biological diversity and encourage international cooperation to supplement such national efforts and, where appropriate and agreed by the States or regional economic integration organizations concerned, to establish joint contingency plans.

Another article of relevance to environmental assessment is Article 8(j) on preservation of knowledge, innovations, and practices of indigenous and local communities embodying traditional lifestyles relevant for the conservation and sustainable use of biological diversity. Their wider application should be promoted with the approval and involvement of the holders of such knowledge. Plans and projects can constitute major threats to biological diversity on which indigenous and local communities depend for their survival. An 'Ad Hoc Open-ended Inter-Sessional Working Group on Article (j) and Related Provisions' of the CBD has drafted, in cooperation with indigenous and local communities, guidelines for the conduct of cultural, environmental, and social impact assessments (CBD, 2004c). The guidelines and recommendations should ensure the participation of indigenous and local communities in the assessment and review process. Environmental legislation in many countries requires the assessment of potential environmental, social, and cultural impacts of proposed developments. However, it is rare that traditional knowledge, technologies, and customary methods are included or required as part of the assessment process. The draft guidelines suggest a framework within which governments, indigenous and local communities, decision makers, and managers of development and planning projects could ensure appropriate participation and involvement of indigenous and local communities and inclusion of their traditional knowledge, technologies, and customary methods as part of environmental, social, and cultural impact assessment processes.

The Ecosystem approach: biodiversity and humankind are inseparable entities

Discussing biodiversity is very much about discussing people's behaviour and interests. The first World Summit on Environment and Development in Rio de Janeiro in 1992 emphasised the importance of biodiversity as the basis of our very existence, to be used wisely and sustainably and conserved for current and future generations. The main threats to global biodiversity are associated with human activities. This strongly anthropocentric view on biodiversity is explicitly followed in this book, based on the simple reasoning that evolution and biodiversity have been around before the emergence of humankind and will probably also be there after its possible disappearance. Or, as Bröring and Wiegleb (2005: 534) put it:

biodiversity is neither static nor given. It is know that global biodiversity easily recovered after each event of mass extinction. Both measurement and evaluation of biodiversity consequently depends on social conventions and agreements.

Their conclusion is that none of the known approaches to value biodiversity *justifies a value of biodiversity per se*. In their reasoning the value of biodiversity is best guaranteed among people by means of a participatory discussion of environmental goals. This observation is in agreement with the ecosystem approach and the approach in this book. So, it is not biodiversity *per se* which is the theme of both this chapter and this book, but it is the reliance of humankind on biodiversity. Conserving biodiversity is not primarily in the interest of biodiversity itself, but moreover in the interest of the human race. Following this reasoning, the use of the term 'intrinsic value' of biodiversity is avoided as much as possible, because a value can only be attributed as an expression of a human interest. In absence of humans, value is a meaningless concept. However, groups of people insist on attributing an intrinsic value to biodiversity. In our approach this group is considered to be a stakeholder, recognising the maintenance of biodiversity as an ecosystem service and assigning a value to this service for the very existence of biodiversity.

The *ecosystem approach* of the CBD explicitly states that humans, with their cultural diversity, are an integral component of many ecosystems. The ecosystem approach was endorsed by the Convention on Biological Diversity in 2000 (Decision V/6). The original document contained 12 principles and additional guidance on the implementation. In 2004 further guidance was provided in a document refining and elaborating the approach, which was based on an assessment of experiences in the implementation of the approach (Decision VII/11). The ecosystem approach is a strategy for the integrated management of land, water, and living resources that promotes conservation and sustainable use in an equitable way (CBD, 2000a and 2004a). The application of the ecosystem approach aims to reach a balance of the three objectives of the Convention. In addition, the ecosystem approach has been recognised by the World Summit on Sustainable Development as an important instrument for enhancing sustainable development and poverty alleviation (CBD, 2004a). People and biodiversity depend on healthy, functioning ecosystems and processes; these have to be assessed in an integrated way and not constrained by artificial boundaries, such as administrative boundaries. The ecosystem approach is participative and requires a long-term perspective based on a biodiversity-based study area. It requires adaptive management to deal with the dynamic nature of ecosystems and the absence of complete understanding of the ways in which ecosystems function.

The ecosystem approach is based on the application of appropriate scientific methodologies focused on levels of biological organisation, which encompass the essential structure, processes, functions, and interactions among organisms and their environment. It recognises that humans, with their cultural diversity, are an integral component of many ecosystems. The approach incorporates three important considerations:

(1) Management of living components is considered alongside economic and social considerations at the *ecosystem level of organisation*, not simply a focus on managing species and habitats;
(2) If management of land, water, and living resources in equitable ways is to be sustainable, it must be integrated, *work within the natural limits*, and utilise the natural functioning of ecosystems;
(3) *Ecosystem management is a social process*. There are many interested communities, which must be involved through the development of efficient and effective structures and processes for decision making and management.

There is no single correct way to achieve the ecosystem approach to management of land, water, and living resources. The underlying principles can be translated flexibly to address management issues in different social contexts. Box 2.4 provides a summary overview of the principles.

Box 2.4. The ecosystem approach principles

Principle 1: *The objectives of management of land, water, and living resources are a matter of societal choice.* Different sectors of society view ecosystems in terms of their own economic, cultural, and societal needs. Both cultural and biological diversity are central components of the ecosystem approach, and management should take this into account. Societal choices should be expressed as clearly as possible. Keywords in the accompanying guidelines refer to the process of decision making: transparency of decision making, accountability, stakeholder interests, equal access to information of all involved, and equitable capacity to be involved (referring to less-privileged groups). The need to include the interests of future generations is highlighted. Good environmental assessment is based on similar principles.

Principle 2: *Management should be decentralised to the lowest appropriate level.* This principle of subsidiarity is well known from various sectors; practical experience stresses the need for a mechanism to coordinate decisions and management actions at different organisational levels. Furthermore, good governance arrangements ask for clear accountabilities. If no appropriate body is available at certain management levels, a new body may be created, an existing body modified, or a different level chosen. Without institutional arrangements that support and coordinate decision-making authorities, their work is worthless.

Principle 3: *Ecosystem managers should consider the effects (actual or potential) of their activities on adjacent and other ecosystems.* Effects of interventions are not confined to the point of impact, and can influence other ecosystems. Time-lags and nonlinear processes are likely to occur. In case of effects elsewhere, relevant stakeholders and technical expertise have to be brought together. Feed-back mechanisms to monitor the effects of interventions should be established.

Principle 4: *Recognizing potential gains from management, there is usually a need to understand and manage the ecosystem in an economic context. Any such ecosystem-management programme should: (a) Reduce those market distortions that adversely affect biological diversity; (b) Align incentives to promote biodiversity conservation and sustainable use; (c) Internalise costs and benefits in the given ecosystem to the extent feasible.* Many ecosystems provide economically valuable goods and services and it is therefore necessary to understand and manage ecosystems in an economic context. Frequently, economic systems do not make provisions for the many, often, intangible values derived from ecological systems. In this regard it should be noted that ecosystem goods and services are frequently undervalued in economic systems. Even when valuation is complete, most environmental goods and services have the characteristic of 'public goods' in an economic sense, which are difficult to incorporate into markets. Deriving economic benefits is not necessarily inconsistent with attaining biodiversity conservation and improvement of environmental quality.

Principle 5: *Conservation of ecosystem structure and functioning, in order to maintain ecosystem services, should be a priority target of the ecosystem approach.* The conservation and, where appropriate, restoration of ecosystem interactions and processes is of greater significance for the long-term maintenance of biological diversity than simply protection of species. Given the complexity of ecosystem functioning, management must focus on maintaining, and where appropriate restoring, the key structures and ecological processes rather than just individual species. However, vulnerable and economically important species have to be monitored to avoid loss of biodiversity. Management of ecosystem processes has to be carried out despite incomplete knowledge of ecosystem functioning.

Principle 6: *Ecosystems must be managed within the limits of their functioning.* There are limits to the level of demand that can be placed on an ecosystem while maintaining its integrity and capacity to continue providing the goods and services that provide the basis for human well-being and environmental sustainability.

Principle 7: *The ecosystem approach should be undertaken at the appropriate spatial and temporal scales.* Failure to take scale into account can result in mismatches between the spatial and time frames of the management and those of the ecosystem being managed.

Principle 8: *Recognizing the varying temporal scales and lag-effects that characterise ecosystem processes, objectives for ecosystem management should be set for the long term.* Management systems tend to operate at relatively short time scales, often much shorter than the timescales for change in ecosystem processes.

Principle 9: *Management must recognise that change is inevitable.* Natural and human-induced change in ecosystems is inevitable; therefore management objectives should not be construed as fixed outcomes but rather the maintenance of natural ecological processes. Traditional knowledge and practice may enable better understanding of ecosystem change and help in developing adaptation measures.

Principle 10: *The ecosystem approach should seek the appropriate balance between, and integration of, conservation and use of biological diversity.* Biological resources provide goods and services on which humanity ultimately depends. There has been a tendency in the past to manage components of biological diversity either as protected or nonprotected. There is a need for a shift to more flexible situations, where conservation and use are seen in context and the full range of measures is applied in a continuum from strictly protected to human-made ecosystems.

Principle 11: The ecosystem approach should consider all forms of relevant information, including scientific and indigenous and local knowledge, innovations, and practices. Information from all sources is critical to arriving at effective ecosystem management strategies. Sharing of information with all stakeholders is equally important.

> **Principle 12:** The ecosystem approach should involve all relevant sectors of society and scientific disciplines. The integrated management of land, water, and living resources requires increased communication and cooperation, (i) between sectors, (ii) at various levels of Government (national, provincial, local), and (iii) among Governments, civil society, and private sector stakeholders.

The consequences of the ecosystem approach for environmental assessment can be summarised under six headings:

(1) *Transparency and stakeholder involvement.* The first principle of the ecosystem approach is of utmost importance to all parties involved in any decision-making process involving biological diversity (or natural resources, in general), because it defines in general terms the 'rules of the game'. The required transparency of decision making and involvement of stakeholders at the earliest possible stages of a project may sometimes be conflicting with confidentiality of strategic information. It should be realised that full involvement of stakeholders provides the best guarantee to avoid problems later on in the process. Potential miscommunications or real conflicts of interest can be identified in an early stage and dealt with. Procedures and mechanisms should be established to ensure effective participation of all relevant stakeholders and actors during the consultation processes and decision making on management goals and actions (Principle 12). Government, industry, and civil society have a shared responsibility to achieve real sustainability.

(2) *SEA can deal with layered responsibilities.* Principle 2 on subsidiarity can be related to the so-called tiering in environmental assessment, where central government develops policies, plans, and programmes subject to SEA, while lower government and private sector proponents perform project-level environmental and social impact assessments. It is in the interest of the project proponents that a mechanism for planning and SEA is in place in order to clearly define accountabilities. Many impacts can be managed at the lowest level, that is, by a proponent itself, but other impacts that either result from or require higher-level involvement are beyond the management responsibility of a proponent.

(3) *Wider horizons in time and space/scale issues.* Principle 3 clearly states that impacts work beyond 'the gate' and may also surpass the lifetime of a project. It is the proponent's responsibility to deal with

such impacts. Environmental assessment (including environmental management plans) is the tool to address these issues, at the project level by a project proponent and at the strategic level by government authorities. Trade-offs between short-term benefits and long-term goals in decision-making processes should be taken into account (Principle 8). A proponent (often private sector) is primarily interested in the lifetime of a project; political decision making has to address long-term objectives that create the boundary conditions for project development. In this respect the private sector has a stake in proactively seeking policy development at government level in order to create clarity in responsibilities between proponent and government. Similarly in cases where the management responsibility of a proponent does not cater for the scale of an ecosystem, responsibilities may be combined with other authorities (Principle 7). The scale at which ecosystems work and may be influenced differ widely. Climate change works at global level, hydrological changes may work at river catchment level, but land clearance may only have local consequences, unless it interferes with pathways of, for example, migratory animals.

(4) *Economic values/stakeholders.* In its fourth principle the convention stressed the economic value of ecosystem services. Social and economic values of ecosystem services should be recognised and incorporated in impact assessment and resource management decisions. The translation of ecosystem services into social and economic values provides a strong tool to identify stakeholders that need to be involved in the decision-making process. Environmental assessment cannot be limited to looking at presence of protected areas or species. Areas with important ecosystem services, not necessarily protected, may also require special management measures (Principle 10). Broad stakeholder consultation is an important tool in identifying important biodiversity-related goods and services.

(5) *Ecosystem processes.* Principle 5 stresses the need to pay more attention to ecosystem structure and processes, in other words, those processes that create, structure, and maintain viable ecosystems. In relatively unknown or complex ecosystems (e.g. tropical rainforests) this provides an effective means to assess potential impacts of activities, without exactly knowing the species composition of such ecosystems, and without losing time and energy on endless species inventories.

(6) *Monitoring, adaptive management, and local knowledge.* Principle 6 stipulates that ecosystems should be managed within the limits of their functioning. This principle is similar to the concept of carrying

capacity. Our current understanding is often insufficient to allow these limits (or capacity) to be precisely defined. In such circumstances a precautionary approach coupled with adaptive management is advised (see Box 2.5 on the precautionary principle). In practice this means that in case of insufficient information, an activity should only be implemented with continuous monitoring; when unexpected impacts occur, management should be adapted. Starting at small scale is recommended. Depending on the rigour of the scoping procedure, impact assessment procedures cater to the precautionary approach; an environmental management plan would have to define the consequences of adaptive management. The notion that maintenance of ecological processes is more important than fixed outcomes (Principle 9) may, in some cases, bear important consequences for the formulation of environmental management plans. Local knowledge can provide relevant clues as local stakeholders may provide important insights in the effect of proposed interventions/decisions (Principle 11). Sharing of knowledge is fundamental for effective stakeholder participation. In some cases the sharing of classified information may pose difficulties, especially in early stages of project development. Nevertheless, it is stressed that active sharing of information and knowledge creates a better basis of trust, a sense of ownership, and overall support for an activity.

Box 2.5. The precautionary principle

In order to protect the environment the Precautionary Principle shall be widely applied by States according to their capabilities. Where there are threats of serious or irreversible damage, lack of full scientific certainty shall not be used as a reason for postponing cost-effective measures to prevent environmental degradation. (Rio Declaration, 1992, Principle 15)

Similar texts can be found in the preamble of the Convention on Biological Diversity (CBD, 1992) and Resolution Conf. 9.24 of the Convention on International Trade in Endangered Species of Wild Fauna and Flora (1994).

Simply stated, if we are not sure what is going to happen as a result of doing something, we should avoid taking any risks. A distinction can be made between *irreversible impacts* on biodiversity, which are impacts that cannot be reversed in time (e.g. the conversion of rainforest into degraded rangeland), and *irreplaceable losses* of biodiversity, being the total loss of biodiversity (e.g. the extinction of a species). An impact leading to irreplaceable loss of biodiversity is, by definition, irreversible (Brownlie *et al.*, 2006).

The Precautionary Principle Project (2005) has drafted guidelines on how to apply this often challenged principle. Apart from the well-known, more general principles on public participation, sharing of information, transparency in decision making, and use of local knowledge, the guidelines provide detailed guidance on the application of the Precautionary Principle. Some excerpts from the document follow.

Be explicit: When decisions are made in situations of uncertainty, it is important to be explicit about the uncertainty that is being responded to, and to be explicit that precautionary measures are being taken.

Be proportionate: A reasonable balance must be struck between the stringency of the precautionary measures, which may have associated costs, and the seriousness and irreversibility of the potential threat. It should be borne in mind that countries, communities or other constituencies may have the right to establish their own chosen level of protection for their own biodiversity and natural resources.

Be adaptive: An adaptive approach is particularly useful in the implementation of the Precautionary Principle as it does not necessarily require having a high level of certainty about the impact of management measures before taking action, but involves taking such measures in the face of uncertainty, as part of a rigorously planned and controlled trial, with careful monitoring and periodic review to provide feedback, and amendment of decisions in the light of new information. Applying the Precautionary Principle may sometimes require strict prohibition of activities.

This is particularly likely in situations where urgent measures are required to avert imminent threats, where the threatened damage is likely to be immediately irreversible (such as the spread of an invasive species), where particularly vulnerable species or ecosystems are concerned, and where other measures are likely to be ineffective. This situation is often the result of a failure to apply more moderate measures at an earlier stage.

Further reading: The Precautionary Principle Project (2005) and Peel (2005).

Identifying relevant biodiversity-related issues

CBD parties must identify activities that are likely to have significant adverse impacts on the conservation and sustainable use of biological diversity and monitor their effects (convention text Article 14). This requirement asks for biodiversity to be treated similar to the way that other environmental issues are treated in environmental assessment. Nevertheless, biodiversity is considered to be a difficult issue in environmental assessment and is treated in a haphazard, inconsistent manner in spite of the availability of very clear and unequivocal definitions provided by the CBD. In this chapter, a possible way is provided to single-out the relevant biodiversity-related issues, based on recent ecological insights, that need to be studied in an environmental assessment study. It aims at limiting a study as much as possible to the essentials, because available time and resources usually are limited in environmental assessment.

This book addresses biodiversity from a human perspective, rather than a species-based approach and goes far beyond conventional nature conservation. Understanding human-induced changes in biodiversity and its impact on humankind, requires understanding of the *goods and services* provided by biodiversity as important contributors to human well-being. The Millennium Ecosystem Assessment (MA) (2003) provided an elaborate conceptual framework using the common denominator *ecosystem services* for the goods and services provided by biodiversity. The MA defines ecosystem services as 'the benefits that people obtain from ecosystems'. Bröring and Wiegleb (2005: 531) stated that one problem with this approach is 'that ecosystem services need not necessarily relate to diversity measures in terms of species richness'. This reasoning, followed by many biologists, is based on the simple assumption that more species is better. This leads to phenomena such as the so-called biodiversity 'hot spots', designed to focus conservation efforts on places on Earth with high species diversity and endemicity under threat (Myers *et al.*, 2000). As the hot spots concept became popular and attracted the bulk of conservation funding, people started realising that many other places may also be important, because even with fewer species 'hot spots' provide crucial services for the maintenance of Earth system. In a shear attempt to refocus attention, biodiversity 'cold-spots' have been proposed by Kareiva and Marvier (2003). They argue that no one strategy is enough; conservationists need a way to make explicit trade-offs: 'Preserving 1,000 species in a "cold spot" like Montana would be more important than preserving 1,000 species in a hot spot like Ecuador because in Montana 1,000 species represents a third of the total, while in Ecuador it represents just 5 percent'.

Ehrlig (2003) provided similar arguments: 'Even if people succeeded in preserving a single viable population of every species on earth, the human race would die out unless it managed to protect the ecosystems that support broader populations of plants, animals and people too'. The approach in this book is based on practical experience showing that recognition and quantification of ecosystem services and the involvement of their stakeholders represent the best opportunity to translate biodiversity in language understood by decisions makers. Biodiversity hot spots in this respect provide a very important service by maintaining high species diversity in certain (limited) areas; other areas may provide important flood storage, water cleaning, or coastal protection services. All of these are necessary to keep this world an inhabitable place; this argument has to be made explicit in the decision making on hundreds of thousands of planning and investment decisions taken yearly around the globe and for which environmental assessments have to be carried out. A focus on maximal species diversity *per se* simply does not provide enough arguments, limits the scope of biodiversity to too few places, and often focuses mitigation efforts on species rescue efforts in stead of real conservation measures. A more scientific line of reasoning is provided by Thompson and Starzomski (2006) who state that there is currently not sufficient scientific evidence to conclude that higher species diversity always positively contributes to the better functioning and the stability of ecosystems (see Box 2.6). In other words, there is not enough scientific evidence to protect species for reason of maintaining functioning ecosystems. (For a critical evaluation of research efforts in soil ecology, see Bengtsson, 1998). The recognition of valued ecosystem services at least provides a focus on ecosystems and the need to maintain or restore processes essential for the creation or maintenance of ecosystems. There is no doubt that the maintenance of ecosystems contributes to the conservation of species, even although what may be adequate in terms of land management for sustaining specific ecosystem services may not match expectations in terms of conservation (Chan, et al., cited in Ghazoul, 2007). Pyke (2007) calculated that the historic trend in prioritising protected areas would lead to uneven distributions of important ecosystem services. This observation highlights the limitations of the present species diversity–based arguments to identify conservation priorities and suggests the need for more comprehensive approaches to planning and prioritizing future land protection. Principle 5 of the

Box 2.6. Species diversity, the insurance hypothesis, stability, and resilience

According to the insurance hypothesis, biodiversity insures ecosystems against declines in their functioning, because many species provide greater guarantees that some will maintain functioning even if others fail. In a theoretical modelling exercise by Yachi and Loreau (1999) species richness showed two major insurance effects: a buffering effect by reducing the temporal variance of productivity and a performance-enhancing effect by increasing productivity. Three factors are of main importance: (i) the way ecosystem productivity is determined by individual species responses to environmental fluctuations, (ii) the degree of asynchronicity of these responses, and (iii) the detailed form of these responses. In particular a greater variance of the species responses contributes to the insurance effects. There is growing evidence that only a few species play a key role in ecosystems, making other species seemingly redundant. However, there is growing evidence that a large pool of species is required to sustain the assembly and functioning of ecosystems in landscapes subject to increasingly intensive land use (Loreau et al., 2001). Species loss may therefore not be important in terms of the role of that species now, but rather in terms of that species being available to fulfil a functional role in a future, changed environment (Thompson and Starzomski, 2006).

Closely linked to this are the terms stability, resilience, and resistance. A stable ecosystem is characterised by the absence of change and a lack of disturbance. Resistance is the ability of a community to maintain structure in the face of disturbance. Resilience is the ability to bounce back after disturbance. There is growing evidence that alternative stable states exist in communities. A community may return to the same configuration after a small perturbation but may shift to a different configuration or equilibrium after a large perturbation (Beisner et al., 2003). The collapse of the Northwest Atlantic cod stock is a well-known example. Even after years of restricted fishing the population does not return; most probably the ecosystem has switched to another stable state and does not bounce back, even when the driver of change (fishing) has been removed.

Walker et al. (2004) provide a conceptual framework to redefine resilience for social-ecological systems. This framework goes beyond the boundaries of traditional ecology and considers *the stability dynamics of all linked systems to humans and nature.* Further elaboration would go beyond the scope of this book, but the approach definitely provides a framework to define research questions for the further implementation of the CBD ecosystem approach. (See, for example, Gunderson *et al.*, 2006). It changes the focus from 'seeking desirable states and maximum sustainable yield to resilience analysis, with a simultaneous focus on adaptive resource management and adaptive governance'.

ecosystem approach firmly states that the conservation and, where appropriate, restoration of ecosystem interactions and processes are of greater significance for the long-term maintenance of biological diversity than protection of species. With respect to species Diaz *et al.* (2006) stated that

in natural systems, if we are to preserve the services that ecosystems provide to humans, we should focus on preserving or restoring their biotic integrity in terms of species composition, relative abundance, functional organisation, and species number (whether inherently species-poor or species-rich), rather than on simply maximising the number of species present.

Five categories of ecosystem services are distinguished (see Box 2.7): (i) provisioning services (harvestable goods), (ii) regulating services (maintaining natural processes and dynamics), (iii) cultural services (source of artistic, aesthetic, spiritual, religious, recreational or scientific enrichment), (iv) carrying services (providing a substrate for human activities), and (v) supporting services necessary for the production of all other ecosystem services. Ecosystem services influence human well-being and thus represent a value for society. *Values* can be expressed, positively and negatively, in economic, social, and ecological terms. Understanding the factors that cause changes in ecosystems and ecosystem services is essential to the design of interventions that enhance positive and minimise negative impacts. Such factors are called *drivers of change* and can be natural or human-induced. Impact assessment is primarily concerned with human-induced drivers of change. Chapter 4 provides an extensive elaboration of the ecosystem services concept and introduces an impact assessment framework to conceptualise human-induced changes in the interdependency of humankind and biodiversity, through the recognition of ecosystem services. The rest of this chapter focuses on certain aspects of biodiversity which assist us in understanding the causal biological mechanisms

---

Box 2.7. Ecosystem services (see Chapter 4 for detailed elaboration)

Provisioning services are products obtained from ecosystems. A distinction is made between natural and joint production, i.e. products harvested from nature with minimal human effort, or produce obtained with human inputs, such as fertilisers, pesticides, or intense resource management. There is no clear-cut differentiation between natural and joint production as many degrees of human intervention occur.

Regulating services are benefits obtained from the regulation of ecosystem processes. Examples are chemical transformation, dilution, sequestration or processing of waste, the dampening of harmful influences from other components such as flood retention, coastal protection, or protection against UV by the ozone layer.

Cultural services are nonmaterial benefits people obtain from ecosystems through spiritual enrichment, cognitive development, reflection, recreation, and aesthetic experiences.

Carrying services are ecosystems that provide space, a substrate, or a backdrop for human activities. This represents a group of services that are not recognised independently by the MA, but yet are an important aspect of the ecosystem services concept. It is best illustrated by river navigation. Navigation needs water as a substrate; if water depth is not sufficient a ship will not be able to proceed. Yet, a ship does, in principle, neither influence the quantity nor quality of water.

Supporting services are those services that are necessary for the production of all other ecosystem services. They differ from all the other services in that their impacts on people are either indirect or occur over a very long time. For example, soil formation processes usually play on a time scale which humans cannot oversee; yet they are closely linked to the provision service of food production.

---

through which drivers of change have an effect on biodiversity and consequently influence ecosystem services provided by biodiversity.

Direct drivers of change are human interventions (activities) resulting in biophysical and social/economic effects with known potential impacts on biodiversity and associated ecosystem services. As also shown in the subglobal assessments of the Millennium Ecosystem Assessment (Capistrano *et al.*, 2005), the probability that such impacts on

Box 2.8. Biophysical effects known to act as potential drivers of change (Slootweg *et al.*, 2006)

- *Land conversion*: the existing habitat is completely removed and replaced by some other form of land use or cover. This is the most important cause of loss of biodiversity.
- *Fragmentation* by linear infrastructure: roads, railways, canals, dikes, powerlines, and so forth affects ecosystem structure by cutting habitats into smaller parts, leading to isolation of populations. A similar effect is created by isolation through surrounding land conversion. Fragmentation is a serious reason for concern in areas where natural habitats are already fragmented.
- *Extraction of living organisms* is usually selective since only few species are of value, and leads to changes in species composition of ecosystems, potentially upsetting the entire system. Forestry and fisheries are common examples.
- *Extraction of minerals, ores, and water* can significantly disturb the area where such extractions take place, often with significant downstream and/or cumulative effects.
- *Wastes (emissions, effluents, solid waste), or other chemical, thermal, radiation, or noise inputs*: human activities can result in liquid, solid, or gaseous wastes affecting air, water or land quality. Point sources (chimneys, drains, underground injections) as well as diffuse emission (agriculture, traffic) have a wide area of impact as the pollutants are carried away by wind, water, or percolation. The range of potential impacts on biodiversity is very broad.
- *Disturbance* of ecosystem composition, structure, or key processes: this will be treated in more detail in the following section.

biodiversity actually occur depends on local circumstances that should be part of an impact assessment study. Box 2.8 lists direct drivers of change.

Some social effects can also be considered to be direct drivers of change as they are known to lead to one of the above-mentioned biophysical changes. A nonexhaustive list of categories of social effects is provided in Box 2.9. An example is provided by UNEP (2006) where areas for the location of refugee camps in Liberia were subjected to a vulnerability assessment – vulnerability was based on the ecosystem services provided

Box 2.9. Social effects known to act as potential drivers of change (based on Schooten *et al.*, 2003, reproduced from Slootweg *et al.*, 2006)

- *Population changes* due to permanent (settlement/resettlement), temporary (temporary workers), seasonal in-migration (tourism), or opportunistic in-migration (job-seekers), usually lead to land occupancy (land conversion), pollution and disturbance, harvest of living organisms, and introduction of nonnative species (especially in relatively undisturbed areas).

- *Conversion or diversification of economic activities*: especially in economic sectors related to land and water, diversification will lead to intensified land use and water use, including the use of pesticides and fertilisers, increased extraction of water, introduction of new crop varieties (and the consequent loss of traditional varieties). Change from subsistence farming to cash crops is an example. Changes to traditional rights or access to biodiversity goods and services fall within this category. Uncertainty or inconsistencies regarding ownership and tenure facilitate unsustainable land use and conversion.

- *Conversion or diversification of land use*: for example, the enhancement of extensive cattle raising includes conversion of natural grassland to managed pastures, application of fertilisers, genetic change of livestock, and increased grazing density. Change to the status, use, or management of protected areas is another example.

- *Enhanced transport infrastructure and services*, and/or enhanced (rural) accessibility; opening up of rural areas will create an influx of people into formerly inaccessible areas.

- *Marginalisation and exclusion* of (groups of) rural people: landless rural poor are forced to put marginal lands into economic use for short term benefit. Such areas may include erosion sensitive soils, where the protective service provided by natural vegetation is destroyed by unsustainable farming practices. Deforestation and land degradation are a result of such practices, created by nonequitable sharing of benefits derived from natural resources.

by the receiving environment, and the resettlement of displaced people was the direct driver of change.

Indirect drivers of change are societal changes which under certain conditions may influence direct drivers of change, ultimately leading to impacts on biodiversity and associated ecosystem services. These drivers will be further elaborated in Chapter 7 on strategic environmental assessment.

Aspects of biodiversity provide focus to describe impacts

When we study the impacts of human interventions on biodiversity, there is a great need to focus studies on the right research questions. Because biodiversity is a complex phenomenon, studies can easily get out of hand and fail to provide the necessary answers. Therefore an expert judgement approach is provided which should guide the impact assessor into the right direction. It is important to realise that in environmental assessment it may not be needed to provide detailed quantitative information on the nature and magnitude of the impacts of human intervention on biodiversity. Often the very fact that important ecosystem services might be affected already leads to the search for alternative solutions to avoid these expected impacts. The probability of such impacts occurring can be assessed by the following sequence of steps: first, define the spatial and temporal range of influence of drivers of change expected to result from a proposed activity; second, identify the ecosystems within this range; third, an ecologist defines for each ecosystem under influence of a driver change, what level of diversity is affected (genetic, species, ecosystem) and whether it affects one of the following aspects of biodiversity:

- **Composition:** what is there and how abundant is it. This is the most commonly known and used aspect of biodiversity. Composition can be described in terms of (i) genetic variability within a species or between populations of a species, (ii) diversity of species expressed in a variety of different diversity indices, and (iii) diversity in types of ecosystems.
- **Structure (or pattern):** how are biological units organised in time and space. Structure is multidimensional and relates to spatial (horizontal and vertical) and temporal patterns in the organisation of biological diversity. Although the structure aspect applies to all three levels of biodiversity, in the practice of environmental assessment structure usually relates to ecosystem level of diversity and in cases where critical species are studied at the population and habitat level. Many concepts from ecology fall under this heading. Some of these will be discussed in

some detail: connectivity, scale, pattern, foodweb, functional groups, and keystone and foundation species.

- **Key processes:** what physical, biological, or human processes are of key importance for the creation and/or maintenance of biodiversity. A key process can be any physical, biological, or human factor of critical importance for the creation and/or maintenance of biodiversity. For each ecosystem one or a limited number of key processes can be defined. When a driver of change affects a key process there is serious reason to believe that such a driver of change will lead to changes in biodiversity and consequently may alter the ecosystem services provided by this biodiversity.

Noss (1990) described three components of biodiversity: composition, structure, and ecosystem function. LeMaitre and Gelderblom (1998) used these components of biodiversity to evaluate the performance of environmental assessments. They concluded that composition was about the only component being studied in assessments. Their review suggested

that many people perceive biodiversity as the diversity of species and ecosystems without appreciating the significance of functional components. For example, transformation of a natural area is commonly perceived to be simply loss of that patch unless rare species or habitat (composition) are located in that patch; the fact that the loss may increase fragmentation or decrease connectivity in natural systems, or disrupt ecosystem services significantly is overlooked.

They suggested to design scoping for biodiversity impacts based on these components of diversity, to be carried out by experts, with peer review to verify the correctness of scoping. The Netherlands Commission for Environmental Assessment (2001) adopted this approach to draft the first CBD guidelines on biodiversity in impact assessment (CBD, 2002), by providing an '*issues table for scoping on biodiversity*' (see Table 2.1).

The table differs from the original publications by Noss (1990) and LeMaitre and Gelderblom (1998) by following the CBD distinction in three levels of diversity, leaving out the landscape level of diversity for reasons explained earlier. The word 'component' is being replaced by 'aspect', because 'component' often leads to confusion with products derived from biodiversity, or sometimes even with the levels of diversity. Furthermore, developments in ecology made it necessary to rename the 'function' aspect the 'key-process' as explained on the next page.

Table 2.1 *Issues table for scoping on biodiversity.*

	Levels of Diversity		
Aspects of diversity	Genetic	Species	Ecosystem
Composition	Minimal viable population (inbreeding/ genetic erosion)	Rarity/abundance Conservation status (threat) Endemic/exotic	Ecosystem diversity
Structure/pattern	Dispersal of natural genetic variability (e.g. spatial division between subspecies)	Alpha, beta, and gamma diversity Minimal area needed Foodweb Functional groups Distribution and abundance Stepping stones for migratory animals Regular (seasonal, lunar, tidal, diurnal) and irregular rhythms	Patchy or continuous (gradients) Connectivity (horizontal) Layers/ stratification (vertical) Linkages with other ecosystems (support services) Regular (seasonal, lunar, tidal, diurnal) and irregular rhythms
Key process	Exchange of genetic material between LMOs and wild ancestors or local cultivars	Keystone species Foundation species	Key processes (different for each ecosystem)

Composition: what there is and how abundant it is (see Box 2.10)

Even although it is in many cases relatively easy to establish a relationship between human activities and their potential impact on biodiversity composition, a major problem remains in the determination of the ultimate effect that this may have on the functioning of an ecosystem as a whole (Thompson and Starzomski 2006, Loreau *et al.*, 2001, Cardinale *et al.*, 2006) and in the ecosystem services provided by these ecosystems. Few empirical studies demonstrate improved functioning of ecosystems at high levels of species richness. Theoretical models predict that species richness beyond the first few species does not typically increase ecosystem stability. The reason for this is that most communities are characterised by strong dominance, such that a few species provide the vast majority of the community biomass. However, due to the rapid turnover of species higher diversity may lead to maximum stability, but this has hardly been investigated (Swartz *et al.* 2000). Nevertheless, when dealing with protected or highly valued species, habitats, or landscapes, the effects of human drivers of change on composition provides a valid entry point to focus impact studies.

Box 2.10. Examples of how human activities affect biodiversity by changing the composition

- *Genetic diversity*: introduction of genetically modified organisms creates the possibility that these organisms interbreed with wild ancestors, thus introducing modified genes into the wild. On 29 January 2000, the Conference of the Parties to the Convention on Biological Diversity adopted a supplementary agreement to the Convention known as the Cartagena Protocol on Biosafety. The Protocol seeks to protect biological diversity from the potential risks posed by living modified organisms resulting from modern biotechnology. It establishes an advance informed agreement procedure for ensuring that countries are provided with the information necessary to make informed decisions before agreeing to the import of such organisms into their territory. The Protocol contains reference to a precautionary approach. The Protocol also establishes a Biosafety Clearing-House to facilitate the exchange of information on living modified organisms and to assist countries in the implementation of the Protocol (CBD, 2000c).

- *Species diversity*: One of the most important impacts on species diversity is caused by the (often deliberate) introduction of alien (nonindigenous) species turning into invasive species, outcompeting indigenous species. The list of examples is without end; some famous examples are the rabbit in Australia, rats on Polynesian islands, South American water hyacinth invading waterways around the world, or the rapid spread of disease organisms by human activities causing, for example, outbreaks of West Nile virus or avian flu. An example of local impact on species diversity is provided by logging, leading to a selective removal of a few species of trees, thus influencing the species composition of the forest. If the ecosystem is a tropical rainforest, characterised by high species diversity, the changed species composition may not have a significant impact on the functioning of the forest ecosystem (unless the exploited species perform key functions), but it may lead to a total disappearance of the exploited species since the remaining specimens may suffer from reproductive isolation. Furthermore, species depending on this tree species will also disappear. If the ecosystem is a species-poor temperate forest, removal of one species alters the composition in such a way that the entire ecosystem is threatened.

- *Ecosystem diversity*: wetlands are among the most rapidly disappearing ecosystems, as these flat and often fertile soils are converted for agriculture (floodplains) or aquaculture (mangroves). Ecosystem services provided by these wetlands for the maintenance of the surrounding area are lost, for example, fish breeding on floodplains to stock the entire river system, or coastal protection by mangroves to protect hinterland from storm surges (or tsunami!). The changed ecosystem diversity has an important impact on the remaining ecosystems.

Structure: how biological units are organised in time and space

Structure is a multifaceted aspect having a spatial and temporal dimension. *Spatial structure* refers to distribution patterns of populations of species, or ecosystems (e.g. patchy or continuous). Apart from this *horizontal* pattern, a very visible expression of structure is the vertical stratification of multilayered forests. *Vertical* structure is related to strong vertical differentiation of physical parameters, such as penetration of light, local temperatures (thermocline), or oxygen (stratification). In primary tropical rainforests up to five layers are distinguished (soil, weeds, shrubs, trees, emergent trees). Each layer has its own communities of plants and animals. Disturbance of this structure by human interventions will lead to a change in physical parameters, which will alter the local habitat conditions. For example, the removal of trees increases the penetration of light to lower layers of a forest, usually leading to a shift in species composition. *Temporal structure* is reflected by regular or irregular patterns in events occurring over time (see Box 2.11). Most species and ecosystems are adapted to cyclic phenomena, such as seasonality (summer–winter or dry–wet season in relation to breeding, flowering, migration, hibernation, etc.), tidal rhythm (mangroves, mudflats), diurnal rhythm (daytime and nocturnal animals/flowers), or lunar cycles (*Chaoborus* mosquitoes appearing at full moon). Some species or ecosystems may even be

---

Box 2.11. Avoiding impacts on biodiversity linked to temporal structure

Proposed dredging activities in the Wadden Sea, an internationally important tidal wetland area in Northwestern Europe, coincided with the reproductive season of oysters and mussels. Both bivalve species are of economic importance for fisheries and of ecological importance because they represent an important source of food for shorebirds. The area is the major stopover for millions of migratory birds along the East Atlantic flyway. In autumn birds fatten here before flying the long stretch to their wintering areas in West Africa. The turbidity caused by the dredging would cause massive death of young bivalves. Rescheduling of the dredging activities to a later season was enough to avoid great ecological and economic damage (Kolhoff, personal communication).

---

dependent on irregular phenomena, such as prolonged drought, erratic floods, or fire (Netherlands Commission for Environmental Assessment, 2001). The ecological concept of *succession* falls under this heading. Ecosystems are often characterised by a gradual, and to a certain extent predictable, change in species composition over time, driven by internal (autogenic succession) or external factors (allogenic succession). Internal processes relate to biological processes that gradually change local conditions. Examples of external processes are sea level rise or volcanic activity. *Primary succession* occurs in areas where no living organisms were present before; *secondary succession* occurs when the original ecosystem has been destroyed or seriously degraded. Succession leads to a stable climax situation, although major disturbances, such as fire, floods, or hurricanes, often put an ecosystem back to an earlier stage of succession.

A structure-related aspect of special importance is *connectivity* (see Box 2.12). With increasing fragmentation of ecosystems and habitats due to human activities, the movement of organisms and exchange of genetic material is increasingly hampered, threatening the survival of many species. Connectivity refers to the arrangement of patches on the landscape and the ability of organisms to use those patches (Lindenmayer, 1994). If a species of wildlife cannot travel between patches, then those patches are considered disconnected. It is difficult to exactly define specifications for connectivity. Connectivity requirements for species may

---

Box 2.12. Restoring connectivity: the National Ecological Network of the Netherlands (Kolhoff and Slootweg, 2005).

The approval of the EU bird directive in 1979 was the occasion that marked the actual start of national planning of biodiversity conservation areas, as 50,000 hectares of important bird areas had to be demarcated. In 1990, the first edition of the National Nature Policy Plan launched the concept of establishing a National Ecological Network (NEN). The aim of the NEN, being a network of protected areas, is to secure the maintenance of biodiversity in the Netherlands. Measures include (i) the enlargement of existing biodiversity conservation areas by converting 280,000 hectares of agricultural lands, (ii) restoring environmental quality and (iii) creating coherence and connectivity through a system of corridors between biodiversity conservation areas. This network has to be realised over a period of 30 years by using the local-level spatial plans. In 2000, the European Union launched a plan for establishing a similar network of conservation areas in the EU, known as Natura 2000.[1]

---

vary from the availability of stepping stones along the flyway of migratory waders or the absence of barriers in the migration route of salmons, spanning large sections of the globe, to connectivity of patches inhabited by soil invertebrates with a range of movement of less than 100 metres. Connectivity in relation to plant dispersal is even more complex and relates to seed dispersal strategies through the action of gravity, wind, water, and/or animals. Consequently, it is difficult to extrapolate from individual species connectivity requirements to general rules. However, it is known that *fragmentation*, the opposite of *connectivity*, is a major cause of loss of biodiversity. In the planning of linear infrastructure or hydro-engineering works a common measure to mitigate fragmentation is to restore connectivity by means of constructed wildlife bypasses ('ecoducts') over or under major roads, fish passages around weirs, and so forth (see Box 2.12).

Scale is a complicating issue in relation to all biodiversity-related studies, in the spatial as well as the temporal sense (see Box 2.13). 'The scale at which an assessment is undertaken significantly influences the problem definition and assessment results, as well as the solutions and

---

[1] http://ec.europa.eu/environment/nature/index_en.htm.

Box 2.13. Species diversity and scale

An interesting phenomenon observed in studies on alpha diversity is the relationship between human presence and species richness. It has been observed that species richness is higher with increased human presence. Explanations range from the diversifying effect of human activities on the environment, the choice of humankind to settle in biodiversity-rich areas, or the increased introduction of exotic species in the neighbourhood of humans. More strikingly the correlation coefficient between human presence and species richness is positively related to study grain and extent (larger areas with larger observation plots show a stronger relation between species diversity and human presence). The correlation becomes negative below a study grain of 1 km and an extent of 10,000 km^2 (on small areas with small observation plots human presence is negatively related to species diversity) (Pautasso, 2007).

Gabriel et al. (2006) studied farming system and species diversity of associated weeds in farmland at three scale levels by comparing different plots in a field, different fields in a region, and different regions. They showed that total observed species richness was mainly explained by beta diversity between fields (37%), to a lesser degree between regions (25%), and less so between plots (16%). In normal language, different agricultural fields in one region show the greatest dissimilarity in associated weeds, thus contributing most to overall species richness.

These two examples show that species diversity in itself is not as simple a measure for biodiversity as often suggested in environmental assessment studies. The first study shows that depending on the grain size or extent of a study, the relation between human presence and species richness can be positive or negative. The second study shows that having fields with many species is not necessarily providing the most diversity; on the contrary having dissimilar fields contributes more to overall species diversity. Selecting the proper method of inventory consequently needs careful consideration.

responses selected' (Capistrano et al., 2005). It is, for example, of no use to study the biodiversity effects of climate change when doing an impact assessment for one thermal powerplant; this is typically an issue when studying the impact of a national energy policy which sets the boundary conditions for individual powerplants to be constructed under this policy.

Global, regional, and local factors influence biodiversity. The impacts of climate change on species distribution may be studied at global or regional level; at local level however, microclimatic conditions probably are of greater influence on the local dispersal of species. In time, similar scale issues can be recognised; ecosystem changes that may have little impact on ecosystem services and human well-being over days or weeks (soil erosion for instance) may have pronounced impacts over years or decades. Assessments need to be conducted at spatial or temporal scale appropriate to the process or phenomenon being examined. Similarly, ecological processes have certain spatial requirements which can work at entirely different scales. De Villiers *et al.* (2005), for example, recognise scales at which ecological processes may work from five hectares for some specialist pollinator relationships to plant–herbivore processes for megaherbivores at a scale up to one million hectares. Scale issues are defined by grain size (finest level at which observation are made) and extent (total surface area and time horizon of studies). Nature has fine grain and large extent. Assessments done over large areas or over a prolonged period of time usually have coarser resolutions, easily overlooking processes at a finer resolution.

Patterns show that composition and structure are very intimately linked; some important aspects, specified below, can hardly be assigned to one or the other category. Watt (1947) in a groundbreaking publication proposed a distinction between process and pattern. Recently, some authors have suggested combining composition and structure again under the heading 'pattern' (LeMaitre, personal communication; Brownlie, 2005a, Brownlie *et al.*, 2005).

Foodweb structure and interactions shape the flow of energy and the distribution of biomass. Changes in the foodweb have immediate repercussions for the functioning of the entire system. In this context Diaz *et al.* (2006) used the term *vertical diversity* to describe the diversity in trophic relations in the food pyramid (trophic levels). (The term 'vertical' is not used here in its literal meaning of a recognisable pattern of layers). Foodweb has direct linkages with *functional groups*. A functional group is a group of species linked to the same ecosystem process, often in relation to food web. They basically have the same 'job', such as decomposing, or nitrogen fixation, and so forth. For example, the introduction of the predatory exotic Nile perch in Lake Victoria has upset the entire ecosystem by eradicating a functional group of specialised cichlid fish species that feed on algae, leading to a turbid and locally deoxygenised lake (Ligtvoet and Witte, 1991; Goldschmidt, 1998). The definition of functional groups and the measurements of functional diversity in an ecosystem are seen by some as the most useful and efficient ways to link biodiversity to the functioning of ecosystems (Bengtsson, 1998; Tilman *et al.*, 1997). Functional groups can also be based on nontrophic relations. Common examples are ecosystem engineers, such as earthworms, termites, or ants (Folgarait, 1998). Hooper *et al.* (2005) in a review of current knowledge concluded that

> ecological experiments, observations and theoretical developments show that ecosystem properties depend greatly on biodiversity in terms of the functional characteristics of organisms present in the ecosystem and the distribution and abundance of those organisms over space and time. Species effects act in concert with the effects of climate, resource availability and disturbance regimes in influencing ecosystem properties.

Some species singularly represent a given functional group. These species are referred to as *keystone species*. Keystone species have a large impact on communities or ecosystems (Bengtsson, 1998).[2] The usual addition is that the species has a disproportionate effect on its environment relative to its abundance, although there is no agreement on the definition of abundance. A limited change in numbers of individuals has disproportional effects on the entire system. Classical examples are the giant sea otter in kelp fields, elephants on African savannahs, starfish in intertidal zones, and beaver in some freshwater habitats. Conceptually closely related to keystone species are *foundation species*, which are dominant primary producers in an ecosystem both in terms of abundance and influence. According to Bengtsonn's definition, both keystone and foundation species are lumped because of a lack of clear definitions. It will be clear that any human action expected to affect such species will trigger the 'biodiversity alarm' during an environmental assessment.

Key processes: processes of key importance for the creation and/or maintenance of ecosystems

In the publications by Noss (1990) and LeMaitre and Gelderblom (1998) reference was made to 'ecosystem function'. However, the concept of ecosystem function has become rather diffuse and definitions are

---

[2] A keystone is the topmost stone in a roman arch holding both sides of the arch in place but receiving the least pressure: the funny thing about a keystone is that it is probably the only stone one can take from an arch and still have a chance the arch will remain (relatively) intact. One can dispute the choice of the word to describe an opposite phenomenon: the collapse of an entire ecosystem when removing the keystone species.

indecisive or even contradictory on whether external and nonbiological processes are part of it or not. The relationship between biodiversity and ecosystem functioning has emerged as a central issue in ecological and environmental sciences during the last decade (Loreau *et al.*, 2001) creating considerable academic dispute and controversy. Noss (1990) defined function as 'ecological and evolutionary processes, including gene flow, disturbance, and nutrient cycling'. This definition leaves room for biological as well as physical processes; physical processes are largely overlooked in the literature on ecosystem function, although Risser (1994) stressed the importance of abiotic interactions as key structuring processes. The 'Ecosystem Valuation' website[3] describes functions as 'the biophysical processes that take place within an ecosystem. These can be characterized apart from any human context (e.g. fish and waterfowl habitat, cycling carbon, trapping nutrients)'.

Fishbase[4] describes ecosystem function as 'an intrinsic ecosystem characteristic related to the set of conditions and processes whereby an ecosystem maintains its integrity (ex: primary productivity, food chain, biogeochemical cycles, etc.)'.

The 'Biodiversity and Ecosystem Function Online' website[5] provides a dual definition: '(i) the collective intraspecific and interspecific interactions of the biota, such as primary and secondary production and mutualistic relationships"; and (ii) "the interactions between organisms and the physical environment, such as nutrient cycling, soil development, water budgeting, and flammability'.

Usher (personal communication) defines two overarching functions: production (starting a food chain with fixation of solar energy) and decomposition (breakdown of all material). All ecosystem functions can be deducted to these two functions. Taking a quick look at experimental research into the linkage between species diversity and ecosystem function there is a narrow focus on the effects of (manipulated) species diversity on aggregate biomass (standing stock) and resource depletion (see, for example, a meta-analysis of 111 experiments by Cardinale *et al.*, 2006) which is not of much use when assessing human influence on biodiversity. Being indecisive on the inclusion of physical processes makes the concept difficult to use; moreover, the essential role of humans

---

[3] www.ecosystemvaluation.org/Indicators/economvalind.htm#introdef (last accessed June 2007).
[4] www.fishbase.org/Glossary/Glossary.cfm?TermEnglish=ecosystem%20function (last accessed June 2007).
[5] www.abdn.ac.uk/ecosystem/bioecofunc/intro.htm (last accessed June 2007).

in the maintenance of certain ecosystems is totally ignored. For example, in Europe the creation and maintenance of alpine meadows, heather land, or nutrient-poor and species-rich grasslands is fully dependent on grazing by livestock, a practice introduced by humans several centuries ago.

In order to avoid a lengthy conceptual discussion on the use of the term ecosystem function, the term *key process* is proposed as a tool to assess how drivers of change may affect biodiversity. In this respect Risser (1994) is followed. Risser calls for

a strategy to simplify the myriad potential relationships between biodiversity and ecosystem function and to identify the most important of these.

Silver *et al.* (1994) provide a similar vision and have suggested the use of the term key function. A key process can be any physical, biological, or human factor of critical importance for the creation and/or maintenance of biodiversity. When a driver of change affects a key process there is serious reason to believe that such driver of change will lead to changes in biodiversity, and consequently may alter the ecosystem services provided by this biodiversity. It is argued that, based on expert or local knowledge, for each ecosystem one or a limited number of critically important key processes can be defined. (See, for example, the outstandingly elaborated ecosystem guidelines for environmental assessment by De Villiers *et al.* (2005), describing 'key ecological drivers' for maintenance of various ecosystems in the Cape Floristic Region). Knowledge of these key processes provides us with the knowledge to predict in a relatively easy manner, potential negative consequences of human activities on biodiversity, also in the case of incomplete knowledge. Box 2.14 provides some elaborate examples; Table 2.2 provides a tentative list of key processes.

The challenge: linking impacts on biodiversity to ecosystem services

The *issues table* for scoping (Table 2.1) disappeared from the 2006 CBD guidelines on biodiversity in impact assessment (CBD, 2006) as it was considered conceptually too complex for a guidelines document. It does however provide a tool to determine which issues need to be studied in an environmental assessment. The table is not intended to expand the workload, but rather to provide a selection mechanism to determine which issues are most relevant to study. For example, if an activity leads

Box 2.14. Examples of key processes

Abiotic/physical process

The damming of a river results in reduced discharge of sediments in the river's estuary. The sediment balance in the estuary is upset, causing massive erosion of the mangrove ecosystem, in its turn reducing the numbers of fish and shellfish that breed in the mangroves, and thus decreasing the numbers of wader birds that prey upon these organisms, and so forth. The physical effect of reduced sediment discharge in the estuary affected the key process of maintaining a delicate balance in sediment deposition and removal in an estuarine mangrove ecosystem.

Biotic process

A man-made wetland in the Netherlands has, unintendedly, become a Ramsar site of international importance, due to the presence of tens of thousands of wintering geese that have stopped the succession of wet reedlands into dry shrubland. By intensive grazing, the shallow open water did not get a chance to grow over, and peat formation largely stopped. The intended conversion of the area into a business park has been cancelled and it has become a national reserve. (The intended biophysical effect, creation of new land, was effectively stopped by geese, thus creating a new ecosystem due to the introduction of a key process.) (See picture at back cover)

Human process

Agriculture has for centuries been one of the key processes in European mountain areas creating new ecosystems, first valued for its productive services (predominantly lifestock) but more recently highly valued for its rich plant and animal diversity. Declining agriculture in these remote areas (economic drivers) results in the loss of this diversity as 'natural' succession turns these areas back into relatively species-poor forested areas. The need for a sustainability assessment was triggered by this perceived loss of biodiversity. (Sheate, 2003; Sheate et al., 2008).

Table 2.2 *Examples of key processes in the formation and/or maintenance of ecosystems (adapted from Koning and Slootweg, 1999).*

Key ecological processes	Relevant for ecosystems
Soil–surface stability and soil processes	Lowland dryland rainforest, montane tropical forest, coniferous montane forest, coastal dunes
Soil erosion patterns due to wind	Coastal dunes, degraded land
Soil erosion patterns due to water	Desert, coastal dunes, degraded land
Erosion patterns of upland area and riverbed	Upper, middle, and lower course of rivers and streams
Erosion patterns of soil and vegetation due to wave action	Rocky coastlines and beaches, freshwater lakes, mangroves, and sea grass beds
Sedimentation patterns	Middle and lower course of rivers, floodplains, estuary, tidal flats, mangrove
Replenishment of sand due to up drift sources	Beaches, tidal flats, mangroves
topography and elevation due to wind erosion	Desert
Local climate (temperatures) determining plant available moisture	Desert, rocky coastline
Seasonal drought/desiccation patterns determining plant available moisture	Deciduous forest, nonforested mountains, savannah, steppe, desert
Seasonal hydrological situation (evaporation, water quantity, water quality, and current/velocity)	Beaches, rivers, and streams, freshwater, saline, or alkaline lakes, reservoirs
Tidal influence (tidal rhythms, tidal range, and tidal prism)	All coastlines, estuary, lagoon, tidal flat, mangrove, sea grass beds,
Permanent waterlogged condition of the soil	Peat swamp
Salinity levels and/or brackish water gradient	Lowland river, saline lakes, estuary, mangrove, sea grass beds, coral reef
Water depth, availability of sunlight, and/or thermocline stability	Freshwater lake and reservoirs, coral reef, coastal sea
Regional groundwater flow and groundwater table (source or sink function of landscape)	Freshwater marsh or swamp, saline, or alkaline lakes
Flooding patterns (frequency, duration)	Tropical flooded forest, floodplain, freshwater swamp or marsh, mangrove
Hydrological processes (vertical convection, currents and drifts, transverse circulation)	Coral reef, coastal sea, open (deep) sea
Biological processes in the root system	All dryland forests

Table 2.2 (*cont.*)

Key ecological processes	Relevant for ecosystems
Protection of soil humus layer by vegetation cover	Lowland tropical rainforest
Canopy density determining light intensity and humidity	Lowland tropical rainforest, deciduous forest
Plant-dependent animal reproduction	Lowland tropical rainforest
Animal-dependent plant reproduction	Lowland tropical rainforest
Grazing patterns by herbivorous mammals	Savannah, steppe (grasslands), tropical flooded forest, floodplain, freshwater swamps or marsh
Grazing patterns by herbivorous birds	Freshwater lake, floodplain, tidal flat
Grazing patterns by herbivorous fish	Freshwater lake, floodplain
Grazing patterns by herbivorous marine mammals	Sea grass beds
Seed dispersal due to water	Mangrove
Seed dispersal by animals (birds, primates)	Lowland tropical rainforest, tropical flooded forest, freshwater swamp or marsh
Pollination due to environmental factors (e.g. wind)	Deciduous forest, mangrove
Pollination by animals (insects, birds, mammals)	Lowland tropical rainforest, montane tropical forest, deciduous forest, mangrove
Production of pelagic and benthic organisms	Saline or alkaline lake or marsh, estuary
Primary production by phytoplankton	Saline or alkaline lake or marsh, coastal sea, open sea
Nutrient inflow due to environmental factors (i.e. water run-off, drainage)	Upper and middle course of rivers, freshwater lake, tropical flooded forest, tidal flat, sea grass bed
Nutrient input by animals	
Nutrient cycling due to water movement/rainfall	Nonforested mountains, lagoon
Nutrient cycling due to fire	Savannah, steppe
Nutrient cycling by juvenile fish	Tidal flat, mangrove
Nutrient cycling by arthropods/insects	Lowland tropical rainforest, savannah, steppe
Nutrient cycling by invertebrates (earthworms, bivalves, starfish, crabs, shrimp)	Montane tropical forest, deciduous forest, coniferous montane forest, rocky coastline, lagoons, tidal flat, mangrove, coastal sea, open sea
Nutrient cycling by fungi and bacteria	Deciduous forest, savannah, steppe
Nutrient cycling by filter feeders	Coral reef

Table 2.2 (*cont.*)

Key ecological processes	Relevant for ecosystems
Gallery forest structure providing shade and nutrient input	Upper course of river
Disruption of vegetation structure due to fire	Lowland tropical rainforest, montane tropical forest, deciduous forest, savannah, steppe, tropical flooded forest, floodplain
Disruption of vegetation structure due to storms/hurricanes/cyclones	Lowland tropical rainforest, deciduous forest, coniferous montane forest, (coconut) beaches, mangrove
Disruption of vegetation structure due to Wave action	(Coconut) beaches, mangrove
Disruption of vegetation structure due to land slides/mud flows	Montane tropical forest, coniferous montane forest, nonforested mountains
Disruption of vegetation structure by animals (herbivores)	Savannah, range land, sylvi-pastoral associations
Peat building by decaying vegetation (accumulation rates vs. decomposition rates)	Peat swamp
Dynamics of sedimentation, accretion, and grazing of the coral skeleton	Coral reefs
Predation of coral polyps by starfish and fish (parrotfish, butterflyfish), and smothering of coral polyps	Coral reefs

to the selective removal of some tree species from a forest the impact will work through the composition aspect. Research effort should go into the distribution of these species in the forest to establish its conservation status and the level of exploitation it can sustain. In some cases structural aspect may have to be taken into account, or the species may act as a foundation or keystone species. Other aspects of diversity do not have to be studied in detail (such as a complete species inventory of the forest). In another case the seasonal flooding of an area may be altered by human intervention; the effect of this driver of change works through temporal structure; flooding will, for some ecosystems, also be a key process. Again, expert judgement can provide a rapid means to focus studies on the relevant issues. Impacts can be identified and described to a level of detail often sufficient for environmental assessment without having a complete description of the biodiversity. If an intervention is expected to result in changes of composition, structure, or key processes, there is a serious reason for concern and further studies can be focussed on this expected change. Especially for areas where available data on biodiversity is limited, this approach had the advantage of focussing costly data collection efforts on the relevant aspect of biodiversity (and thus avoiding lengthy descriptive studies of all biodiversity aspects in the intervention area). It will be clear that the issues Table 2.1 is a first attempt to produce the ultimate table. The ecological literature is presently expanding at a rapid pace, providing relevant information for most biomes in the world. One of the main challenges is to describe the processes of key importance for the maintenance of ecosystems. This will provide a strong tool to reduce uncertainty on the probability of significant impact occurring as a result of human interventions.

One of the biggest challenges in ecology nowadays is to find the precise linkages between biodiversity and ecosystem services. The Millennium Ecosystem Assessment already stressed the fact 'that the impacts of biodiversity changes on ecosystem services are still poorly understood. Even where knowledge is better, there are almost no studies documenting the trends over time' (Hassan *et al.*, 2005). (For similar arguments, see Carpenter *et al.*, 2007.) From the perspective of environmental assessment it is a positive fact that on lower spatial scales often much more is known on the relation between biodiversity and the provision of ecosystem services (Capistrano *et al.*, 2005). In the subglobal assessments of the Millennium Ecosystem Assessment, identification of indicators for provisioning services did not give much problems as production statistics on, for example, food, timber, and water are readily available; the same applies to water quality data. Supporting services were more complex and required, for example, net primary production data and complex calculations of nutrient transport and balances. Indicators for regulating services were expressed as the conditions ecosystems are required to have in order to provide such services. For cultural services many new techniques were required to describe the condition of these services. So, even without perfect knowledge the subglobal assessments did provide relevant information. For many environmental assessment studies it may not even be necessary to have absolute quantitative data. Environmental assessment often is about avoiding negative impacts and/or enhancing positive impacts. By comparing the impacts of various alternatives even a relative estimate of impacts – that is, one alternative has more serious consequences compared to the other – already provides relevant information for decision making. Expert judgement with semiquantitative information – that is, an impact is small or large – can be used to provide reliable and relevant information.

By affecting the magnitude, pace and temporal continuity by which energy and materials are circulated through ecosystems, genetic, species, and ecosystem diversity influences the provision of ecosystem services. According to Diaz *et al.*, 2006),

the most dramatic changes in ecosystem services are likely to come from altered functional compositions of communities and from the loss, within the same trophic level, of locally abundant species rather than from the loss of already rare species.

The authors provide one of the first attempts to link main aspects of biodiversity to ecosystem services. Because of the profound impact of the Millennium Ecosystem Assessment, this area of research is rapidly developing; probably by the moment of printing of this book research will have produced significant new outputs. This is an encouraging development, because it will provide the environmental assessment community with much better information to predict impacts on biodiversity, translate these impacts in terms of changing ecosystem services, and provide decision makers with relevant information on the (un)sustainability of proposed plans and projects.

3 · Biodiversity conservation and development: challenges for impact assessment

Asha Rajvanshi and Vinod B. Mathur

Introduction

The literature linking conservation and development presents a number of different perspectives on the relationship between biodiversity conservation and socioeconomic progress. These differences in perspectives are rooted in the differences in objectives of 'development' (which is seen as the intended modification of the biosphere and the application of human, financial, living, and nonliving resources to satisfy human needs and improve the quality of life (IUCN, UNEP, and WWF, 1980)) and of 'conservation' (which is generally taken to mean the management of human use of the biosphere through approaches of land use and renewable natural resource management with the objective to yield the greatest sustainable benefits to present and future generations).

Most of the conservation-oriented literature also actually reflects that the local community welfare and development are directly conflicting with the objectives and practice of biodiversity conservation and regard development as the main causal agent of biodiversity loss. According to Sanderson (2002), development and conservation are altogether separate goals as clearly illustrated in his expression:

If development has ignored conservation, conservation has paid too little attention to development. Economic policymakers have concentrated on growth, developers on the distribution of the benefits of growth, and conservationists on the costs and consequences of growth for nature and the environment. The result has been an agreement to disagree, with the growth, development, and conservation communities proceeding down separate paths.

According to Sanderson and Redford (2003), the achievement of economic development goals for half of the number of poor people 'will either mark the true beginning of sustainability or the end of biodiversity'. Swanson (2006) also mirrors the earlier held views of apparent incompatibility between biodiversity and development in his expression that 'states with high material wealth have low biodiversity wealth and vice versa'.

Development perspectives, on the other hand, identify conservation as a threat to human welfare and condemn the 'fortress conservation' approach that alienates the local people from biodiversity-rich areas and deny them the rights to resources that supplement sustenance and livelihoods. It is on this argument that the development community is urging the conservation organisations to take poverty issues onboard while formulating conservation strategies and policies.

Biodiversity and development: conflicting goals or complementary objectives?

The journey through time has portrayed many different caricatures of conservation and development. The outcome has been the mixed opinions resulting in conservation versus development debates at times (Box 3.1) and the appreciation of complementarity between conservation and development at other times.

As early as the 1940s, Adams (2004) held the view that economics and human well-being and biodiversity conservation are linked. The linkages were largely seen in conservation providing the revenue-generating opportunities (e.g. trade and tourism) that could contribute to local economic development in poor countries. Later, in 1980, the World Conservation Strategy first articulated the link between conservation and development and stressed the need to promote development that provides real improvements in the quality of human life and, at the same time, conserves the vitality and diversity of the Earth. The World Conservation Strategy conveyed the important message that the sustainable development depended on the conservation and sustainable use of living organisms and ecosystems. The most often-quoted definition of sustainable development is development that 'meets the needs of the present without compromising the ability of future generations to meet their own needs'. WCED (1987) emphasises the need to adopt patterns of resource use that aim to ensure their use not only in the present but in the indefinite future.

The 1992 Earth Summit put forward the 'triple bottom line' concept of sustainable development that encompassed ecological, social, and economic sustainability. The categorical use of the term 'poverty' stressing human deprivation in conservation literature, however, is a more

---

Box 3.1. Examples of controversial projects

The Silent Valley Hydroelectric Project in India, proposed way back in the 1980s, was opposed by the ecology movement on the grounds of it being a threat to the rare and threatened flora and fauna of the Western Ghats ecosystem due to the impoundment of the only remnant of primal rain forest in India. The consequent decision was to withdraw the proposal for dam construction (Shiva, 1991; Rangachari et al., 2002).

At Lake Cowal in New South Wales (Australia), approval for North's deep open-cut gold mine intruding into a section of the lake came after a second Commission of inquiry was conducted to evaluate a revised proposal with lesser cyanide levels; power lines relocated away from bird flight paths; and other modifications. The initial development application was rejected as unacceptable during the 1996 Brisbane Ramsar Convention meeting (Reeves, 1999).

Damming of the Nu River in China sparked a series of activist confrontations with the Chinese authorities to protect China's astonishingly rich biological resources threatened by the dam project proposed in the heart of the Hengduan Mountain Range. This mountain range is a part of the eastern wing of the Himalaya, an area now known to the world as the Three Parallel Rivers (Sanjiang bingliu) World Heritage site (Litzinger, 2008)

Local and international opposition mounted protest against the Pascua Lama project, straddling the border between Chile and Argentina. This project that seeks to start extracting gold, silver, and copper in the year 2009 is likely to move three Andean glaciers that presently sustain the mountain and valley ecosystems and agriculture production in the fertile valley bottom (Anonymous, 2008a).

---

recent development (World Bank, 1999) and is linked to the priority that development agencies have given to poverty reduction as a prerequisite for ensuring sustainable development.

Although it is evident from the literature that the opinions have been divided on whether biodiversity conservation would be a failure if it does not take into account the human well-being aspects or would have better prospects of success if it does (Brockington, 2003), there is growing evidence of a major paradigm shift in conservation thinking during the past two decades. Blaikie and Jeanrenaud (1997) believed that contemporary conservation policy and practice is undergoing rapid transformation and opined that policies which earlier saw people as a threat have subsequently started recognising people as potential partners in sustainable development strategies. Hulme and Murphree (1999) coined the term 'new conservation' to distinguish the traditional conservation approaches from those that mainstream participation of communities in conservation.

The view put forth by Sachs and Reid (2006) that a disaggregated approach to conservation and development neither serves the interests of the conservation community nor the developers has perhaps provided the strongest impetus for re-examining conservation and development as complementary goals.

Some newer models of conservation through development are emerging to foster the synergistic relationship that can help bridge the long-standing, conservation-development divide (Anonymous, 2002a). These newer approaches to conservation stress complementarities and trade-offs rather than conflicts between conservation and development (Salafsky and Wollenberg, 2000). From the strengths of such a model, many conservation agencies are encouraging business practices that can reduce industry's ecological footprint, contribute to conservation, and create value for the companies that adopt them (Anonymous, 2002a).

IUCN's Business and Biodiversity Initiative (http://biodiversityeconomics.org/business), launched in September 2001, firmly rests its activities on the CBD's three objectives – conservation, sustainable use, and equitable benefit sharing. The Energy and Biodiversity Initiative (www.theebi.org), operating under the slogan 'making capitalism work for conservation', aims at facilitating the emergence of three interconnected instruments: corporate biodiversity plans, business plans for nature, and biodiversity business investments. It is a consortium of four major energy companies and five leading conservation organisations that aims to develop tools and guidelines for integrating biodiversity into oil and gas development

The International Institute for Environment and Development (IIED) likewise is involved in a number of corporate sector projects to promote partnerships for natural resource management with local communities, developing instruments for sustainable forestry, creating markets for environmental services for poverty reduction, and encouraging sustainable agriculture and rural livelihoods. The Mining, Minerals and Sustainable Development project (MMSD) was one of the most impressive of the IIED projects. One of MMSD's most important elements was a partnerships established in four of the world's principal mineral-producing and mineral-consuming regions (Australia, North

America, South America, and Southern Africa) to assess the global mining and minerals sector in terms of the transition to sustainable development (www.iied.org/mmsd/what_is_mmsd.html). The project was an independent two-year process of consultation and research with the objective of understanding how to maximise the contribution of the mining and minerals sector to sustainable development at the global, national, regional, and local levels.

The International Finance Corporation (IFC) of The World Bank has been involved, for instance, in establishing the first biodiversity business investment fund in Latin America. The objective of such a fund is to promote commercially viable and environmentally sustainable enterprises, such as sustainable agriculture, forestry, and ecotourism, in Latin American countries that have ratified the Convention on Biological Diversity.

Developers are also finding new mechanisms by which they can realise the economic values created by good stewardship of biodiversity. Examples of biodiversity-related business initiatives drawn from business cases (Box 3.2) from around the world (Anonymous, 1999; Bishop *et al.*, 2008) amply demonstrate how the integration of biodiversity priorities into business planning can provide multiple benefits including managing risks, capitalising on opportunities, and meeting corporate social responsibilities.

The intent of the chapter is thus to discourage the discord that pose biodiversity and development as a question of 'either/or,' but to identify synergies that can help strike a better balance between the two. It aims to explore the linkages between conservation and economic development, more specifically the linkages between the key indicators of economic prosperity and environmental health including, in particular, the role that biodiversity plays in meeting the global development goals, alleviating poverty, improving livelihoods, and addressing climate change. The chapter also tries to explore when, how, and why some types of human development are compatible with biodiversity conservation and others are not. Finally, it presents new challenges for impact assessment practitioner for integrating biodiversity conservation in development decisions.

Conserving biodiversity for sustainable development: priorities and challenges

It is abundantly clear from the contents of Chapter 2 that the term 'biodiversity' represents both a popular notion and a widely used scientific terminology for the natural biological wealth that influences

---

Box 3.2. Examples of biodiversity-related business initiatives

Due to excessive deforestation, the Atlantic Forest of Brazil has been reduced to less than 10 percent of its original size. The Guaraqueçaba Climate Action Project has sought to regenerate and restore natural forest and pastureland. Companies such as American Electric Power Corporation, General Motors, and Chevron-Texaco have invested US\$18.4 million to buy carbon emission offset credits from the approximately 8.4 million metric tons of carbon dioxide that the project is expected to sequester during its lifespan. The project has initiated sustainable development activities both within and outside the project boundary, including ecotourism, organic agriculture, medicinal plant production, and a community craft network. The project has made significant contributions towards enhancing biodiversity in the area, creating economic opportunities for local people (such as jobs), restoring the local watershed, and substantially mitigating climate change (Anonymous, 1999)

The IPIECA/OGP Biodiversity Working Group is an industry-led joint initiative established in 2002 to develop technical guidance and promote good practice of biodiversity management in the oil and gas industry (see www.ipieca.org). The working group also provides a forum for members to exchange information and discuss how the industry can improve its biodiversity performance.

---

environmental security, human life, and well-being. The importance of global commitment towards the conservation of biodiversity is therefore no longer questioned. Instead, there is ample support for its conservation, sustainable use, and equitable sharing of the produce and benefits of bioresources and ecosystem functions (CBD). Because of the universal recognition of biodiversity as a fundamental component for achieving sustainable development and poverty eradication, its conservation is seen as an overriding global priority (Pisupati and Warner, 2003). As a counter to this well meaning objective, there is, however, a threat from many different factors, such as globalisation, poverty, land diversion for fast pace economic development, extraction of living resources (Fahrig, 1997; Forman, 2000; Wilcove et al., 1998; Miller et al., 1992), and human migration to pristine areas that is eroding biological diversity.

This unabated rate of loss of plant and animal species and the predicted impacts of this loss of germplasm on humankind is well-documented (Hoyt, 1988; Wilson and Peter, 1988; Bower, 1989; Brockelman, 1989; Bunting, 1990; Abelson, 1991; Beattie, 1991; Ehrlich and Wilson, 1991; Loesch, 1991; Solbrig, 1991; Pimm and Raven 2000; Novacek and Cleland 2001; Brooks *et al.*, 2002; Singh, 2002;)

Reversing the biodiversity loss induced by poor planning of developmental projects has become a daunting challenge because the benefits of conserving biodiversity and ecosystem tend to be considered as long-term, indirect, and diffuse, while the benefits of development actions that destroy or degrade biodiversity tend to be immediate, direct, and easily captured by individuals (Kiss, 2004). The most common examples that represent this short-term approach to quick material gains is the iron ore extraction from areas especially designated for protection of rare and threatened biodiversity of the tropical evergreen ecosystem in the Western Ghats, a biodiversity hotspot area (CES, 2001; Krishnaswamy and Mehta, 2003) and the construction of giant multilateral schemes in the Mekong and Narmada river basins in South East and South Asia for harnessing of hydropower at the expense of long-term implications of massive ecological dislocations.

Only in situations where economic benefits from extractive use of biological products are high, there may be sometimes incentives to contribute to biodiversity conservation. For example, in Hawaii, artifacts made of *Acacia koa* wood sell for thousands of dollars, leading some landowners to allow land previously converted to sugarcane or pasture to return to natural *A. koa* woodland to tap the lucrative market (Kiss, 2004).

As pressure is mounting to ensure the compatibility between economic development and conservation of world biodiversity, Environmental Impact Assessment (EIA) is being widely acknowledged as a powerful decision support tool and a potent mechanism for implementing principles of sustainability and 'wise use'. Principle 17 of the Rio Declaration on Environment and Development endorses the universal application of EIA 'as a national instrument' (McNeely, 1994) and assigns a clear role for EIA in the implementation of National Sustainable Development Strategies (NSDSs) (Sadler, 1993). Similarly, there is increasing recognition of the value and importance of Strategic Environmental Assessment (SEA) for mainstreaming and upstreaming environmental sustainability in higher level decision making on policies, plans, and programmes (OECD, 2006a).

Challenges for impact assessment

The current knowledge of how the drivers of biodiversity change are interfering with essential ecological functions and threatening the different dimensions of human society (including security and well-being) is rapidly expanding (Bridgewater and Arico, 2002). The Millennium Ecosystem Assessment report based on the global project commissioned by the United Nations in 2000 provides the most comprehensive and scientifically defensible assessment of the causes and consequences of ecosystem change for human well-being. The conclusions of the Millennium Ecosystem Assessment (2005) provide a stark warning about the direct drivers of ecosystem degradation (see Millennium Ecosystem Assessment 2005a). The predictions are that the progressive negative trends in ecosystem degradation in the next fifty years are likely to have disastrous effects on human behaviour and the ability of ecosystems to provide essential services to people around the world. Chapter 2 summarises the existing understanding on the various biophysical and societal drivers of biodiversity and ecosystem changes.

In practice, however, this knowledge of drivers of biodiversity loss is still somewhat disparate and disconnected and can hardly deliver good EIA outcomes for decision making. There is clearly an emergent need to develop a new framework for impact assessment that can capture conservation–development links and promote economically profitable, socially acceptable, scientifically sound, politically feasible, and environmentally sustainable development.

The Millennium Ecosystem Assessment (2005a) explicitly clarifies that the key drivers of biodiversity loss, being large-scale exploitation of resources, habitat loss and/or alteration, and introduction of invasive alien species, are all linked to economic development. It is therefore vitally important to establish the relationship between biodiversity and development for each proposed policy, plan, programme, or project, and to also include these as part of the information flow for EIA. It also stands to reason that use of valuation approaches taking into account ecological, social, and economic values should be an inseparable part of impact assessment. More sophisticated approaches are needed at this point of time for undertaking impact assessment that can harmonise conservation priorities with those of achieving social development and economic security. Such integrated assessments should be able to give the sense of what the real priorities must be – conservation or development; development at what price (by forgoing what conservation values?), and

---

Box 3.3. Biodiversity is a development issue

Kenya's Tana River Delta is inhabited by 350 species of birds, lions, elephants, rare sharks, and reptiles. Kenya's National Environment Management Authority, NEMA, has approved a proposal by the Mumias Sugar Company, a publicly traded company based in Nairobi, to convert 2,000 km^2 of the pristine delta into irrigated sugarcane plantations for the production of ethanol. The EIA was hurriedly produced and ignored the assessment of the delta's ecological benefits and the irreversible loss of ecosystem services, such as flood prevention, the storage of greenhouse gases, and the provision of medicines and food. The conversion of the delta will bring about loss of biodiversity, affect the cultural value and the revenue generated from tourism, fishing, farming, and other lost livelihoods worth US$59 million. An independent report commissioned in May 2008 by Nature Kenya and the Royal Society for the Protection of Birds found deficiencies in the calculation of ecosystem benefits presented by Mumias and has called for a fresh EIA (Anonymous, 2008b).

---

how much cost for protection of ecosystems (ecological cost of protecting ecosystem services and economic costs for impact avoidance and clean up) should be enough to justify economic benefit. The challenge for EIA practitioners is to develop an impact assessment framework that takes into account the interrelationships between social, economic, and ecological environments.

Presently, the fundamental shortfall of impact assessment in delivering win–win outcomes for both conservation community and development planners is the failure to recognise that biodiversity conservation is also a key development issue. Experience suggests that the lack of appreciation of the linkages between biodiversity conservation and human security finds its basis in the failure to properly include ecological, social, and economic features as part of the preproject baseline (or autonomous development) description. The consequence of such a pitfall has been a neglect of ecological and well-being issues within the subsequent steps of impact assessment. (Box 3.3). The outcome is an assessment that constrains the very objective of impact assessment in finding common causes and common solutions to the problems associated with proposed development.

EIA that can organise, interlink, and synthesise information for a meaningful output therefore requires a more exhaustive understanding of the linkages between biodiversity and human and well-being.

Biodiversity, human security, and the Millennium Development Goals (MDGs)

Human security covers a wide range of issues including basic elements, such as food to eat, homes to live, good health, education, freedom from violence, safety during natural and human-caused disasters, democracy, good governance, and respect for human rights (see www.unescap. org/esid/GAD/Issues/Humansecurity/index.asp). As a great deal of human security is tied to peoples' access to natural resources and vulnerabilities to environmental change, the relationship between biodiversity and human well-being is close and complex. The understanding of the interlinkages between the two is critical to develop policy prescriptions and appropriate impact assessment tools to mainstream all the essential elements of sustainability of the development process: biodiversity conservation, livelihood security, poverty eradication, and environmental stability.

In September 2000, at the United Nations Millennium Summit, 189 world leaders agreed to a set of time-bound and measurable goals for combating poverty, hunger, disease, illiteracy, environmental degradation, and discrimination against women (United Nations, 2007). These Millennium Development Goals (MDGs) focus the efforts of the world community on achieving significant improvements in people's lives by the year 2015. It is for this reason that some policy advocates even recognise the MDGs more as Millennium Security Goals, because they concentrate primarily on achieving sustainable security rather than sustainable development (Khargam et al., 2003) and consider priorities such as 'reducing biodiversity loss' as environmental security and not as sustainable development goals.

In September 2005, a statement from the Secretariats of the five biodiversity conventions argued that biodiversity underpins all MDGs and could 'be the basis for ensuring freedom and equity for all'. Links between biodiversity and MDGs have also been explored by Department of International Development (DFID), European Commission (EC), United Nations Development Programme (UNDP), the World Bank (DFID, EC, UNDP, and World Bank, 2002) and Roe (2004). It will be worthwhile to attempt a quick review of the direct and indirect linkages

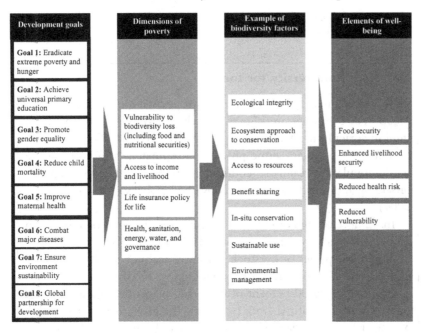

Development goals	Dimensions of poverty	Example of biodiversity factors	Elements of well-being
Goal 1: Eradicate extreme poverty and hunger			
Goal 2: Achieve universal primary education		Ecological integrity	
Goal 3: Promote gender equality	Vulnerability to biodiversity loss (including food and nutritional securities)	Ecosystem approach to conservation	Food security
Goal 4: Reduce child mortality	Access to income and livelihood	Access to resources	Enhanced livelihood security
Goal 5: Improve maternal health	Life insurance policy for life	Benefit sharing	Reduced health risk
Goal 6: Combat major diseases	Health, sanitation, energy, water, and governance	In-situ conservation	Reduced vulnerability
Goal 7: Ensure environment sustainability		Sustainable use	
Goal 8: Global partnership for development		Environmental management	

Figure 3.1 Contributions of biodiversity in the achievement of Millennium Development Goals.

between biodiversity and each of the Millennium Development Goals. Figure 3.1 illustrates how biodiversity can make a critical contribution to the achievement of the MDGs.

MDG-1: Eradicating extreme poverty and hunger

The links between poverty alleviation and natural ecosystems is well understood (Wunder, 2001; Adams *et al.*, 2004; Roe and Elliot 2004; Timmer and Juma, 2005). Natural ecosystems widely serve as 'safety nets' for the rural poor and provide essential resources to combat malnutrition, hunger and poverty and offers key options for sustainable livelihoods in most developing countries. For poor people with limited capacity to buy food, access to a diverse range of locally produced food including nutrient supplements is vital for maintaining a balanced diet. Undomesticated biodiversity not only underpins many aspects of domestic food production systems, through maintaining soil structure and fertility, nitrogen fixation, pollination, and natural pest control but also provides material factors essential for well-being, such as shelter, clothing, food, medicine, and livelihood (Box 3.4).

Box 3.4. Biodiversity for food

Of the approximately 270,000 known species of higher plants, 10,000–15,000 edible species are known, of which around 7,000 have been used in agriculture (Ash and Jenkins, 2007; Zedan, 2007)

Thirty crop species alone provide an estimated 90 percent of the world population's calorific requirements, with wheat, rice, and maize providing about half the calories consumed globally (Ash and Jenkins, 2007).

In 62 developing countries, wild meat and fish provide more than 20 percent of all protein (Bennett and Robinson 2000).

About 14 species of livestock currently account for 90 percent of global livestock production (Ash and Jenkins, 2007).

Approximately 90 percent of flowering plants overall, and at least one-third of agricultural crops (including three-quarters of the world's principal crops) are dependent on insects and other animals for their pollination (Ash and Jenkins, 2007).

One of the factors responsible for hunger and poverty today is the unprecedented loss of biodiversity associated with ecological deterioration as a consequence of development projects. The result is the downward spiral caused by increasing deprivation of human communities in a scenario of development-induced biodiversity losses. The causal relationship between poverty and biodiversity actually runs in both directions (Figure 3.2). Poverty can force people to deplete natural resources in their surrounds, thus destroying the resource base upon which their livelihoods and incomes rely. Conversely, persistent natural resource degradation can contribute to poverty due to deprivation of resources and benefits particularly to subsistence-based community.

An estimated 350 million poor people presently rely on forests as safety nets or for supplemental income in developing countries (Scherr, 2003). Simply letting these 'natural safety nets' disappear could push many people to even greater poverty and undermine many of the development agencies' broader agendas of providing economic prosperity and securing human well-being. Restoring ecosystem services and biodiversity will be essential in many regions to meet the international imperatives

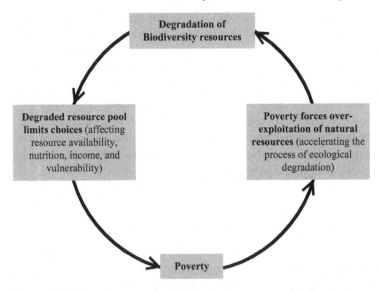

Figure 3.2 Effect of resource degradation and poverty on biodiversity.

of the new millennium – an imperative stressed by the Millennium Development Goals (MDGs), the 2002 World Summit on Sustainable Development, and the more recent 2005 UN World Summit. Biodiversity-driven impact assessment could possibly guide the development process to avoid and reduce impacts on ecosystem services.

MDG-2: Achieve universal primary education

Increased dependence on biodiversity for sustenance is likely to even engage children in the collection of water, fuel wood, and other resources and, in the process, deny them the opportunity of education in most developing countries. This is particularly worrying when there is sufficient evidence of the investments in education, especially for girls leading to better healthcare and nutrition, declining birth rates, poverty reduction, and improved economic performance at both the family level and beyond. Natural disasters linked to biodiversity loss and impairment of ecosystem services are also known to have limited children's available time and access to education opportunities. This has severely hampered the prospects that under normal circumstances could possibly improve the human resources capital, encourage alternative income options, and thereby reduce dependence on biodiversity for subsistence living.

Added to these risks are the risks of disruption of formal educational activities resulting from development-induced displacements and the concomitant reduction of access to basic ecosystem benefits and services (Mathur and Marsden, 1998; Cernea and McDowell, 2000). Empirical studies show that a number of displaced children never return to school and instead engage in unsustainable harvesting of resources to contribute to subsistence needs of the family. Failure to mitigate or avoid these risks and associated consequences may degrade natural resource base over the long term and generate what Cernea (1996) refers to as 'new poverty', as opposed to the 'old poverty'.

MDG-3: Promote gender equality and empower women

This goal is essentially to ensure that women have equal contributions in accessing resources, availing public and ecological services, and in influencing decision making. When biodiversity and ecosystem services are degraded or destroyed and secured access and rights to resources is denied, both the travel time and physical burden increases for women for collection of water, fuel wood, fodder, and other resources (Box 3.5).

Box 3.5. Estimates of travel time for women for fetching water and other resources

An average household spends 1 hour and 40 minutes each day collecting water in Kenya, Uganda and Tanzania (IIED, 2002).

Girls and women in Africa spend 40 billion hours a year hauling water to their homes from sources up to 15 km away (Maurits la Riviere, 1989; UNICEF, 1997).

In the Gujarat State of India, women spend up to six hours per day hauling water from distant sources to their homes (UNDP, no date).

Similarly, average time spent on collecting fuel wood and fodder by women in the hill areas in India is about four to six hours per day, respectively (Saksena et al., 1995).

Studies in Burkina Faso, Uganda, and Zambia indicate that women and girls could save hundreds of hours a year for education and generating alternative incomes if walking to sources of fuel wood and water were reduced to 30 minutes or less (Pisupati and Warner, 2003).

This provides the opportunity of generating surplus to eventually escape poverty but limits the opportunities for education and literacy for taking up alternative income-generating activities.

Many development-induced impacts are also not gender-neutral (Box 3.6) as women and men have well-demarcated gender roles in indigenous communities.

Box 3.6. Examples of gendered impacts of development projects

The Social and gendered consequences of involuntary resettlement of the Orang Asli displaced by dam projects in Malaysia, namely the Temenggor Dam in Upper Perak and the Sungai Selangor Dam in Kuala Kubu Bahru, led to a greater loss of access and rights to land for women than for men (Lin, 2006).

The gender segregated impacts of displacement and rehabilitation induced by coal mining in the Jharkhand region of India are well studied. Iron ore deposits are spread over large tracts of the rich and diverse forest of the Jharkhand region. Coal India Limited has been extracting iron ore from these forests through mechanised open-cast mining. In recent years, the ever-expanding and encroaching coal mines have been responsible for the displacement of a significant number of indigenous populations in India and their impacts have been disastrous, particularly on women. In a survey for assessing impacts of mining on local communities, women respondents listed as many as 30 items that they had previously collected from the forest and which provided them both food security and additional monetary benefits for the family before mining of coal was initiated. These items included leaves, flowers, and roots used for food and medicine, wild fruits, bamboo, stick to make brooms, wood for house building, and fuel wood. Most important was the Mahua tree (*Madhuca indica*) whose flowers were used for food and whose seeds were used for extracting cooking oil. The surplus resources collected from the forests or their products were sold for generating income. These resource rich forests were lost to open-cast mining. As a consequence, women were relocated to places away from the forest, which restricted their access to most resources that fetched money but also limited their choices of alternatives for food and livelihoods. (Ahmad and Lahri-Dutt, 2006).

The gender division of labour makes women still more dependent on productive systems and resources from the environment for collection of the essentials for subsistence: food, fodder, fuel, and water (Agarwal, 1992). It is therefore important that if gender equality was listed as MDG-3, a more pragmatic approach of impact assessment should ensure mainstreaming impacts of development projects on gender along with those on biodiversity to help empowerment of women. This empowerment would not only help women to secure rights and access to biodiversity resources but would also help them contribute a fair share of the benefits from their knowledge of biodiversity to improving biodiversity dependent livelihoods. If the opportunities of education and skill development become an inherent part of development process, women's empowerment may also help them reduce their dependence on biodiversity for a living.

MDG-4: Reduce child mortality

Under nutrition, unhealthy environment and agents of disease (malaria, dengue fever, and other insect-borne and water-borne diseases) are the underlying causes of child mortality that have links with degraded ecosystems. A wide range of wild foods and the resulting dietary diversity contributes to nutritional well-being of children. It has been well-established that development-induced displacements pose the biggest risk of impoverishment and nutritional deficiencies among the most vulnerable groups including children (Cernea, 1996).

MDG-5: Improve maternal health

The loss or impoverishment of biodiversity has a direct impact on maternal and infant health. Functioning ecosystems serve as a wealthy source of plant and animal foods, which can supplement and improve diets that are dependent largely on one or two staple crops. These sources of food become critical when productivity declines because of drought, floods, or land degradation. Healthy ecosystems filter and clean water and provide access to fuel wood and fodder, reducing the amount of time women and girls must spend in fetching these essential supplies. As a consequence of development-induced displacement, women have been found to be more vulnerable to impoverishment due to lack of access to common property resources (Cernea, 1996).

Box 3.7. Importance of biodiversity in medicines

Scientists have identified more than 2,000 tropical plants as having anticancer properties (GEF, 2005).

According to WHO's Traditional Medicine Strategy (2002–2005), about 80 percent of the African population, 70 percent in India and around 40 percent in China, rely on traditional health care systems that use traditional medicines that are derived from plants and animals (WHO, 2002).

Some 50,000 of the world's plant species are used in traditional medicine. Of these, 7,500 species are being used medicinally in India (WWF, 2004).

Many countries, such as Thailand, Sri Lanka, Mexico, China, and India, have integrated traditional medicine into their national health care systems. About 85 percent of traditional medicine involves the use of plant extracts from about 50,000 species, including almost 20 percent of the Chinese flora, around 7,000 species in India, and some 10 percent of Indonesia's flora. Estimates of the number of marine species used for medicinal purposes ranges from a few hundred to a few thousand, the use of which is mainly confined to Asia (Ash and Jenkins, 2007).

An estimated 4,160 to 10,000 medicinal plants are endangered by habitat loss or overexploitation (Hamilton, 2004).

MDG-6: Combat HIV/AIDS, malaria, and other diseases

Biodiversity plays a crucial role not only in providing medicines to deal with issues of health and nutrition (Box 3.7) but is also a constituent of healthy ecosystems that play a significant role in dealing with diseases, such as malaria and others (Chivian, 2002). Some diseases are known to flare up in ecological systems which have their regulation component altered or impaired by irrigation projects, dams, construction sites, standing water, and poorly drained areas. For example, it is estimated that the deforestation and consequent immigration of people into the Brazilian interior increased malaria prevalence in the region by 500 percent (Smith, 2002).

Biodiversity buffers humans from organisms and agents that cause disease. By diluting the pool of virus targets and hosts, biodiversity reduces their impact on humans, and provides a form of global health insurance. The intrusion into the world's areas of high biodiversity for meeting the development agenda can disturbs these biological reservoirs and exposes people to new forms of more virulent disease organisms, including SARS, Ebola, malaria, and the HIV pandemic.

Some plants have been found to significantly boost the immune system without the side effects of expensive antiviral drugs. The widespread reliance of the poor on natural sources for building immune systems and drawing medicines is met largely through the use of locally harvested plant extracts. Brazil has the world's largest reserve of biodiversity, a very promising source of material for the discovery of new medicines. The market has been increasing at a rate of 20 percent per year, with good possibilities for natural supplements for vitamins and herbs (see www.moiti.org/pdf/Brazil%20Health%20Care%20System.pdf).

MDG-7: Ensure environmental sustainability

The real challenge for development is to ensure that the benefits from goods and services provided by biodiversity are optimised for human development. Ecosystem depletion and species extinction can reduce the capacity to respond to future stresses, such as climate change and degradation of resources, such as water and soil. Averting environmental problems and managing risks associated with unplanned development will have to be a priority agenda for planning developments that ensure ecological sustainability. MDG 7 in particular, encourages participating countries and organisations to ensure environmental sustainability through good environmental management practices for the long-term success of development and for the overall achievement of the MDGs (United Nations 2007). Furthermore, MDG 7 requires that principles of sustainable development are made an integral part of policies and programmes, and that the environmental factors are considered while making decisions. Finally, MDG-7 indicates that the most promising way for this to occur is through the use of environmental assessment (OECD 2006b).

The role of environmental impact assessment tools become more significant and the mechanisms for clean development become relevant for determining the development that is most acceptable. Also, SEA directly supports the requirement of MDG-7 to 'integrate the principles of sustainable development into country policies and programmes and reverse loss of environmental resources'.

The MDG-7 poses sound business arguments for companies as well to be proactively engaged in action oriented towards efficient production processes and greening of the financial market dynamics with the aim of improving environmental stewardship (Nelson and Prescott, 2003).

MDG-8: Develop a global partnership for development

Conversion of land rich in biodiversity for industries and large scale infrastructure projects can erode the resource base for food, medicines, and livelihoods, pollute the environment, increase health risks and affect livelihoods of communities challenged by poverty. Destroying habitats, which support wildlife, therefore undermines the capacity of governments to generate income from tourism and support projects which could eradicate poverty, improve maternal health, and reduce child mortality. Maintaining biodiversity and the integrity of critical ecosystems will require partnerships among all stakeholders. Mainstreaming biodiversity conservation in business is already being encouraged for bridging the conservation–development divide and for striking global partnerships for development (Bishop *et al.*, 2008; Gutman and Davidson, 2007).

The role of impact assessment in meeting the MDGs
From a policy perspective, the impact assessment process must be strengthened for mainstreaming human well-being factors into development planning. From the action-oriented perspective, this calls for evolving an impact assessment framework for mainstreaming various human security factors in development choices. Such an impact assessment framework should be able to focus on the identification of threats and barriers to achieving the Millennium Development Goals. The concerns can be systematically captured by linking the project objectives to the human well-being objectives as represented by the MDGs, and their direct linkages to biodiversity and ecosystems. Table 3.1 presents a generic guidance for aiding in the development of project specific assessment approaches to incorporate human security objectives in development planning.

Linking biodiversity and climate change

It will not be out of place to also review the relationship between development interventions and climate change impacts, because the links between climate issues and sustainable development are manifold

Table 3.1 *Integration of human security goals in impact assessment.*

MDGs	Key considerations in impact assessment	Impacts on biodiversity and ecosystem services
Goal 1: Eradicate poverty and hunger	• How many poor households are likely to benefit from the project, and how many of these are headed by men/women? • What income-producing opportunities are linked to biodiversity resources collected by men/women? • Will there be a change in income of the men/women displaced from the project site? • What corporate social measures are to be incorporated for augmenting income?	• Increased dependence on biodiversity due to lack of alternative livelihood options. • Reduced dependence on biodiversity due to financial security through better employment prospects under the project
Goal 2: Achieve universal primary education	• Will the project influence opportunities of education by supporting increasing enrollment and attendance of boys/girls in primary school?	• Project-induced increase in pressures on biodiversity by idle children.
Goal 3: Promote gender equality	• What might be the effect of on off-farm employment on the income of women? • Will there be an overall increase in women's income linked to the sale of biodiversity resources and products? • How will the project influence the time spent by women in household activities? • What will be the effect of the project on the total daily work load of women? • Will the project encourage the decision-making power of women in household, community, or government? • Will there be opportunities for literacy and skills training for men/women?	• Increased access to common property resources and risk of unsustainable harvesting of critically scarce resources for food, medicines, shelter, and other subsistence needs. • Lack of purchasing power for food and nutrition supplements may induce new pressures on bioresources.
Goal 4: Reduce child mortality	• Will there be a change in the dietary habits of children? • What might be the effect on health conditions?	• Literacy, awareness, and empowerment may reduce dependence on renewable resources and encourage mothers to use healthy food supplements for children

Goal	Questions	Considerations
Goal 5: Improve maternal health	• Will the project reduce physical burden and exposure to indoor air pollution? • Will the project-induced displacement lead to changes in the distance traveled by women for collecting forest food, fuel, wood, water, and other nonwood products? • Will the project ensure the provision of health clinic facilities and services for the affected population? • Will there be hardship that exclusively affects women's health and reduces her time for collection of forest-based resources and child care?	• Project-induced increase in pressures on biodiversity for meeting subsistence needs from pristine areas in adversity and increased burdens. • Provision of houses in relocation site may reduce pressure on forest for wood and thatch earlier needed for construction of shelters.
Goal 6: Combat major diseases	• Will the living conditions improve after the displacement under the project? • Is there likely to be any new risk of disease in the relocation site? • Are there provisions for health monitoring facilities and services for treatment of disease?	• Reduced ecological footprint of the project due to benefits from biodiversity.
Goal 7: Ensure environment sustainability	• What will be the effect on the use of wood? • How will the transition to modern fuels effect carbon dioxide emissions? • Will the project ensure access to clean water and sanitation? • Is there a well conceived plan for the reclamation of eroded forest and agricultural land? • Do the project objectives ensure promoting water harvesting and de-silting of water sources under its environmental management strategies?	• Protection of forest land • Improved benefits from healthy ecosystems if project ensures sustainable use of resources. • Stressed ecosystems due to added burden of demand of resource (e.g. water and wood) induced by the project. • Long-term resource securities through appropriate strategies for resource management and conservation.
Goal 8: Promote global partnership	• Is there enough evidence of transparent participation in decision making?	• State of well-being may motivate people to support conservation. • Alienation and deprivation may lead to opposing conservation actions.

(Robinson and Herbert, 2001; Reid *et al.*, 2004). A stable climate is a global public good. Rapid climate change puts stress on the ability of ecosystems and in turn on human society to adapt to these changes. Climate change is one of the major environmental effects of economic activity (IPCC 2001; Hansen, 2004), and one of the most difficult to handle because of the 'softness' of the knowledge and the broad scale on which the impacts become visible. The achievement of MDG targets will depend on effective planning for managing climate risks. The need to identify the appropriate points at which to introduce climate change adaptation into development activities has been clearly perceived in many policy formulation initiatives (OECD, 2006b).

The climate science community urges caution about development decisions based on current uncertainty which may lead to 'maladaptations' or future 'dangerous development'. Environmental impact assessment can be one important tool for mainstreaming both climate change mitigation and adaptation. Like the link between poverty and biodiversity, the link between climate change and biodiversity also runs both ways: while biodiversity can be threatened by climate change (Table 3.2), conservation of biodiversity can reduce the impacts of climate change (Figure 3.3).

The state of the art of impact assessment can be significantly improved by drawing on climate change factors and impacts linked to development proposals for cobenefiting biodiversity and development goals. The final outcome of good EIAs must deliver biodiversity-based adaptive and mitigative strategies through which the resilience of ecosystems can be enhanced and the threats to human well-being can be reduced.

Conclusions

It is concluded that impact assessment can serve as a potent tool to guide sustainable development by tightening the actions and development processes that can serve the 'common good'. This can however only happen if society at large, development planners, and conservation community forge common agendas.

Having clearly understood the relationship between biodiversity and the pursuit of sustainable development, the EIA practitioners cannot ignore the need to retool impact assessment for mainstreaming biodiversity in environmental decision making. Impact assessment must provide clear understanding about the answers to the following questions.

Table 3.2 *Consequences of climate change on ecological functions and biodiversity.*

Increase in air temperature	Increase in the number of hot days	• Increase in heat-related stress on biological diversity • Increased exposure to diseases and pests • Drying of wetlands
	Increase in water temperature	• Decrease in dissolved oxygen • Coral mortality
	Reduction of ice sheets	• Decrease in sediment deposition
	Glaciers retreat	• Change in hydrological regimes
Change in rainfall patterns	Increase of droughts in the dry seasons	• Desertification • Loss of soil biodiversity • Increase in fire risks • Drying of wetlands
	Increase of floods in the wet seasons	• Soil erosion • Risks linked to waterborne diseases • Change in natural flow of rivers
Increase in sea level	Salt water intrusion in costal wetlands	• Habitat modification
	Coastal floods	• Coastal erosion
Increase in frequency and intensity of extreme climatic events	Decrease of ecosystems resistance and resilience	• Reduction of productivity • Increased mortality of certain organisms
	Increased height of storms waves	• Perturbation and loss of habitat
	Increased frequency	• Less time to recover from perturbations

What is to be sustained (e.g. the perpetuity of benefits from biodiversity and ecosystems within project area?)

It is a fundamentally acknowledged fact that ecosystems provide a range of services essential to humanity, which in short can be described as supporting life, supplying materials and energy, and absorbing waste products. As these services are encoded in biodiversity, the importance of ensuring their perpetuity and protection of their natural variability to help sustain and fulfil human life must be clearly reflected in development-planning processes that depend on utilization of

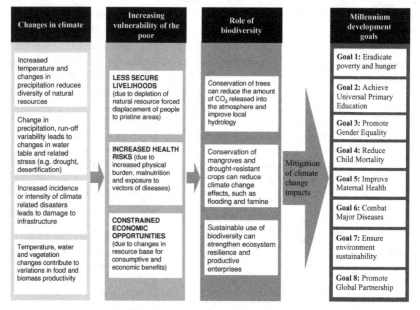

Figure 3.3 Role of biodiversity in mitigating climate change for achieving human well-being.

biodiversity resources. The risks threatening the ecosystem benefits of biodiversity and the measures to ensure their perpetuity should be rendered sufficient importance in assessment of impacts of development projects.

What is to be developed (e.g. economic benefits?)

Biodiversity and ecosystem services, particularly food production, timber, and marine fisheries contribute significantly to global GDP. Development activities that may lead to the depletion and degradation of these ecosystem benefits would represent a direct loss of a capital assets. For impact assessments to be really meaningful, capacities need to be developed to actually improve the process of assessment to predict how changes in ecosystems benefits associated with specific development activities would typically yield economic benefits for some people and impose costs on others through deprivation of access to resources or livelihoods. Such an assessment approach would be extremely useful in guiding decisions on whether to pursue or not to pursue certain development initiatives at the cost of long-term sustainability of economic benefits of biodiversity.

How is the sustainability of resources linked to the processes that define development?

Some people consider the term 'sustainable development' as closely linked with continued material production, and prefer to use these terms to actually indicate the efficiency of resource flow for the intended development. The impact assessment approaches that help to perceive the risk of resource consumption at a rate higher than that of its replenishment can actually help to assess resource securities linked with production processes and the various technologies involved in development processes. Such assessments can better help in making informed choices for permitting best development alternatives based on assurances of ecological security and economic sustainability.

Changes are thus needed in the conventional EIA approach to retool impact assessment for harmonizing development with human well-being objectives. This rationale underpins the broader scope of the book and assigns a more responsible role to ecologists and ecological economists in providing much greater inputs in EIA and SEA.

Part II

Assessment tools

4 · The impact assessment framework

Roel Slootweg and Peter P. Mollinga

Introduction

The Brundtland Report (WCED, 1987) introduced the concept of sustainable development, which has been further elaborated by the biodiversity, climate change, and antidesertification conventions created at the UNCED conference, or Earth Summit, in Rio de Janeiro in 1992. 'Sustainable development' as a concept is based on the idea that environment and development are intimately connected and coevolve. As a policy notion it also suggests that a positive connection between and coevolution of environment and development is possible. This was made more explicit by the adoption of the Millennium Development Goals in Johannesburg in 2000 (United Nations, 2000).

Sustainability is usually described in terms of its three pillars: (1) economy, (ii) environment, and (iii) society. This interpretation of sustainability is translated into different vocabularies, such as the 'triple bottom line' in sustainability appraisal, or the triple 'P' of *people, planet, profit* in corporate social responsibility. In a sense, even the objectives of the Convention on Biological Diversity (see Chapter 2) can be translated in these terms: conservation addresses *planet*, sustainable use relates to *profit*, and equitable sharing is about *people*.

The consequence of the desire to integrate environmental, social, and economic aspects in assessments of projects, plans, programmes, and policies is the need for an integrating framework. In practice, the worlds of environmental impact assessment (in its strict meaning of assessing biophysical impacts only), social impact assessment, and economic cost–benefit analysis largely continue to operate in their separate realms and experience great difficulty in working in a multidisciplinary environment (Slootweg *et al.*, 2001). As Treweek (1999) indicated, the inconsistency of methodologies and the inconsistency of reporting on methodologies and results have, among other reasons, seriously hampered the accumulation of one body of relevant knowledge in the prediction of impacts that human activities have on biological diversity. This chapter introduces such a framework in an attempt to create more consistency in addressing issues related to biological diversity.

Because Environmental Impact Assessment (EIA), and progressively more Strategic Environmental Assessment (SEA), are well-developed instruments backed by legal and/or procedural frameworks in most parts of the world, these instruments are increasingly used to also assess the social and economic impacts of planned interventions. The conceptual framework presented in this chapter provides an integrated way of thinking to assist in the identification of potential environmental, social and economic impacts of planned human interventions. The framework is designed to cover all imaginable effects of human interventions in the biophysical and/or societal environment and goes far beyond a conservation-oriented interpretation of biodiversity. It relates biodiversity to human well-being and provides insight into, and understanding of, the complex cause–effect chains that may lead to desired or undesired effects of human interventions.

The framework is based on the translation of biodiversity into functions for human society, often referred to as function evaluation. Since the early sixties the functions concept is a recurrent theme in the environmental literature (W. T. de Groot, 1992: 229). In the Netherlands the classifications of functions of the environment grew into a scientific tradition in the 1970s, ultimately resulting in an influential book on Functions of Nature by R. S. de Groot (1992). Functions of nature bear great similarity to environmental goods and services that figure in other publications. In 2003, the Millennium Ecosystem Assessment (MA, 2003) and many publications following) provided a worldwide platform to at least harmonise vocabulary: the term 'ecosystem services' has become generally accepted, encompassing the concepts of functions of nature and environmental goods and services. The MA also provided a conceptual framework, which is extensively discussed in this chapter.

The framework considers social as well as biophysical mechanisms through which impacts occur, including impacts on biological diversity. The framework is comprehensive in the sense that it can be applied to any imaginable impact, including those on biological diversity. It provides an integration framework for impact assessment studies, potentially encompassing environmental impact assessment, health impact assessment, social impact assessment, and strategic environmental assessment, to name some of the more commonly applied assessment instruments. In this book the framework will be elaborated from a biodiversity point of view. To avoid unnecessary confusion the vocabulary is harmonised with the Millennium Ecosystem Assessment (2003). Part of the material presented here has appeared in earlier publications; the conceptual background largely remains intact, but because of the harmonisation with the MA, the terminology in this book may deviate from the earlier publications. An overview of these earlier publications is provided below.

The conceptual framework presented in this chapter is not intended to be a fixed procedure, nor is it intended to be a predictive model. It is a way of thinking to assist in the clarification of the issues that may need to be studied in any environmental assessment and to assist in the communication among and within multidisciplinary teams of experts and stakeholders. It can be used in an iterative way, for example, by first qualitatively identifying the issues at stake, and later quantitatively when the issues are actually studied. Using boundary crossing as a metaphor for interdisciplinary work (Klein, 1996), the literature on interdisciplinarity has developed the notions of 'boundary concepts' and 'boundary objects' as the cognitive and practical instruments used for effective interaction at the interfaces of scientific disciplines, research and policy, and research and society (Star and Griesemer, 1989). Environmental assessment and the question how biodiversity has to be treated in environmental assessment has all characteristics of interdisciplinarity and the use and deployment of boundary concepts and objects. In the last section of this chapter, the framework will be positioned as a 'boundary object', that is, as a device that allows us to act in situations with incomplete knowledge and divergent interests.

History of the framework

The foundation for the framework was the development of a methodology for the assessment of functions and values of wetlands, published as an informal document by The Wetland Group Foundation (Koudstaal *et al.*, 1994). The earliest version of the present framework was developed between 1997 and 2000 to provide a conceptual basis for a computerised instrument to assist in the identification (qualitatively) of potential impacts of proposed water resources development projects. For that purpose, the authors were forced to create a formalised and unequivocal conceptual framework. In doing so, it was realised that the analytical dimension of environmental and social impact assessment practice could be strengthened. Very often, implicit assumptions are used in both EIA and SIA without this being acknowledged. For example, in many terms of reference for EIA studies, impacts on water quality are considered negative impacts, without any statement of the reasons why this should be the case. Implicitly, water is assumed to provide ecosystem services, such as water supply for irrigation or to act as a medium for fisheries. Water quality would probably not be an issue if the water would drain into an uninhabited area without any living organisms. Slootweg *et al.* (2001) provided the first formal publication on the framework. Schooten *et al.* (2003) and Slootweg *et al.* (2003) enriched the framework from a social impact assessment perspective by providing a conceptualisation of social change processes and social impacts – a distinction which was new to the world of social impact assessment.

The framework was further elaborated in a proposed conceptual and procedural framework for the integration of biological diversity considerations within national systems for impact assessment (Netherlands Commission for Impact Assessment, 2001; conceptual background published in Slootweg and Kolhoff, 2003). The purpose of this elaboration was to assist the Convention on Biological Diversity (CBD) in providing guidance on the implementation of Convention article 14 on impact assessment. The procedural part of this work appears in Decision 7A of the sixth Conference of Parties of the CBD on further development of guidelines for incorporating biodiversity-related issues into environmental impact assessment legislation or processes and in strategic impact assessment (CBD, 2002). Upon request by the CBD the same framework has been further elaborated for Strategic Environmental Assessment, which resulted in Decision 27 of the 8th Conference of Parties providing voluntary guidelines on biodiversity-inclusive impact assessment (CBD, 2006). Again, the conceptual background text does not appear in the convention text but has been published in Slootweg, *et al.* (2006).

Upon invitation by the World Bank, the framework has been used to create a tool to facilitate integrated and participatory planning in the irrigation and drainage sector (Abdel-Dayem *et al.*, 2004). Two pilot trials in Egypt and one in Pakistan have resulted in the development of two different modules: one for a strategic analysis of development opportunities and constraints in a defined area, and one for the assessment of potential impacts of proposed irrigation and drainage interventions, including participatory negotiation on various alternatives (Slootweg *et al.*, 2007). The oil industry's Energy and Biodiversity Initiative has adopted the framework in its guidelines to integrate biodiversity in ESIA (Energy and Biodiversity Initiative, 2004).

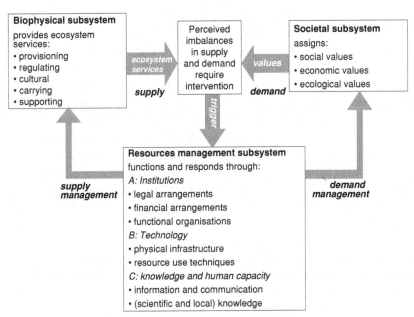

Figure 4.1 Linkages and interactions of three subsystems.

Three subsystems

The framework aims to provide insight into the relationship between human society and the biophysical environment, and the way in which both biophysical environment and human society are being influenced and managed. In order to better understand this relationship, the world is split up in three different but intensely interacting subsystems: (i) the biophysical subsystem, (ii) the societal subsystem, and (iii) the resources management subsystem, the latter being a combination of institutions, technology, knowledge, and human capacity. Figure 4.1 illustrates the way in which these three settings interact.

The core element of the conceptual approach presented in this chapter is the characterisation and classification of ecosystem services provided by the biophysical environment and the assessment of their value for sustaining human livelihoods. In this respect the focus of attention lies on the biophysical and societal subsystems. The resources management subsystem is the initiator of policies, plans, programmes, and projects that are subject of environmental assessment at project or at strategic level. In this chapter we will not go into any further detail on the resources-management subsystem.

Human society utilises products and services that are provided by the biophysical environment. In economic terms, society constitutes the demand side, and the biophysical environment constitutes the supply side. The demand for ecosystem services may surpass the available supply, leading to a present or expected future problem (e.g. overexploitation of fish stocks or degradation of soils). Reversely, the potential supply of ecosystem services can be larger than the actual demand; in this situation one can speak of an opportunity for sustainable development (e.g. sustainable exploitation of forests and groundwater aquifers, tourism development). Perceived imbalances thus include both threats *and* opportunities for human development. Simply stated, sustainability deals with the equilibrium in supply and demand, now and in the future![1] Imbalances in the supply of and demand for ecosystem services trigger the resources management system to act by managing either the supply of ecosystem services (through hydraulic engineering, agriculture, forestry, etc.) or the demand from society (through tax incentives, setting of quota, trade negotiations, etc.).

Delineation of boundaries within which a study needs to be carried out can be difficult, because the boundaries for these three subsystems are often different. The *biophysical subsystem* boundaries are generally defined in terms of specific geographic areas, which can be outlined on maps. These areas can be referred to as supply areas of ecosystem services. Boundaries to be considered for the societal subsystem may differ from ecologically defined areas. These so-called demand areas encompass the main users of the ecosystem services of the supply areas. The *societal system* is less specifically geographically defined and boundaries may even differ for different types of demand. In many cases of environmental assessment, a demand area may be defined as the area where the majority of the people live and work which use, either for their living or for their economic activities, the goods and services produced by the supply area. Below we provide arguments why this simple approach will in many instances not work. For the management subsystem administrative units in many cases provide a convenient base for the delineation. Administrative units have jurisdiction over the ecosystems and their services and encompass the main users and economic activities. The administrative level involved very much depends on the scope of the policy, plan, programme, or

---

[1] We are aware that both equilibrium and sustainability are contested concepts. However, for our purposes, the exact meaning is not crucial. The central point is the dynamics: action triggered by *perceived* imbalances and unsustainability.

project under study. It can range from supranational bodies down to local municipalities. Then again, for management subsystems administrative boundaries may, in many instances, not suffice.

The assumption that supply, demand and management areas are well-demarcated geographical areas may in many environmental assessments be valid, especially where it concerns EIA's of projects, but it definitely does not apply to all cases. Users of goods and services from a certain supply area (say a soybean producing area in Brazil or a wild coffee cultivation area in Ethiopia) may be spread across the globe. In addition to 'areas' it is necessary to conceive of 'networks' as the organising forms of the supply and demand of ecosystem services. In water management studies this is captured in the concept of 'problemshed' – as an alternative to 'watershed' (Viessman, 1998: 5). In a watershed (catchment) perspective, boundaries are predefined spatially, sectorally, and analytically through the primacy of 'water'. The latter seems unwise given the complexity and multidimensionality of water management problems. For example, an issue involving wetlands and migratory waterfowl may involve environment agencies and NGOs from outside of the watershed and even the country, whereas saving water through conversion from water-consumptive crops to other crops might require linkages with infrastructure agencies and the private sector to develop markets for alternative crops. In a problemshed perspective, the question of the boundaries of a given water management issue, in space, in time, and in a society, is treated as an open, empirical question. This avoids confining the scope of analysis to a hydrologically defined unit. In terms of strategic analysis the problemshed can be regarded as an 'issue network' (see Van Waarden, 1992; Howlett and Ramesh, 1995, 1998). It is open to what constitutes the 'issue' and which actors, processes, and mechanisms influence it, rather than using preconceived ideas of the structure of the action arena, for example, by confining it to the river basin area and the actors directly involved in water use and management.

In the biological diversity 'arena' similar concepts can be found in the ecosystem approach, defined by the CBD as a strategy for integrated management of land, water, and living resources (CBD, 2000a and 2004a). It considers management of living components, alongside social and economic considerations, at the ecosystem level of organisation. It considers ecosystem management as a social process that has to work within natural boundaries. Recognising the potential gains of ecosystem management also necessitates understanding and managing ecosystems in an economic context. Furthermore it stresses to address issues at the right scale, because the failure to take scale into account can result in mismatches between the spatial and time frames of the management and those of the ecosystem being managed.

The biophysical subsystem: supply of ecosystem services

The Millennium Ecosystem Assessment (2003) has set the standard for terminology to translate the natural environment into terms understandable for human society. It has defined ecosystem services[2] as 'the benefits people obtain from ecosystems'. These include *provisioning services* (such as food and water), *regulating services* (such as regulation of floods, drought, land degradation, and disease), *cultural services* (such as recreational, spiritual and other nonmaterial benefits), and *supporting services* to maintain the other services (such as soil formation and nutrient cycling). The framework for assessment of the MA has solved the problem of 'tiered services', which appeared in earlier literature on functions of nature. Some functions directly provide goods or services for human society, while other functions are needed to maintain these direct services.[3] The MA therefore made a distinction between supporting services and the other three categories. This also solved the problem of double counting in economic valuation of ecosystem services; supporting services are not taken into account and only the final, directly exploited service is valued (Hein *et al.*, 2006).

Yet, the MA missed out an important category of services, which has been described extensively in earlier literature, that is, the carrier or carrying functions.[4] R. S. de Groot (1992) defined carrying functions as 'providing space and substrate'. All living organisms need a certain amount of space in accordance with their particular environmental requirements. Although all organisms need space, humans show a wide differentiation in their appreciations of space. Requirements for the construction of houses, roads or ports, to make use of rivers for

---

[2] Norman Myers (1996) argued that *environmental* services was a better term as it embraces large-scale services such as the albedo of Amazonia, being a region too large to conform to the conventional understanding of ecosystem. However, with the flexible definition of ecosystem provided by the ecosystem approach, the MA could stick to the term 'ecosystem services'.

[3] A summary of conceptual difficulties around these 'conditional' and other functions can be found in W. T. de Groot, 1992: 229–36.

[4] Both W. T. de Groot and R. S. de Groot use the terms 'carrier' and 'carrying functions'. We will stick to 'carrying functions'.

navigation, or to develop tourism differ widely. Consequently, the carrying functions of ecosystems can be multiple but very different for each ecosystem.

A subtle differentiation in functions introduced by W. T. de Groot (1992) not followed by the MA, is the differentiation between joint and natural production functions. Natural production functions are part of the provisioning services of the MA and relate to the products derived from natural production (such as fish). Joint production functions relate to products that partly depend on natural productivity, but where human decisions and inputs are a dominant factor (e.g. agricultural crops, aquaculture products, and plantation crops).

Some have highlighted the problem of double counting with the present classification of ecosystem services (Hein *et al.*, 2006 Wallace, 2007). Depending on the local conditions, services may underpin other services (e.g. water purification partly or entirely supports provision of drinking water); by taking both services into account, double counting may occur in an assessment. Hein *et al.* solved this by only taking the locally defined, end variable into account; Wallace proposed an entirely new classification system. This classification system is less developed and more complex to understand and is therefore less useful for interdisciplinary work in environmental assessment. This is the reason we stick rather closely to the mainstream ecosystem services classification.

Summarising the above we arrive at five categories of ecosystem services, closely following the MA framework, but with some additions as pointed out above:

(1) *Provisioning services*: products obtained from ecosystems. For each of the products obtained from ecosystems a distinction can be made between natural and joint production, that is, products harvested from nature with minimal human effort, or produce obtained with human inputs such as fertilisers, pesticides, or intense resource management. There is no clear-cut differentiation between natural and joint production as many degrees of human intervention occur. It is important to realise that provisioning services *always*, although to varying degrees, depend on biodiversity.

(2) *Regulating services*: benefits obtained from the regulation of ecosystem processes. In earlier literature a distinction was made between processing and regulation, where processing referred to chemical transformation, dilution, sequestration, or processing of waste, and regulation to the dampening of harmful influences from other

components (W. T. de Groot, 1992). In practice however, it is very difficult to make a proper distinction between these two categories. R. S. de Groot (1992) and the Millennium Ecosystem Assessment (2003) have consistently maintained these as regulating services.

(3) *Cultural services*: These are nonmaterial benefits people obtain from ecosystems through spiritual enrichment, cognitive development, reflection, recreation, and aesthetic experiences. In earlier publications the terms 'signification functions' or 'information functions' have been used for the same group of services.

(4) *Carrying services*: Ecosystems provide space, a substrate, or backdrop for human activities. This is best illustrated by river navigation. Navigation needs water as a substrate; if water depth is not sufficient a ship will not be able to proceed. Yet, a ship does, in principle, neither influence the quantity nor the quality of water. The service is often considered as being for free, until water shortage occurs (e.g. by extraction of water by upstream irrigation works to provision agricultural production), the economic damage becomes apparent, and, for instance, dredging becomes an option at a certain cost.

(5) *Supporting services*: These are the services that are necessary for the production of all other ecosystem services. They differ from all the other services in that their impacts on people are either indirect or occur over a very long time. For example, soil formation processes usually play on a time scale which humans cannot oversee; yet they are closely linked to the provision service of food production. Biodiversity is said to provide an 'insurance' service as the very diversity itself insures ecosystems against declines in their functioning (e.g. Loreau *et al.*, 2001; Thompson and Starzomski, 2006; see Box 2.6 in Chapter 2). We consider this being a supporting service, because it guarantees the provision of all other services.

A detailed list of ecosystem services can be found as an appendix at the end of this chapter.

Water is an interesting example showing the strength of the ecosystem services concept. Water is often considered as one natural resource, but in terms of ecosystem services it constitutes a multiple resource. Depending on the type of ecosystems providing these water-related ecosystem services, the set of services will be differently composed. Some examples: the ecosystem can be harvested for public water supply or irrigation as a provisioning service, water bodies regulate water quality (regulation service), and water plays an important role in many religious ceremonies as well as recreational activities (cultural services), it provides a substrate for navigation (carrying service), and it plays a major role in climate regulation (supporting service). Exploitation of one service often goes at the cost of another service. Obviously, for the assessment of impacts of any water resources project a multisectoral approach is necessary, as also stated by the Millennium Ecosystem Assessment (2003: 60).[5] Economic sectors are usually constituted around a single service; the multiple services provided by water usually go far beyond sectoral boundaries, thus constituting the problem of 'integration'.

The description of water resources in terms of ecosystem services provided by water also provides the means to define the relevant units of measurement. The description of ecosystem services is done in terms linked to the service and not in terms of the values attached to the service by stakeholders (for an explanation of 'values', see the next section). For example, water quality parameters are a major concern for public water supply, while the provision of irrigation water is more focussed on quantity. For river navigation only, width and depth of the river may be of interest. Religious and recreational services provided by water probably have to be expressed in more qualitative terms. Timber production can be expressed in cubic meters of harvestable wood for various tree species. This can be done in a relatively unequivocal manner. According to the Millennium Ecosystem Assessment (2003), ecosystem services have to be described in terms of stock (how much is there), flow (how much can sustainably be harvested), and condition (an ecosystem's capacity to continue providing a service). Other important parameters are variability, resilience, and thresholds (see the Millennium Ecosystem Assessment (2003: Chapter 2).

The societal subsystem: stakeholders and the demand for ecosystem services

In the terminology of the MA, ecosystem services contribute to human well-being by having an influence on multiple constituents of human well-being: security, access to basic material for a good life, health, social relations, and freedom of choice and action. How well-being, ill-being, or poverty are experienced and expressed depends on context

---

[5] An elaboration of the ecosystem services concept for the water sector can be found in Abdel-Dayem et al., 2004.

and situation, reflecting local physical, social, and personal factors, such as geography, environment, age, gender, and culture (Millennium Ecosystem Assessment, 2003). In all contexts, however, ecosystems are essential for human well-being through their services. Although the approach of our framework is fundamentally similar, we do not use the 'constituents of human well-being' terminology as used by the MA. The information needs in environmental assessment differ on a case-by-case basis and may strongly deviate from the MA, which is a worldwide scenario evaluation. For decision making on any plan or project it is more practical to look at the values of affected ecosystems services for stakeholders in society. In our approach values are expressed in three broadly defined categories, more fitting with decision-makers' language, that is, social, economic, and ecological values. The precise description of values differs for each environmental assessment, as these values need to be identified and valued in consultation with groups 'having a stake' in any of these values. Values may change among different stakeholders; for example, indigenous people will value forest resources (provisioning service) differently as compared to an international wood processing company. Similarly, regional authorities or a chamber of commerce may be interested in the contribution of forestry to the regional economy, local inhabitants may be more interested in generating employment, and indigenous people will value the forest as their very basis of life.

Stakeholders

The societal subsystem consists of individuals or groups of people that value the goods and services provided by the biophysical subsystem. In economic terms society constitutes the demand side for ecosystem services. By putting a value to one or more ecosystem services, individuals or groups of people automatically become a stakeholder in any policy, plan, programme, or project which may affect these ecosystem services. The opposite also applies – it can be argued that an ecosystem service does not exist for society as long as the service does not have stakeholders. For environmental assessment it is therefore of utmost importance to identify stakeholders in order to be able to identify, describe, and quantify the values of ecosystems services for society. In this respect the term 'stakeholder' is interpreted in its widest possible sense, and includes those speaking on behalf of future generations. Four categories of stakeholders can be identified (see Figure 4.2):

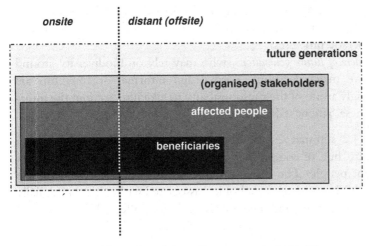

Figure 4.2 Stakeholders (adapted from Slootweg *et al.*, 2006: 37).

- *Beneficiaries* of a policy, plan, programme or project – target groups making use of ecosystems services which are purposefully enhanced. A distinction can be made between direct, or onsite, beneficiaries, such as fisherfolk (service: productivity of aquatic resources) or farmers (service: soil productivity), or distant beneficiaries, such as foreign tourists or urban inhabitants dependent on water supply from wetlands elsewhere;
- *Affected people*, that is, nontarget groups that experience intended or unintended changes in ecosystem services that they value;
- *General stakeholders*, representing a wide variety of formal and informal organisations and groups, including:
  - *National or local government agencies* having a formal government responsibility with respect to the management of defined areas (town and country planning departments, etc.) or the management of ecosystem services (fisheries, forestry, water supply, coastal defence, etc.);
  - *Organisations representing affected people* such as established governmental or nongovernmental organisations or spontaneous initiatives (water boards, trade unions, consumer organisations, civil rights movements, ad hoc citizens committees, etc.);
  - *Organisations representing (the intrinsic value of) biodiversity itself* (nongovernmental nature conservation organisations, park management committees, scientific panels, etc.);

- *The general public* that wants to be informed on new developments in their direct or indirect environment (linked to transparency of democratic processes).
- *Stakeholders of future generations*, who may rely on biodiversity around which we make decisions. Formal and informal organisations are increasingly aware of their responsibility to take into account the interests of these 'absent stakeholders'.

Figure 4.2 summarises the above. It puts future stakeholders in a separate box, but, of course, only present *real* stakeholders can represent nonexistent people. The figure provides a distinction between on site stakeholders, that is, those stakeholders associated with what traditionally is called the *project site*, and stakeholders indirectly affected by the project through biophysical changes with distant impacts.

Values

The value of something is its worth, merit, or importance. Different categorisations of values exist. Distinctions are often made between use and nonuse values, where use values are further subdivided in direct use values of harvestable products (i.e. linked to provisioning services) and indirect use values, linked to the other categories of ecosystem services. Nonuse values can be further subdivided in optional values, bequest values, existence, and intrinsic values. Chapter 8 will discuss this in greater detail. In order to identify the values that ecosystem services represent for society we propose three broad categories of values (Koudstaal *et al.*, 1994; Netherlands Commission for Environmental Assessment, 2001; Slootweg *et al.*, 2001):

- *Social values*: These refer to the quality of life in its broadest sense and can be expressed in many different units, depending on the social context and cultural background. For example, health and safety can be expressed in the numbers of people protected from forces of nature by coastal mangroves (regulation service), or prevalence of disease; human wellbeing can be related to the level of food (in-kind income) obtained directly in nonmonetary societies. Schooten *et al.* (2003) provide an extensive overview of social values; many of these can be directly linked to ecosystem services.
- *Financial and economic values*: These are related to both direct consumption (e.g. fish, timber) and the inputs to the production of other goods and services (e.g. water for irrigation, wetlands store flood water to reduce downstream flood damage). Examples of expressing economic

values are monetary value assigned to individual economic activities (agriculture, industries, construction, etc.), household income as an overall expression of the financial conditions of the population, or per capita gross regional or domestic products as an overall expression of the income of society as a whole.

- *Ecological values*: These refer to the value that society places on or derives from the maintenance of the earth's life support systems. They come in two forms:
  - *Temporal ecological values* refer to the potential future benefits that can be derived from biological diversity (genetic, species, and ecosystems diversity) and key ecological processes that maintain life-support systems for future generations. Examples include the unknown potential of genetic diversity to be exploited in biotechnology, but also the simple maintenance of production and decomposition processes on which all life on earth depends.
  - *Spatial ecological values* refer to the interactions of ecosystems with other systems, providing services for the maintenance of other ecosystems. Examples include coastal lagoons and mangroves which serve as breeding grounds for marine fish, supporting an economic activity elsewhere (fisheries); wintering areas for migratory birds linking West Africa with Northwestern Europe and Siberia; or floodplains acting as flood buffer for flood protection of downstream areas, or acting as a silt trap preventing reservoirs from silting up.

Values are linked to stakeholders, and one ecosystem service can have different stakeholders. Consequently, different and even opposing values can be attached to one ecosystems service. For example, in Bangladesh farmers think very negatively about prolonged floods, while such floods are considered a blessing for fishermen. Measures to reduce floods enhance agricultural production (provisioning service) at the cost of flood attenuation (regulation service), resulting in local loss of fisheries income (provisioning service) and increased floods downstream (Oliemans *et al.*, 2003).

The impact assessment framework

Linkages to the MA framework

The impact assessment framework has been developed to further elaborate on the cause–effect chains which impact upon biodiversity and ultimately affect human well-being. To understand the strengths and limitations of the impact assessment framework, we first explain some of the concepts introduced by the MA. The MA was a four-year international work programme designed to meet the needs of decision makers for scientific information on the links between ecosystem change and human well-being. It was launched by United Nations Secretary General Kofi Annan in June 2001. Leading scientists from more than 100 nations have contributed. The first product of the MA was a conceptual framework providing the thinking behind all ongoing work (Millennium Ecosystem Assessment, 2003). The MA conceptual framework is fully consistent with the CBD Ecosystem Approach (CBD, 2000a and 2004a). An important feature of the MA is the translation of biodiversity into ecosystem services, which contribute to human well-being and poverty reduction. Ultimately, humanity is fully dependent on the flow of ecosystem services. The degradation of ecosystems places a growing burden on human well-being and economic development.

The performance of ecosystems to provide ecosystem services can be influenced by drivers of change. In the MA, a 'driver' is any factor that changes an aspect of an ecosystem. A direct driver unequivocally influences ecosystem processes and can therefore be identified and measured to differing degrees of accuracy. An indirect driver operates more diffusely, often by altering one or more direct drivers, and its influence is established by understanding its effect on a direct driver. Demographic, economic, sociopolitical, cultural, and technological processes can be indirect drivers of change. Actors can have influence on some drivers (endogenous driver), but others may be beyond the control of a particular actor or decision maker (exogenous drivers). Strategies and interventions can affect a driver of change at geographical scales varying from local to global and may work at widely different time scales. Consequently, the organisational scale at which to best address a driver of change needs to be assessed for each situation.

The Impact Assessment framework introduced in this chapter provides a structure to describe direct drivers of change that result from human interventions in more detail. It establishes linkages between biophysical and societal effects of interventions and provides insight in how interventions may lead to impacts, either through biophysical interventions or through societal interventions. It makes a clear distinction between 'facts' and 'values'. 'Facts' are analytical understandings of physical and social reality that can be tested through established research methodologies. We may or may not like the 'facts' that we find in our analyses, but that does not alter their factuality. 'Values' are, crudely put, what we like or do not like about reality, what importance, merit, and appreciation we lend to it from the perspective of our particular position and worldview. The kinds of values we hold regarding ecosystems were outlined above (also see Chapter 8). These values are the basis of how 'impact' is understood and felt.

The framework is a conceptual basis for impact assessment at levels where interventions in the social and biophysical environment are (more or less) known, at project level but also at the level of strategic assessment for regional or sectoral plans. The Millennium Ecosystem Assessment is not developed for such types of impact assessment but aims at providing information for natural resources management policies. The MA concepts are largely similar to the Impact Assessment Framework, but serve the highest level of strategic assessment where interventions are not precisely known. The notion of indirect drivers of change, or in other words, diffuse societal processes that influence or even govern direct drivers of change, provides a concept to coherently describe chains of cause and effect at higher policy level. Although conceptually similar, the two frameworks have been developed for different settings and can be considered as complementary. Chapter 7 provides more detail on the combined use of the MA framework and ours for strategic environmental assessment at the level of policies, plans, and programmes.

The MA framework largely overlooks that social processes can also be considered direct drivers of change. For example, the creation of employment in a relatively uninhabited area will attract migrants that settle in the vicinity of the facility where people are employed, occupying formerly uninhabited areas. There is nothing diffuse or indirect to this as it is a planned activity with predictable consequences.

Activities and effects: direct drivers of change

The first step in the impact assessment framework is the description of direct drivers of change, that is, activities and their effects that change ecosystems and services provided by these ecosystems. (Activities can consist of biophysical (Figure 4.3, part 1) as well as societal interventions (Figure 4.3, part 2) Biophysical interventions lead to biophysical effects being defined as changes in the characteristics of the recipient media soil, water, air, flora, and fauna (Figure 4.3, part 3).

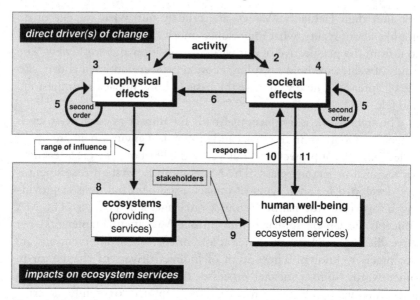

Figure 4.3 Impact assessment framework (adapted from Slootweg *et al.*, 2001).

Example: A new mining activity will physically alter the site of the concession (clearing of site, relocation of streams, creation of dumpsites (physical interventions), operation of the mine, use of processing water, etc.). The resulting biophysical effects may be a change in the quality and quantity of surface waters and groundwater aquifers, clearing of vegetated areas, noise and dust production, and so forth.

Each direct biophysical effect can result in a chain of second-order and higher-order biophysical effects (Figure 4.3, part 5).

Example: A change in the groundwater table in the mining concession area may alter the groundwater table in a surrounding forested area. Change in the hydrology of a local stream may alter the flooding regime of wetlands located downstream.

Activities may also lead to societal effects (Figure 4.3, part 4) being defined as changes in the characteristics of components of society (individuals, families, functional groups, or a society as a whole) and the relations between these individuals and groups; the nature of these characteristics and relations can be demographic, economic, sociocultural, emancipatory, institutional, land use, and so forth (van Schooten *et al.*, 2003).

Example: If the mining concession is located in a rural area, the mining company may have to attract additional labour, leading to an increase in number of inhabitants, often with a skewed sex ratio and a skewed age distribution.

Each direct societal effect can lead to second-order and higher-order effects (Figure 4.3, part 5).

Example: The influx of new inhabitants leads to an increased demand from public services such as health and educational facilities.

A change in the characteristics of a community (societal effect) may lead to biophysical effects (Figure 4.3, part 6).

Example: An influx of migrant labourers, will lead to increased land encroachment for housing and public facilities, food production, and leisure activities.

Biophysical and societal effects are independent of the particular way a specific analyst or stakeholder may choose to understand them. This means that, in principle, it is possible to produce relatively clear 'facts' about the physical and societal effects that would result from a certain intervention. We produce 'facts' for use in an assessment – this is both a physical and a social reality. The processes happening in physical reality are independent of how and whether we think about them – they also happen when there are no human beings, and are in a fundamental sense independent of human thought and presence. This is not the case with social reality, which is – of course – *constituted* by human beings, their presence, actions, and thoughts. This complicates matters considerably as regards the behaviour of social systems and the ease with and the kind of 'facts' that can be produced about them. The view taken by the present authors is that there are no 'social laws' that can explain the behaviour of social systems in the same way as 'natural laws' can help to explain physical system behaviour. Nevertheless, it is quite possible to produce 'facts' about social systems, although these are always historical and contextual. However, in our view these 'facts' do not, in principle, depend on the 'values' that the analyst may hold and the 'meaning' that s/he attributes to them. That is, social analysis and social 'facts' are subject to the same procedural rules as physical 'facts' and analysis, although the nature of the object is different, and testing social 'facts' is often a much more hazardous affair than testing physical 'facts' (see Box 4.1; Van Schooten *et al.*, 2003, are the first to publish on this idea within

---

Box 4.1. 'Facts' in critical realism

The philosophical position taken here is that of critical realism (Bhaskar, 1979; Sayer, 1992). In this perspective, 'facts' are always socially constructed, and therefore the relationship between 'facts' and reality is always a question, although sometimes much more than other times. However, this does not preclude a realist ontology for both physical and social objects. Despite the epistemological problems related to the idea of 'objectivity', something that can be at best approached, we are still able to produce 'practically adequate' knowledge about physical as well as social objects – although not always very easily. The confusing part of the analysis of social systems, at least from a natural science perspective, may be that 'meanings' are both present in the object of research and in the perspective from which the analysis of the object is undertaken. Critical realism considers this as following from the nature of social objects and human beings (societies are defined by the presence of meaning; people cannot observe and analyse other than by employing certain structures of meaning) rather than as a 'problem' necessitating the expulsion of 'meaning' from scientific analysis.

---

the context of social impact assessment, although they use a somewhat different vocabulary).

The magnitude of the effects and the direction of change are determined by the combined characteristics of the intervention and the recipient involved (see Table 4.1).

Example: Dams (intervention) change the hydrology of watercourses (recipient). Forced migration (intervention) will change the demographic characteristics of a population (recipient).

As we pointed out biophysical and social effects can be identified independently and relatively objectively (e.g. by external experts). This does not mean that local knowledge shouldn't be used when describing and/or quantifying these changes. Local knowledge may provide important clues about the dynamics of the physical and social systems at hand, and it is part of the reality that is investigated. However, local opinion, like any other opinion, needs to be separated from the identification of the structures and mechanisms at work in the physical and social systems

Table 4.1 *'Recipients' or 'carrying media' of change with some characteristics that may be affected.*

Recipient	Characteristics
Soil: quality	Further described in: change in soil chemistry (may be further detailed in salts, acidity, naturally poisonous elements), structure, texture, moisture, fertility, man-induced pollution.
Soil: quantity	Sedimentation and scouring of stream beds, susceptibility to wind erosion, water erosion, landslides, subsidence.
Water: quality (surface, groundwater, run-off water).	Salt/freshwater balance (sea-land interface), sediment load, turbidity, acidity, man-induced chemical pollution, poisonous elements in groundwater, oxygen contents, nutrient contents, temperature, stratification
Water: quantity surface, groundwater, run-off water.	Regime of peak flow, base flow, and flooding, change in water level or water level dynamics of surface and groundwater reservoirs, flow velocity, stream profile (wet section).
Air	Microclimatic and macroclimatic change (complex of factors related to temperature, humidity, force and frequency of weather phenomena); airborne solid particles (dust, asbestos), odours, noise level, chemical pollution, greenhouse gasses.
Flora	Removal of vegetation (clearing, felling), infestation with terrestrial or aquatic weeds, algal bloom, plant diseases, invasion of exotic species, replacement of traditional plant varieties or cultivars by high-yielding varieties.
Fauna	Removal of indigenous species (hunting), breeding of disease transmitting animals, outburst of pests (nematodes, insects), damage by animals (rodents, birds), invasion of exotic species, replacement of traditional animal breeds, breeding of pathogenous organisms.

under analysis. According to Principle 11 of the ecosystem approach, the use of local knowledge should be incorporated as much as possible in any environmental assessment, as long as local knowledge is separated from value judgements.

So, while biophysical and social effects can be defined independently and in a relatively objective manner, the impacts resulting from these changes can *never* be defined independently from the affected stakeholders. Stakeholders perceive impacts within their system of norms, values and beliefs, and give value to impacted ecosystem services.[6] This is why we distinguish between an 'effect' (by external – nonstakeholder – experts measurable and to some extent predictable changes) and an 'impact' (the consequences of interventions as they are perceived or 'felt' by affected individuals, groups, or society as a whole).

Impacts on ecosystem services and human well-being

Biophysical effects resulting either directly from biophysical interventions, or indirectly through societal interventions, are pivotal to the impact assessment framework as these are the actual agents of change in ecosystems and in the services provided by these ecosystems.

Example: A change in groundwater table will change the lumber production of an exploited forest, and it may change the capacity of the forest to maintain its biological diversity.

An important characteristic of biophysical effects is their spatial as well as temporal *range of influence* (Figure 4.3, part 7). Each biophysical effect has a geographical range of influence (area of influence: where and how far away), and a time range (when and for how long, permanent or temporary) (Slootweg, 2005). Experts can model biophysical effects or use empirical evidence to predict where and when an effect will be noticeable. Local stakeholders with their day-to-day experience can provide relevant information.

By knowing the range of influence of each biophysical effect (e.g. drawn on a map), the ecosystems and land use types (sometimes referred to as man-made ecosystems) under the influence of biophysical effects

---

[6] Modern societies have developed legislation (e.g. protected areas) and formal environmental norm systems which are intended to represent the values and perceptions of society as a whole. Having such formalised rules and norms greatly facilitates environmental assessment as on many issues stakeholders don't have to be asked for their opinion. The disadvantage is that environmental assessment in many instances has become a technocratic exercise where impacts are only compared to norms. Thinking stops and stakeholder involvement is minimised. Furthermore, one cannot assume that norms as codified in law, regulations etc. are shared by everyone. 'Society as a whole' is not an actor. Norms at that level may be the norms of a dominant group or class or cultural elite. They may obscure divisions and differences. For stakeholder analysis it is exactly important to look beyond 'society as a whole'.

Table 4.2 'Recipients' or 'carrying media' for biophysical effects and their range of influence (not exhaustive).

Recipient	Range of influence
Soil	Predominantly local changes and vertically through soil layers.
Air: noise	Local and neighbouring areas following noise contour lines.
Air: dust, odour	Local and neighbouring areas following direction of prevailing wind.
Open surface water	Local and downstream, permanently or temporarily submerged areas connected to the water course.
Run-off water	Terrestrial and aquatic areas downhill from where the activity is carried out.
Groundwater	Local and neighbouring areas sharing the same underground aquifer.
Aquatic organisms	Upstream and downstream, permanently or temporarily submerged areas connected to the water course.
Disease-transmitting flying insects	Settlements of man and domestic animals within flying range of insects (approx. 2–5 km).
Disease-transmitting aquatic snails	Aquatic habitats used by people immediately downstream of snail breeding area (approx. 2 km).

can be identified (Figure 4.3, part 8). Each biophysical effect can have a different range of influence. Examples are the range of influence of changes in air (downwind), surface water (downstream), and groundwater (aquifer) (see Table 4.2). Only when the biophysical effects have been identified and their range of influence established can the potential area of influence of an activity be established. This reasoning shows that the idea of making a baseline description at the start of an assessment, an idea promoted and automatically copied in many EIA frameworks, in fact is impossible. One first has to have a clear picture of the type of activities and their range of influence through biophysical and social effects before one can have an idea of the area that needs to be studied. Starting with a baseline study often leads to nonfocussed, wasteful data collection efforts and consequently a waste of time and resources.

If the area of influence is known, the type of land use and/or ecosystems under the influence of each biophysical effect can be described. Each type of land use or ecosystem provides a unique set of ecosystem services.

Biophysical effects influence the composition or structure of biodiversity, or affect key processes that create or maintain biodiversity (see Chapter 2 for an extensive elaboration). Through these biophysical effects, the ecosystem services provided by biodiversity may change. As different ecosystems or land use types will respond differently to these biophysical effects, the impact on the ecosystems services has to be determined for each individual ecosystem or type of land use.

A change in the services that are provided by the natural environment will lead to a change in their value for human society (Figure 4.3, part 9), and has an impact on human well-being. The ecosystem services concept is anthropocentric, as it translates nature into services for human society. Society puts a value on these services. Values can be expressed in economic, social, or ecological terms, providing valuable information for decision making on proposed (alternative) activities.

Example: A change in the lumber productivity of an exploited forest, caused by lowering groundwater, may lead to loss of income and jobs, and to marginalisation of rural villages and a reduced Gross Domestic Product for the region (values expressed in economic and social terms); similarly the drop in groundwater level may affect the forests capacity to maintain certain levels of biological diversity (value expressed in ecological terms).

Different (groups of) stakeholders can value the effects on ecosystem services differently. Some types of ecosystem services may not even be recognised at all among certain groups of stakeholders. For example, wetlands near Alexandria (Egypt) have a strong potential for recreational use by the city's five million plus inhabitants, thus creating an incentive for the cleaning up of these highly polluted wetlands. However, none of the participating formal stakeholders from the water sector was willing to recognise this ecosystem service identified by an external expert in a pilot SEA for improved water management in the region (IPTRID, 2005). In Western Europe recreation and quality of living environment is one of the most important arguments for wetland restoration and water quality improvement because of the strong demand from inhabitants. This example shows that the recognition of ecosystem services is context dependent; one has to know the exact nature of the ecosystem or land use type where biophysical effects occur and one has to know the use that a local society makes of these services (including people's perception of these services). This relates to the norms and values system of a society, represented by its laws, regulations (customary rules or formalised legislation), and other normative structures. The important consequence of this notion of context dependency is that impacts cannot be determined by external experts only, but that representatives of the local society have to be consulted. The importance of broadly identifying stakeholders when looking for ways to put biodiversity on the agenda in environmental assessment is apparent. More strongly stated, it can be argued that ecosystem services without stakeholders will go unnoticed in environmental assessment, and as a consequence biodiversity in general will receive less or no attention.

Social effects and social impacts

To make the impact assessment framework complete we also have to address the direct social impacts resulting from an activity. Depending on the characteristics of the existing community (and on availability of mitigation measures), social effects may cause social impacts (Figure 4.3, part 11).

Example: In a situation of relatively low local availability of labour, the creation of new job opportunities by an activity will attract migrant labour thus changing the demography of the area (social effect). This may cause stress on the existing social services (school, health) or, when coming from a different cultural background, may lead to conflicting cultural values (social impact).

As human beings or society as a whole are, contrary to the biophysical world, able to respond actively to impacts, the experience of social impacts in some cases leads to induced social effects (Figure 4.3, part 10) (van Schooten *et al.*, 2003).

Example: The marginalisation of rural villages (impact) forces people to migrate to urban areas (induced social effect).

This last relation of induced social changes in theory creates endless loops in the framework. From a theoretical point of views this is correct, as the world will never stop changing; any human activity will trigger changes, which in their turn will result in new changes. From a practical point of view it is of course not desirable to have endless analytical exercises to describe potential impact mechanisms in as much detail as possible. This is exactly the reason why scoping in environmental assessment is so important. The framework can be used in an iterative manner to provide a rapid overview of the potential chains of cause and effect. In consultation with experts and stakeholders, the magnitude and relevance of each of this long list of potential impacts can be determined, including the extent to which induced effects have to be taken into account. By providing insight in the nature of impact mechanisms, the process of weighing the relevance of impacts can be kept transparent, provided that institutional arrangements are in place where relevant stakeholders can be part of the weighing process on equal terms. A further advantage is the framework's separation of 'facts' and 'values'; stakeholders can possibly agree on the major physical and social changes that are happening, even when they may value these changes differently. Environmental assessment is often carried out as a technocratic exercise with minimal involvement of society. Technocratic information does not necessarily represent the diversity of perceptions in society (very convincingly shown by Stolp, 2003, 2006).

The framework as a boundary object

Complex problems and boundary crossing

What work can a framework as presented in this chapter do? It is not intended to be a 'standardised procedure' to be mechanically applied in any new situation – like a checklist for instance. Neither is it intended to be a predictive analytical model that can conclusively answer 'what-if' questions and make authoritative predictions. It is a device to be utilised for the facilitation and systematising of the interaction between those involved in decision making on human interventions with social and environmental impacts. The framework is an instrument that mobilises available societal and environmental knowledge regarding a certain issue or situation, and provides a structure, process, and language for learning and decision making. Frameworks of this kind are mediators (cf. Morgan and Morrison, 1999). They mediate between different types of knowledge: natural and social science knowledge, lay and expert knowledge, knowledge about facts and knowledge about values, and so forth. They also mediate between interest groups and the individuals representing these in the process of negotiating knowledge in order to make sensible decisions. Such frameworks thus do not produce or predict 'solutions' by themselves, but their active use by those involved in a certain problem situation can help to find sensible and feasible ways forward.

In the science and technology studies literature devices performing such functions are called 'boundary objects' (Star and Griesemer, 1989). This is because effectively addressing complex problems requires the mobilisation of different types of knowledge and involvement of different interest groups. Managing biodiversity is typically such a complex problem, whether addressed at the level of a miniwatershed or the earth system. The knowledge required for addressing such complex problems is developed and organised in different disciplines and organisations, while different interest groups not only have different but often also conflicting interests. This constitutes serious problems in the communication, interaction, integration and required decision making. Addressing biodiversity management issues requires crossing boundaries in knowledge and interest for joint problem solving. To facilitate and systematise the process of boundary crossing, devices are needed that assist in this, so called boundary objects. The framework presented in this book is meant to operate as a boundary object linking the worlds and knowledge of procedurally oriented environmental assessment experts, contents-oriented social and ecological sciences experts, and result-oriented decision makers (see Figure 4.4). In this section we discuss the conceptual ideas underlying the notion of the framework as a boundary object.

Knowledge: salience, credibility, and legitimacy

One of the boundaries to be crossed to make science and technology usefully contribute to environmental assessment is that between the domains of research and policy, the concern being how knowledge can be translated into action. According to Cash et al., (2003) information requires three attributes to successfully cross the boundary from research to policy: *salience*, *credibility*, and *legitimacy*. The problem is that actors in different domains perceive and value these attributes differently. 'Traditionally, scientists, managers and scholars of science, technology and policy have focused on credibility – how to create authoritative, believable, and trusted information'. (Cash et al., 2003: 2) Many, particularly scientists, tend to assume that 'good science' more or less automatically leads to 'good decisions'. In policy practice, the recurrently experienced problem is that 'decisions makers [are] not getting information that they need and scientists [are] producing information that is not used'. (Cash et al., 2003: 1) This suggests that more is needed than credibility (scientific plausibility and technical adequacy) of information: salience and legitimacy. Salience is about the relevance of the information/knowledge

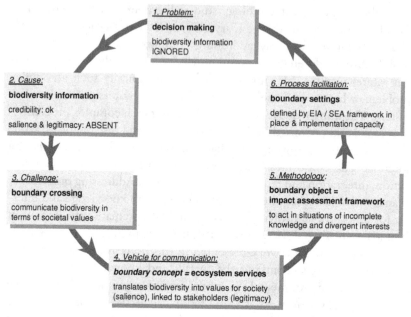

Figure 4.4 Biodiversity information for decision–making: a border-crossing problem.

for the different actors involved in the process (receiving the right kind of knowledge at the right time). Legitimacy is about political and procedural fairness – whether the information/knowledge sufficiently speaks to the different concerns of the multiple actors involved.[7] What Cash *et al.* conclude is that information/knowledge has to be salient, credible, and legitimate *simultaneously* to become actively used in decision making. (Cash *et al.*, 2003: 5) For successful translation of knowledge into action a process is required that strikes a balance between the three attributes of salience, credibility and legitimacy. This balancing act is an interactive social process of negotiation, in which boundary objects like the framework presented in this chapter can be usefully deployed.

---

[7] 'Paying too little attention to salience is exemplified in the case of the Global Biodiversity Assessment, in which the primary intended audience (parties to the Convention on Biological diversity) had little interest in the kinds of questions that were being asked by the assessors. Information relevant to their decision making was not produced and the assessment was largely ignored by the intended audience (Raustiala and Victor, 1996). Likewise, too little focus on legitimacy is seen in the early stages of the IPCC, in which developing country participants began to question the lack of Third World scientists and perspectives in the assessment process'(Agrawala, 1998)' Cash *et al.*, 2003:2)

Interdisciplinarity: boundary concepts, boundary objects, and boundary settings

Problems of environmental management and sustainable development (including biodiversity management) are inherently complex phenomena in the sense of having multiple dimensions. Because 'society has problems; universities have departments' (Abelson, 1997) the production and deployment of knowledge for addressing such problems requires the integration of different disciplinary contributions. In the literature on this topic the most common typology of forms of integration is to distinguish between *multidisciplinarity*, *interdisciplinarity*, and *transdisciplinarity* (Klein, 1990 and 1996; Pohl, 2005). Multidisciplinarity refers to collaboration in which different disciplines study different aspects of a problem without influencing each other's theory and method – an additive or synoptic strategy. Interdisciplinarity refers to those forms of collaborative research in which there is a joint problem definition/research question and in which the answer to the question depends on all contributions – an integrative strategy. In interdisciplinarity the different disciplines have to influence each other because, as a minimum, they need a common language for the joint question addressed. Transdisciplinarity refers to the type of interdisciplinary research that involves different interest groups associated with the problem under investigation in the design and implementation of the research – an integrative plus democratising strategy. Although interdisciplinarity can still be 'science driven', transdisciplinarity is driven by societal problems and takes the different knowledges regarding the problem focussed on very seriously. In situations where interdisciplinarity or transdisciplinarity are required for effective analysis and decision making, as is the case in most environmental assessment situations, the problem of boundary crossing presents itself. Boundaries have to be crossed among three main domains: (i) research, (ii) policy, (iii) and society. Within these domains internal boundary crossing problems exist.

Effective boundary crossing requires the following three elements:[8]

(1) The development of *boundary concepts* that allow us to think, that is, conceptually communicate about the multidimensionality of the issues that we study and address.

---

[8] A fuller discussion of boundary concepts, objects and settings can be found in Mollinga, (2008).

(2) The configuration of *boundary objects* as devices and methods that allow us to act in situations of incomplete knowledge, nonlinearity, and divergent interests – characteristics of most concrete natural resource management situations.

(3) The shaping of *boundary settings* in which these concepts, devices, and methods can be fruitfully developed and effectively put to work.

Put differently, there is a cognitive, an instrumental, and an institutional dimension to boundary crossing.

Boundary concepts

Boundary concepts are words that operate as concepts in different disciplines or perspectives, referring to the same object, phenomenon, process or quality of these. Boundary concepts are often, certainly initially, 'loose concepts' (Löwy, 1992). 'Loose concepts' are not very precise, openly defined concepts and provide the conceptual space to explore and elaborate the different meanings and dimensions of that concept. It is more important to actively explore the interconnections than to impose an authoritative definition, which often represents a particular perspective. Imposing or pushing a certain understanding can easily become an act of power to establish dominance of one perspective over others. An example of a boundary concept is the concept of 'water control' as used in different segments of water resources studies: as hydraulic/technical control, as managerial/organisational control, and as sociopolitical and economic control of water use (Mollinga, 2003). In this case the concept, although it is the same word, carries very different meanings and refers to the different dimensions of water use. 'Water control' becomes a true boundary concept when it starts to be recognised that it is a single multidimensional phenomenon, and the different disciplines or perspectives start asking questions about how these different dimensions are interconnected. Such a process has happened with regard to a central boundary concept in this book, that of 'ecosystem services'. This concept summarises the multidimensionality of the three objectives of the biodiversity convention: (i) the conservation, (ii) sustainable use, and (iii) equitable sharing of biodiversity. It brings together the scientific worlds of ecology, economy, and social sciences in a normative policy context. With the MA the legitimacy of the concept has been enhanced considerably, and the concept 'codified', where salience and credibility were already largely established.[9]

Boundary objects

Boundary objects are the main object of this chapter as we have declared the impact assessment framework to be a boundary object. Ecosystems are complex systems that exhibit for example nonlinear behaviour and have many external linkages. Social mechanisms at work have the additional property that they can change. Social systems are open systems in which human actors can learn to accept or otherwise decide to change the structural properties of the system. So, situations to be studied in environmental assessment are not only location specific, but also historical, that is, variable in both space and time, and their development is inherently unpredictable. We are unlikely ever to reach the point of sufficient knowledge – given the evolving, changing nature of the systems concerned. Therefore 'shortcuts to progress' are needed. Science has to find ways to contribute useful and usable knowledge to decision-making processes in situations structurally characterised by incomplete, and sometimes unreliable, data, uncertainty, nonlinearity, and unpredictability. We suggest there are, in practice, three different ways to 'cut through' this problem, trying to adequately respond to society's demand for useful and usable knowledge to address complex natural resources management problems. We call these three 'routes', as they are ongoing journeys that could possibly (maybe even preferably) also converge.

The *analytical route* is the route that attempts the comprehensive modelling of the behaviour of real complex systems, which lend themselves for use as decision support systems (DSS). In spite of a huge amount of (disciplinary as well as interdisciplinary) research, the track record of such science-driven decision support tools is rather weak, that is, very few make it to actual active use, at least in less-developed countries, but probably in industrialised countries too.[10] DSS approaches

---

[9] Salience through the rapid loss of biodiversity and consequent proliferation of environmental and social problems, policy, and civil society attention for these (exemplified, for example, by the Biodiversity Convention and an active NGO community), and the persistence of environmental problems. Credibility because the concept captures the complexity of socioecological systems quite well.

[10] For the reasons for poor uptake of crop simulation models for agricultural decision making by farmers and agricultural extension agents, see Stephens and Middleton (2002).

tend to have a strong 'good science – good decisions' perspective and tend to underemphasise the salience and legitimacy dimensions of information/knowledge.

The *assessment route*, in which the development of 'frameworks' is the central element, is at present probably the most common strategy in the field of natural resources management to achieve integration and interdisciplinarity in decision-making contexts. With 'framework' we refer to a conceptual construct with limited theoretical (explanatory) ambition as such but which is mainly oriented towards bringing together different pieces of knowledge in a 'workable' manner. Frameworks are simplified, generic conceptual models with practical purposes, ranging from ordering data to be collected to assisting decision making. They are simplifications and abstractions of complex realities, summarising the main features, factors and mechanisms in relation to a given problem in a manner that allows deriving practical conclusions for action from it.

'Frameworks' are typical examples of boundary objects, building connections between the worlds of science and that of policy, and between different knowledge domains. Although developed for decision making and deriving guidance for action, (assessment) frameworks do not always address the process dimension of their use as an explicit part of the framework. The framework presented in this chapter is content oriented, as it is intended to be used in existing procedural frameworks, for example, in Environmental Impact Assessment (EIA). EIA may be mandatory as defined in law and policy, and the process therefore may already be defined. In the case of strategic environment assessment, the existing planning mechanism usually prescribes the process aspects of the assessment. The environmental assessment framework developed in this chapter consequently is flexible enough to be applied in procedurally different settings. It therefore purposefully does not prescribe process aspects.

The *participatory route*: processes as boundary objects. In many natural resources management situations in developing countries (and industrialised countries to a lesser, but certainly not negligible degree) basic information necessary for understanding even only physical behaviour of a resource concerned has not been collected. Historical data is often not available or not reliable, and cannot be recreated. Researchers (and decision makers) often have to accept they will have to deal with absence or partial and fragmented availability of data and information. This is not only a historical problem. It is very unlikely that there will ever be sufficient resources available to do the fine-grained data collection in all relevant places that is necessary for precise analysis. What some have concluded from this is that approaches for analysis and assessment have to be designed in such a way that information generation and iterative learning-by-doing are central features of the approach. Moreover, natural resources management situations can be highly heterogeneous. Biophysical heterogeneity is evident in some landscapes, such as steep mountain areas with very different ecosystems at very close distance, but landscapes that are characteristic of many areas. Social heterogeneity can take many shapes and forms. In terms of knowledge, different groups of people may be the carriers of different kinds of knowledge, and there may be large social barriers for sharing that knowledge across groups. Knowledge held by state and scientific organisations may be very poorly accessible to marginalised communities or to the public at large. The existence and value of local knowledge may not find acknowledgement by state and scientific organisations. Different groups may also have very different understandings of apparently the same problem. Underlying such heterogeneities and divisions are often broader social relations of class, gender, caste, and other categories, and regional, organisational, or other vested interests – knowledge is indeed power in many situations.

There is increasing pressure to enhance 'stakeholder involvement' in natural resources management, governance, and policy-making processes. However, in many contexts a process may not be defined at all and many interest groups excluded from the processes of decision making. The environmental assessment process then becomes a vehicle for increasing inclusiveness of the decision-making process. The framework presented in this chapter prominently includes the identification of relevant stakeholders, with the explicit intention to include such stakeholders in the assessment process. This combination of an assessment framework with participatory approaches to decision making, leads to the third component of boundary crossing.

Boundary settings

Boundary settings are the institutional arrangements in which suitable boundary concepts can be fruitfully developed and explored and in which adequate boundary objects can be designed and deployed. Boundary settings are of two kinds, those internal to the assessment process, and those that determine how that process is embedded in society. The first set of boundary settings refers to the issues related to collaborative work within teams of individuals or consortia of organisations. This includes how a research project or programme organises itself – how it creates subunits to implement the research work, nowadays often called 'work packages', which data sharing procedures it adopts, what funds allocation procedures are used within the project, how it organises communication among partners, how quality control is assured, on what criteria staff working in the project is recruited, what frameworks for internal learning are created, and so forth. Problems in interdisciplinary research projects and teams can be divided into three kinds (Mollinga, 2008):

(1) Syntactic or language and communication problems;
(2) Semantic problems or differences in approaches and paradigms;
(3) Pragmatic problems or problems related to incentives and institutions.

The second, 'external' boundary settings may be thought to have a 'structural' and a 'process' aspect. The structural aspect refers to the institutional arrangements within which activities take place. This may be the legal, administrative rules and regulations of for instance conducting EIAs, as already mentioned above, statutory requirements for stakeholder consultation in regional planning, or legal provisions regulating the (public) access to information. There is enormous variations in the structure of institutional arrangements that enable (and constrain) environmental assessment. The process aspect of external boundary settings refers to the mechanisms through which teams communicate with external actors and organisations. This involves the way dissemination of findings and recommendations is done, the way accountability and auditing is organised, the (participatory) methodologies for 'stakeholder involvement', the strategies for influencing decision- making processes, and so forth. Chapters 5, 6, and 7 will provide more detail on the way in which environmental assessment processes can be institutionally embedded, and what approaches can be used to enhance biodiversity as an area requiring more attention.

Conclusion: the importance of boundary work

The past 10–15 years of problem solving oriented interdisciplinary research is starting to produce a literature that documents and assesses the experiences with 'what works and what does not work' in collaborative research. Telling titles of recent publications on this front are *Managing the interface* (Moll and Zander 2006) and *Design principles for interdisciplinary research* (Pohl and Hirsch Hadorn (2006). The literature on interdisciplinary and transdisciplinary practice shows that a lot of 'boundary work' and 'boundary management' is necessary to align the different views, interests, approaches, and so forth into a joint endeavour, and to get the structure and process right to achieve the objective of interdisciplinary analysis and action. There are many different kinds of boundary objects for different kinds of boundary crossing, ranging from repositories (such as databases), expert systems, joint protocols, models of different kinds (physical, mathematical, simulation, etc.), assessment frameworks, devices/technologies like computer-aided design (CAD) software (see Carlile, 2002), and also people can function as boundary objects (or, rather, perhaps, subjects) (see Frost *et al.*, 2002).

For biodiversity information in environmental assessment to be taken into account in decision making, it needs to have enough salience (and probably legitimacy). In other words, the information needs to be translated into decision maker's language (see Figure 4.4). The border between biological sciences and the decision-making arena has to be crossed. For this purpose the concept of ecosystem services serves as a boundary concept, translating biodiversity into values for society, linked to interest groups. In this chapter we have developed an assessment framework as a boundary object aiming at better integration of biodiversity in environmental assessment, which requires an interdisciplinary or transdisciplinary approach. The external boundary settings are largely defined by the environmental assessment frameworks in place; internal boundary settings are determined by environmental assessment capacity in place (in terms of quantity and quality of available expertise). The rest of this book is an attempt to get both internal and external boundary settings right.

Appendix to Chapter 4: indicative list of ecosystem services

All services listed below can be further detailed depending on the area under analysis and the nature of the product or service obtained. In principle the list is endless because valuation of services by society is constantly changing and new types of services may emerge. For example, before the discovery of the hole in the ozone layer, the protection service provided by ozone was never recognised. Similarly, carbon sequestration in biomass has only recently emerged as an important service to combat climate change. The prospect of new potential uses of genetic material is inconceivable. The list below is an updated version of Slootweg *et al.*, 2006).

Provisioning services: harvestable goods
Natural production

- timber
- firewood
- grasses (construction and artisanal use)
- fodder and manure
- harvestable peat
- secondary (minor) products
- harvestable bush meat
- fish and shellfish
- drinking water supply
- supply of water for irrigation and industry
- water supply for hydroelectricity
- supply of surface water for other landscapes
- supply of groundwater for other landscapes
- genetic material

Nature-based human production

- crop productivity
- tree plantations productivity
- managed forest productivity
- rangeland/livestock productivity
- aquaculture productivity (freshwater)
- mariculture productivity (brackish/saltwater)

Regulating services responsible for maintaining natural processes and dynamics
Biodiversity-related regulating services

- maintenance of genetic, species and ecosystem composition
- maintenance of ecosystem structure
- maintenance of key ecosystem processes for creating or maintaining biodiversity

Land-based regulating services

- decomposition of organic material
- natural desalination of soils

- development/prevention of acid sulphate soils
- biological control mechanisms
- pollination of crops
- seasonal cleansing of soils
- soil water storage capacity
- coastal protection against floods
- coastal stabilisation (against accretion/erosion)
- soil protection

Water-related regulating services

- water filtering
- dilution of pollutants
- discharge of pollutants
- flushing/cleansing
- biochemical/physical purification of water
- storage of pollutants
- flow regulation for flood control
- river base flow regulation
- water storage capacity
- ground water recharge capacity
- regulation of water balance
- sedimentation/retention capacity
- protection against water erosion
- protection against wave action
- prevention of saline groundwater intrusion
- prevention of saline surface water intrusion
- transmission of diseases
- suitability for navigation
- suitability for leisure and tourism activities
- suitability for nature conservation

Air-related regulating services

- filtering of air
- carry off by air to other areas
- photochemical air processing (smog)
- wind breaks
- transmission of diseases
- carbon storage
- protection against cosmic radiation (ozone layer)
- climate regulation

Carrying services

- suitability for human settlement
- suitability for leisure and tourism activities
- suitability for nature conservation
- suitability for infrastructure

Cultural services providing a source of artistic, aesthetic, spiritual, religious, recreational, or scientific enrichment or nonmaterial benefits.

Supporting services necessary for the production of all other ecosystem services

- soil formation
- nutrients cycling
- primary production
- evolutionary processes

5 · Environmental assessment

Arend Kolhoff, Bobbi Schijf, Rob Verheem, and Roel Slootweg

Introduction

While the biodiversity assessment framework that this book promotes is relatively new, the process into which this framework should be integrated is not. Environmental assessment has been around for more than 40 years and is practiced in some form or another in most countries around the world. The principle behind environmental assessment is deceptively simple: it directs decision makers to 'look before they leap'. An environmental assessment should bring into focus what the likely environmental effects of a project or plan could be, before decisions on that project or plan are made. When there is a clear insight into the environmental consequences, decision makers are in a better position to direct development into a more sustainable course. Of course, decision makers do not direct development on their own. Most plans or projects concern a range of actors, from governments to the business sector and the public arena. For this reason, environmental assessment does not merely provide information but brings the various parties together to discuss this information. It provides a process for them to come to a shared understanding of the possible effects and to determine what this knowledge should mean for the plan or project at hand.

Since its early beginnings, the field of environmental assessment has expanded, both in scope and in application. Practitioners now recognise two levels of environmental assessment: (i) Environmental Impact Assessment (EIA) that is applied at the level of individual projects, and (ii) Strategic Environmental Assessment (SEA) which is applied to policies, plans, and programmes (see Box 5.1). First, the origins and early development of EIA and SEA are briefly described in this chapter. It then continues to set out, for EIA and then for SEA, some basic concepts, as well as a selection of best practice principles that have been drawn from practice and from academic research into the effectiveness of these tools. Subsequently this chapter discusses recent trends in thinking about

---

Box 5.1.   Defining policies, plans, and programmes

Policy: A general course of action or proposed overall direction that a government is or will be pursuing and that
guides ongoing decision making.

Plan: A purposeful forward looking strategy or design,
often with coordinated priorities, options, and
measures that elaborate and implement policy.

Programme: A coherent, organised agenda or schedule of commitments, proposals, instruments, and/or activities that
elaborate and implement policy.

Sources: Sadler and Verheem (1996) and OECD (2006a)

---

and the application of both EIA and SEA. The practise of EIA and SEA
benefits from an active professional community and a healthy publication
record. It is beyond the purpose of this book to summarise this literature.
Instead the authors have drawn selectively from printed sources, as well as
their own experience, to put this chapter together. This chapter provides
general views on EIA and SEA, and as such is 'poor in biodiversity'.
When fitting, linkages to biodiversity will be highlighted through the
impact assessment framework. SEA case examples are presented in the
Annex of this book, all dealing with biodiversity issues.

Origins and early development of EIA and SEA

The history of EIA is easier to trace than the beginnings of SEA. Generally, the U.S. National Environmental Policy Act (NEPA) is credited
with first institutionalising EIA. It did so in 1969, in response to the
growing concern about environmental degradation that was building at
that time. Several other countries followed suit in the 1970s, predominantly those in the Western world. Then, in the 1980s, EIA application
began to spread more extensively. The European Union instituted EIA
legislation in member states, and EIA became part of the World Bank
operations. In the 1990s, other international finance institutes, such as
the Asian Development Bank and the European Bank for Reconstruction and Development, adopted EIA, and EIA became legally embedded
in developing countries as well. By 1997, more than 100 countries had
an EIA system in place (Wood, 2003; Sadler, 1996). Since then, there

has not been an updated count, but in the authors' estimation, there are probably no more than ten countries where EIA has not been introduced, and these include countries at war as well as some of the small island states.

In comparison to EIA, SEA is less widespread at present, but its application is rapidly catching up. If SEA is seen as a successor to EIA, as it often is in the literature (Bina, 2007), the lag is easily explained. Several years of practice with EIA showed that cumulative and large-scale effects could not be addressed adequately at the project level, and so a new instrument was needed to assess such effects at the appropriate strategic level: that of policies, plans, and programmes. However, it can also be argued convincingly that EIA and SEA actually originated at the same time. The U.S. NEPA, which is credited as EIA's birthplace, did not differentiate between the project level or strategic level; it intended all levels of decision making to be supported by environmental assessments (Bina, 2007). That SEA practice initially did not take off at the same rate as EIA is probably attributable to the more complex nature of strategic assessment, the long-standing preoccupation with economic priorities in strategic decision making, and the perception that sufficient assessment tools already existed. In any case, by the 1980s a distinct SEA practice was gaining momentum. Canada, New Zealand, and the Netherlands were amongst the first countries to develop a regulatory basis for SEA (Dalal-Clayton and Sadler, 2005). In the 1990s many more developed countries embedded SEA into regulation. SEA practice was also starting to emerge in South Africa (Dalal-Clayton and Sadler, 2005). In 2005, Dalal-Clayton and Sadler estimated that more than 25 countries had SEA regulation in place.

A very recent expansion of the application of SEA is the European Union SEA directive, which came into effect in 2006 (European Council, 2001). All 25 EU member states are now faced with the legal obligation to apply SEA to plans and programmes. If a member state's own regulatory framework for SEA is not fully developed, then the EU SEA Directive applies directly. Non-EU European countries Iceland and Norway also have SEA legislation in place. In Asia, SEA has now been regulated in China, Taiwan, and Vietnam (Liou and Yu, 2004; World Bank, 2006). Sri Lanka has also recently adopted an SEA regulation. Add to this list the early implementers of SEA, and the current total of countries with SEA legislation is brought to 35. However, the adoption of SEA regulation is spreading so rapidly, that this number is likely to be dated the moment this book goes to print. Countries that are in the process of developing their own SEA systems include Indonesia, the Philippines (Briffett *et al.*, 2003), Japan (which already has SEA practice at the local planning level) (World Bank, 2006), Turkey (Innanen, 2004), and the southern and eastern European countries that are introducing SEA in preparation for EU membership (including Croatia, Bosnia, Serbia, Montenegro, and Albania). Lebanon, Jordan, Tunisia, and Egypt are reportedly also working on SEA frameworks (Chaker *et al.*, 2006, El-Fadl and El Fadel, 2004). South Africa has had a steady SEA practise for more than ten years, despite the weak legal basis. In Latin America the SEA track record is patchy but advanced in places.

One of the driving forces behind this growth in application is the United Nations Economic Commission for Europe (2003) Protocol on Strategic Environmental Assessment to the Convention on Environmental Impact Assessment in a Transboundary Context (the 'Kiev Protocol'). The protocol is open to all UN members and was signed by 38 countries in July 2008. Interest in SEA is also sparked by the call for more holistic, integrated, and balanced strategic decision making made in influential initiatives, such as the 2002 World Summit on Sustainable Development, and the Millennium Development Goals. Millennium Development Goal 7, environmental sustainability, is supported by a target that reads like an SEA mission, namely 'integration of the principles of sustainable development into country policies and programmes to help reverse the loss of environmental resources'. International financing institutions and cooperation organisations, such as the World Bank and CIDA, have played an important role in introducing SEA to developing countries by initiating and funding many SEA case studies.

The early developments of EIA and SEA share this ever-expanding scope of application, both globally and in terms of the types of projects and plans to which they are applied. Both have also evolved from a more narrow biophysical focus to a broader inclusion of different types of effects. Social and economic impact assessments have been increasingly gathered under the environmental umbrella. This opened the field up to new categories of information on possible impacts, as well as bringing new methods into the range applied within environmental assessment. Other expansions include the assessment of effects on human health, transboundary effects, and applications in postconflict or postdisaster situations. That is not to say that all, or even most, impact assessment reports produced consider such a wide range of effects, but it is becoming the standard of 'good practise' within assessment to look beyond the effects on the separate components of the direct biophysical environment.

Aside from these commonalities, however, EIA and SEA also have major conceptual differences that will become clear when each is described separately below.

Environmental Impact Assessment (EIA)

Generally accepted procedural framework, with existing variations

Environmental Impact Assessment (EIA) can be defined as (IAIA, 1999) 'the process of identifying, predicting, evaluating and mitigating the biophysical, social, and other relevant effects of development proposals prior to major decisions being taken and commitments made.' The objectives of 'good practice' EIA are: (i) to ensure that environmental considerations are explicitly addressed and incorporated into the development decision–making process in a participatory way; (ii) to anticipate and avoid, minimise or offset the adverse significant biophysical, social. and other relevant effects of development proposals; (iii) to protect the productivity and capacity of natural systems and the ecological processes which maintain their functions; and (iv) to promote development that is sustainable and optimises resource use and management opportunities. EIA is a process of evaluating the likely environmental impacts of a proposed project or development, taking into account interrelated socioeconomic, cultural, and human health impacts, both beneficial and adverse. Participation of relevant stakeholders, including indigenous and local communities, is considered as a precondition for a successful EIA. Although legislation and practice vary around the world, there is little discussion about the fundamental components of an EIA, which would necessarily involve the following stages:

(1) *Screening* is used to determine which proposals should be subject to EIA, to exclude those unlikely to have harmful environmental impacts and to indicate the level of assessment required. Types of existing screening mechanisms include:

(i) Positive lists identifying projects requiring EIA (inclusion lists). A disadvantage of this approach is that the significance of impacts of projects varies substantially depending on the nature of the receiving environment, which is not taken into account.

(ii) A few countries use (or have used) negative lists, identifying those projects not subject to EIA (exclusion lists).

(iii) Lists identifying sensitive geographical areas in which projects would require EIA. The advantage of this approach is that the emphasis is on the sensitivity of the receiving environment rather than on the type of project.

(iv) Expert judgement (with or without a limited study, sometimes referred to as 'initial environmental examination' or 'preliminary environmental assessment').

(v) A combination of a list plus expert judgement to determine the need for an EIA.

(2) *Scoping* to identify which potential impacts are relevant to assess (based on legislative requirements, international conventions, expert knowledge, and public involvement), to identify alternative solutions that avoid, mitigate or compensate adverse impacts (including the option of not proceeding with the development, finding alternative designs or sites which avoid the impacts, incorporating safeguards in the design of the project, or providing compensation for adverse impacts), and finally to derive terms of reference for the impact assessment study. Scoping also enables the competent authority (or EIA professionals in countries where scoping is voluntary) to:

(i) Guide study teams on significant issues and alternatives to be assessed, clarify how they should be examined (methods of prediction and analysis, depth of analysis), and according to which guidelines and criteria;

(ii) Provide an opportunity for stakeholders to have their interests taken into account in the EIA;

(iii) Ensure that the resulting Environmental Impact Statement (or environmental impact assessment report) is useful to the decision maker and is understandable to the public.

(3) *Assessment and evaluation of impacts*: The actual study phase to describe and possibly quantify the likely environmental impacts of a proposed project or development, including the detailed elaboration of alternatives. EIA should be an iterative process of assessing impacts, redesigning alternatives and comparison. Assessing impacts usually involves a detailed analysis of their nature, magnitude, extent, and duration and a judgement of their significance, that is, whether the impacts are acceptable to stakeholders and society as a whole, require mitigation and/or compensation, or do not comply with formal norms or standards. The main tasks of impact analysis and assessment are:

(i) Refinement of the understanding of the nature of the potential impacts identified during screening and scoping and described in the terms of reference. This includes the identification of indirect and cumulative impacts and of the likely cause–effect chains;

(ii) Review and redesign of alternatives; consideration of mitigation and enhancement measures, as well as compensation of residual impacts; planning of impact management; evaluation of impacts; and comparison of the alternatives;

(iii) Identification of the remaining gaps in knowledge and information and an assessment of risks due to these gaps; and

(iv) Reporting of study results in an environmental impact statement (EIS) or EIA report.

(4) *Reporting*: the environmental impact statement (EIS) or EIA report consists of a technical report with annexes; an environmental management plan, providing detailed information on how measures to avoid, mitigate, or compensate for expected impacts are to be implemented, managed, and monitored; and a nontechnical summary for the general public. The environmental impact statement is designed to assist:

(i) The proponent to plan, design, and implement the proposal in a way that eliminates or minimises the negative effect on the biophysical and socioeconomic environments and maximises the benefits to all parties in the most cost-effective manner;

(ii) The Government or responsible authority to decide whether a proposal should be approved and the terms and conditions that should be applied; and

(iii) The public to understand the proposal and its impacts on the community and environment, and provide an opportunity for comments on the proposed action for consideration by decision makers. Some adverse impacts may be wide ranging and have effects beyond the limits of particular habitats/ecosystems or national boundaries. Therefore, environmental management plans and strategies contained in the environmental impact statement should consider regional and transboundary impacts.

(5) *Review* of the environmental impact statement, based on the terms of reference (scoping) and public (including authority) participation. The purpose of the review is to ensure that the information for decision makers is sufficient, focussed on the key issues, and is scientifically and technically accurate. In addition, the review should evaluate whether:

(i) The likely impacts would be acceptable from an environmental viewpoint;

(ii) The design complies with relevant standards and policies or standards of good practice where official standards do not exist;

(iii) All of the relevant impacts, including indirect and cumulative impacts, of a proposed activity have been identified and adequately addressed in the EIA. To this end, experts should be called upon for the review and information on official standards and/or standards for good practice to be compiled and disseminated.

(iv) The concerns and comments of all stakeholders are adequately considered and included in the final report presented to decision makers. The process establishes local ownership of the proposal and promotes a better understanding of relevant issues and concerns.

(v) The effectiveness of the review process depends on the quality of the terms of reference defining the issues to be included in the study and approval by the authority. Scoping and review are therefore complementary stages.

(6) *Decision making* on whether to formally approve the project or not, and under what conditions. Decision making takes place throughout the process of EIA in an incremental way to move from the screening and scoping stages to decisions during data collection and analysis and impact prediction, to making choices between alternatives and mitigation measures, and finally to the decision to either refuse or authorise the project.

(7) *Monitoring, compliance, enforcement, and environmental auditing*: EIA does not stop with the production of a report and a decision on the proposed project. Activities that have to make sure the recommendations from EIS or EMP are implemented are commonly grouped under the heading of 'EIA follow-up'. They may include activities related to monitoring, compliance, enforcement, and environmental auditing. Roles and responsibilities with respect to these are variable and depend on regulatory frameworks and performance by the responsible organisations.

Effectiveness of EIA

In theory EIA is a tool with an enormous potential to influence decision making. This potential is utilised maximally when the good practice principles developed by IAIA (1999) are fully applied. A minimum

Table 5.1 *Minimum and maximum ambition levels in EIA, simplified (sources: EU EIA directive (European Council, 1985; 1997) for minimum variant and IAIA (1999) for maximum variant).*

Aspects of EIA system	Minimum EIA variant	Maximum EIA variant
Objectives	Environmental protection	Environmental protection, sustainable development, well-informed, and participatory decision making
Scope of study	Environmental aspects	Environmental, social/health, and economic aspects
Environmental Protection	Mitigation measures	Alternatives, mitigating, and compensatory measures
Involvement of civil society	Low	Ambitious
Transparency, accountability	Limited	Complete
Quality assurance mechanism	Limited	Advanced

variant as such is not described in the literature, but a number of basic conditions should be fulfilled in order to justify the term 'EIA' for a decision support tool. The EU EIA directive (European Council, 1985, 1997) can be considered as an example of the minimum variant. EU member countries can build on this directive to develop their own more ambitious EIA regulatory framework. Table 5.1 lists the minimum and maximum ambition levels for various aspects of the EIA system; in principle, the EIA systems of all countries and institutes in the world can be positioned in this table. A country can have different ambition levels for different aspects.

One could assume that EIA must be an effective tool, achieving its objectives, because so many countries and international institutions have adopted it. So, the question arises: how effective is EIA in reality? A growing number of evaluation studies confirm that EIA is effective in developed countries, but that EIA is not or hardly effective in developing countries. Sadler (1996) executed an international landmark study on the effectiveness of EIA, mainly based on experiences of EIA practices in industrialised Western countries. He concluded that EIA is effective when it:

- improved project design and site selection;
- led to more informed decision making;
- resulted in more environmentally sensitive decisions;
- increased accountability and transparency during the development process;
- improved the integration of projects into their environmental and social setting;
- reduced environmental damage;
- created more effective projects in terms of meeting their financial and socioeconomic objectives; and
- contributed positively towards achieving a more sustainable development.

More studies have shown that EIA is achieving its objectives to a certain extent in industrialised countries: the Netherlands (Heuvelhof and Nauta 1996), Germany (Wende, 2002), Canada (Gibson, 2002), Hong Kong (Wood and Coppell, 1999), Denmark (Christensen *et al.*, 2003). However, without exception such evaluation studies also state that there still is opportunity to further increase the effectiveness of EIA (Doelle and Sinclair, 2006).

An extensive study in Denmark (Christensen *et al.*, 2003) found that EIA effectively influences the design of project proposals at three moments in the EIA process. The first influential moment is prior to the formal start of the EIA process (also found for the Netherlands by Heuvelhof and Nauta, 1996). The very fact of having an EIA process in place apparently forces proponents to adapt their project plans. Sometimes a proponent may decide to abandon a project altogether, if it becomes clear that environmental standards will not be met. This is known as the 'prevention effect' that only occurs in countries where a 'rule of law' is applied. The EIA process provides the second influential moment, leading to the 'dialogue and transparency effect'. From the formal start of the procedure until final decision making and licensing, the dialogue between proponent and government, in combination with public participation, was considered to have the most important effect on project design. The third influential moment is linked to the actual result of the EIA, leading to requested conditions in the license relating to, for example, an alternative measure or the implementation of mitigation measures (i.e. the 'conditionality effect').

The above does not mean, however, that all of the 56 high-income countries (World Bank, 2006) have an effective EIA system in place.

In 2002 the EU reviewed the application of the EU EIA Directive, adopted in 1985 and amended in 1997. It was concluded that none of the member countries had implemented the directive completely. Without mentioning 1 of the 15 member states by name it was concluded that some practised best practice EIA whilst others still had to remedy many weaknesses. Recent studies on the effectiveness of EIA in Greece (Androulakis and Karakassis, 2006) and Italy (Fischer and Gazzola, 2006) confirm the results of the EU study. The effectiveness of EIA in these countries is weak due to weak enforcement and insufficiently transparent project planning processes due to the highly politicised character of the process. Other high income countries, such as South Korea and Japan (World Bank, 2006), do not have fully effective EIA systems in place either, whilst some Middle Eastern states, such as the United Arab Emirates, do not even have EIA legislation.

The majority of EIA evaluation studies in low-, low middle-, and upper middle-income countries come to the conclusion that EIA is only modestly effective, mainly because the performance of the regulatory EIA framework and compliance by the responsible organisations is weak. These organisations have limited capacities, and their autonomy is often not respected by influential decision makers in a country. A review of EIA regulations in 12 East and Southeast Asian countries by the World Bank (2006) concluded that there is a sharp gap between the existing EIA/SEA legal system on paper and the poor level of implementation. Weak enforcement is considered as a major problem, reflected by EIAs executed when decisions have been taken already, a limited study of alternatives, a lack of information disclosure and weak public consultation. A comparative study of 21 selected Middle Eastern and North African countries concluded that performance of the EIA system is weak (El Fadl and El Fadel, 2004). Espinoza and Alzina (2001) concluded in a comparative review of EIA in 26 selected countries in Latin America and the Caribbean that failures persist in defining the coverage and scope of EIA studies, standardising review methods, monitoring environmental management plans, and involving the local community in all stages of the process. An extensive evaluation study of EIA in Tanzania (Mwalyosi and Hughes, 1998) concluded that EIA had very little impact on decision making. The main causes mentioned were: EIAs were late in starting and underresourced, compliance with EIA recommendations has been the exception rather than the rule, and consideration of alternative project options was often absent or extremely weak. A review of EIA in the Southern African countries (SAIEA, 2003) confirmed that the conclusions of the review study executed in Tanzania are applicable for the countries in Southern Africa.

Concluding one can state that EIA has a great potential to contribute to well-informed decision making and protection of the environment. Currently this potential is achieved in a small number of (predominantly) high-income, democratic countries. In many other countries there is a strong belief in the potential of EIA and the willingness to improve the effectiveness. However, there are also examples where the decision makers are not genuinely interested in accountable and transparent decision making. Nevertheless, EIA is considered a powerful environmental management tool all over the world. Only in a small number of Western countries is the potential of the tool utilised towards a reasonable extent. EIA does influence decision making and contributes to environmental protection, but even in these countries the full potential has not been achieved yet. The effectiveness of EIA can be regarded as a continuum with full potential at one side and no use of the potential at the other side. The question arises what factors contribute to a successful application of EIA.

State of the art: what is needed for effective EIA?

EIA has a legal basis in the majority of countries but seems to meets its objectives predominantly in Western democratic countries. Twenty years of capacity development programmes have learned that copying Western systems does not work or is even counterproductive (Cherp and Antypas, 2003). Three main factors seem to be of major importance: first, the capacities of the EIA system, including the regulatory framework; second, the capacities of the environmental compliance system; and third, the country specific context in which an EIA system functions. The latter determines the enabling environment of the first two factors as it influences the opportunities and constraints of EIA compliance and environmental compliance system performance. The EIA system is the organisational and administrative structure to implement EIA (Espinoza and Alzina, 2001). Performance of the EIA system is determined by its capacities and external factors. The availability of scientifically sound information is a requirement for good quality EIA studies and consequently well-informed decision making. Ideally information should systematically be gathered, analysed, and made accessible for third parties such as the civil society. When this requirement is not or only partly fulfilled the execution of EIA studies will be hampered and EIA cannot utilise its full potential as a tool to provide environmental information for decision making. This illustrates the linkages between the EIA system and its context, in this case the knowledge infrastructure in a country (Kolhoff et al., in press).

The output of an EIA system is an EIA report that is used for decision making and licensing. Whether the conditions set in the environmental licence are met in practice depends on the performance of the environmental compliance system. This performance is again the result of the capacities of the responsible organisations and the performance of the proponent and external factors, such as influential business agents and politicians. In many developing countries there is an imbalance in the focus of capacity development efforts, often emphasising the EIA capacity needs and largely neglecting the environmental compliance system. Capacities of both systems should be developed in parallel.

There is a growing insight that the national context, such as the political system and the socioeconomic structure of a society, has a major influence on EIA implementation and compliance. The checks and balances in a political system determine to what extent the formal autonomy of EIA and compliance implementing organisations is respected. In countries where division of powers between legislative, executive and judiciary are weak there is significant risk of corruption undermining the role of EIA. The ability and capacity of society to act as a counterveiling power is to a great extend determined by the political system. Even when political systems become more transparent, it can take a generation before society is able to participate effectively in EIA (Cherp, 2001; Purnama, 2003). This context should be taken as a starting point for developing EIA systems that do work in developing countries. Ghana is an example of a country that has developed a balanced EIA and environmental compliance system whilst taking the national context as a starting point (UNECA, 2005).

Strategic Environmental Assessment

Current thinking

The practice of SEA is less easily demarcated than that of EIA. Some countries, especially those with a strong planning tradition, had SEA-like assessment even before the term itself came into use. There are now a large number of assessment tools in planning that do not necessarily carry the label SEA, but have strong similarities. A few years ago, Sadler and Dalal-Clayton identified at least 15 different acronyms for SEA-type approaches (Dalal-Clayton and Sadler, 2005) (examples include integrated assessment, country environmental analysis, integrated trade assessment, poverty and social analysis, sustainability appraisal, strategic environmental analysis, etc.). There is no indication that this amount will grow smaller in the near future. A newcomer to this field could easily be dissuaded by the multiple definition and conceptualisations. However, the fundamental differences between approaches are fewer than might be assumed from existing publications. In 2006, the SEA Task Force of the OECD Development Assistance Committee brought together represen-tatives of a wide range of countries and international organisation with SEA experience. In the process of drafting SEA guidance, this diverse group also adopted a shared definition of SEA, which states that SEA is 'a family of tools that identifies and addresses the environmental conse-quences and stakeholder concerns in the development of policies, plans, programmes, and other high-level initiatives'(OECD, 2006a).

There is no generally agreed SEA procedure as such, no 'one-size-fits-all' approach. As planning processes vary greatly from context to context, and even case to case, SEA needs to be applied flexibly. Even when SEA is captured in a formal procedure in legislation (e.g. the SEA Directive of the European Union) there will still be great differences in how the SEA activities are undertaken, when and with whom, as illustrated in the description of SEA cases in the Annex. However, there is general agreement about the activities that make up an SEA process (OECD, 2006a). There is a logical sequence to these activities, but logic is certainly not the only, nor necessarily the dominant, principle governing a given planning process. Realistically then, the activities outlined here may take more or less effort, may follow each other sequentially or not, and some may be repeated or combined.

First phase: creating transparency and joint objective setting:

• Announce the start of the SEA and assure that relevant stakeholders are aware that the process is starting.
• Bring stakeholders to develop a shared vision on (environmental) problems, objectives, and alternative actions to achieve these.
• Check in cooperation with all agencies whether objectives of the new policy or plan are in line with those in existing policies, including environmental objectives (consistency analysis).

Second phase: technical assessment:

- Make clear terms of reference for the technical assessment, based on the results of stakeholder consultation and consistency analysis.
- Carry out a proper assessment, document its results, and make these accessible for all.
- Organise effective quality assurance of both SEA information and process.

Third phase: use information in decision making:

- Bring stakeholders together to discuss results and make recommendation to decision makers.
- Make sure any final decision is motivated in writing in light of the assessment results.

Fourth phase: postdecision monitoring and evaluation:

- Monitor the implementation of the adopted policy or plan and discuss the need for follow up action (OECD, 2006a).

Parallel to or integrated within a planning process?

A key factor that determines what an SEA process will actually look like for a given application is the degree of integration of the SEA activities into the planning process itself. Traditionally, SEA was often applied as a stand alone series of activities, in parallel to planning (left hand diagram in Figure 5.1). This might be a good way to build experience with SEA, but it is less effective in influencing the plan, programme, or policy. By the time results of the assessment are provided to the planning team and the public, a plan strategy has already been developed. Time and resources have been committed to this strategy, and major deviations from it are not likely to be welcome. This is not to say that a stand alone SEA cannot improve a plan at all. Even when decisions have already been taken, SEA can play a meaningful role, for example, to decide on necessary monitoring measures and mitigating actions, as a reference to compare plan implementation outcomes against, or to set the agenda for future policies and plans. However, when SEA is more integrated, it has better scope to influence planning (middle diagram in Figure 5.1). In this form of integration the planning and SEA activities are distinct, but each feeds the other at different stages in the process. Taking integration further, the

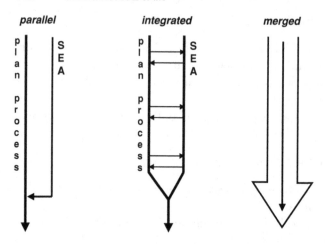

Figure 5.1 Combinations of SEA and planning process.

SEA and planning process could also become merged (right most diagram in Figure 5.1). The key SEA activities are then no longer separate from the planning process, but an integral part of it. This means that critical environmental issues are still identified, and alternative strategies will be developed and assessed, but there will not be an SEA team or SEA report as such. Instead, the assessment work continually informs the planning process, and communication and participation are organised at key moments around assessment and planning issues in combination. A merged SEA may well be the ideal, but this approach does require willing planners that already consider environmental concerns a clear priority. It is also more difficult to regulate, that is, to apply checks and balances that ensure a minimum assessment level.

One of the key questions for the integration of SEA and planning is when the assessment process should start. The answer to this depends on whether SEA is seen as a plan development tool or more as an impact assessment tool. If it is seen as a plan development tool, the SEA process needs to start early on, before the policy proposals exist. SEA then assists in the analysis of the problems that the plan needs to solve, and contributes to the development of proposals. If the SEA starts after the policy proposals have been developed, there is a stronger focus on impact assessment and the reactive identification of alternatives to avoid the negative impact identified. Examples from both approaches can be found in practice, but the assessment tool approach is probably more common. Again, while an SEA will be more effective with an earlier start, this is difficult to regulate. This stage of the planning process is often an inhouse phase, which is difficult to access unless planners themselves choose to make use of SEA at that time.

SEA is not EIA

It is important to point out that SEA is not EIA, because it is necessarily different in nature. At project level, decision making is about a concrete set of activities that makes up a specific development proposal. EIA, then, concentrates on the activity–effects relationships (Herrera, 2007). Strategic decision making is less about the concrete activities that will follow from the plan, as it is about identifying and assessing and comparing the different ways in which the plan can achieve its objectives. Also, the process of designing and approving a project is more amenable to linear structuring and simplification than the commonly more changeable and politically charged development of a plan or policy. As EIA aims at better projects, SEA aims at better strategies, ranging from legislation and countrywide development policies to more concrete sector and spatial plans. Ideally, SEA is applied at each planning tier, and higher level SEAs inform those at a less- strategic level so that there is no overlap in the assessments. In theory, if not always in practice, the process starts with a policy broadly describing objectives and setting the context for proposed actions, usually with a sectoral or geographic scope. Policy objectives are then translated into an action plan, further operationalised in programmes, and actual implementation is done through projects. Figure 5.2 shows an example of such tiering for the flood management planning in the Netherlands.

A national policy and strategy on water management provided the framework for decision making in the water sector. SEA was conducted for the National plan: space for rivers. An EIA was conducted for the design of interventions for the River Meuse. If it is undertaken in this tiered manner upstream from project considerations, SEA can help streamline EIA processes. The SEAs will consider broader environmental issues likely to be common to multiple project initiatives in a sector or in a region and is able to look at cumulative effects. Allowing the subsequent EIA processes to concentrate on impacts specific to individual proposals improves the efficiency and the effectiveness of the overall process.

Another aspect in which SEA is very different from EIA is the more expansive spatial and temporal horizons that are addressed. Where EIA

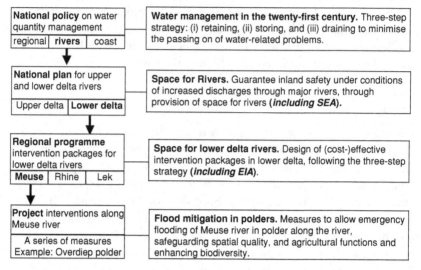

Figure 5.2 Hierarchy of policies, plans, and programmes: an example from the Netherlands.

(and earlier SEA approaches) address issues that could be expected in the intervention area within a relatively short time, the upstreaming of environmental assessment in the decision-making hierarchy leads to more and broader spatial and time horizons, that is, looking at effects elsewhere (such as in other countries) and later (effects on future generations). Table 5.2 sets out the key difference between SEA and EIA.

Effectiveness of SEA

Concretely, then, SEA improves planning by (i) structuring the public and government debate in the preparation of policies, plans and programs; (ii) feeding this debate through a robust assessment of the environmental consequences and their interrelationships with social and economic aspects; and (iii) ensuring that the results of assessment and debate are taken into account during decision making and implementation (Netherlands Commission for Environmental Assessment, not dated). By doing this, SEA promises to improve planning by increasing the quality of the plans, the credibility of decision making, and thereby the support for the plan implementation (See Box 5.2). However, while the practice of SEA has expanded rapidly, research into this practice has been much slower to develop. Seminal studies that took a broad-based review of SEA

Table 5.2 *Characteristics of SEA and EIA compared.*

SEA	EIA
Takes place at earlier stages of the decision-making cycle	Takes place at the end of the decision-making cycle
Considers broad range of potential alternatives	Considers limited number of feasible alternatives
Early warning of cumulative effects	Limited review of cumulative effects
Emphasis on meeting objectives and maintaining systems	Emphasis on mitigating and minimising impacts
Broader perspective and lower level of detail to provide a vision and overall framework	Narrower perspective and higher level of detail
Multistage process, continuing and iterative, overlapping components	Well-defined process with clear beginning and end
Focusses on sustainability agenda and sources of environmental deterioration	Focusses on environmental standards and norms and symptoms of environmental deterioration

---

Box 5.2. Advantages of SEA to decision makers (CBD, 2006)

SEA can offer the following advantages to decision making:
• Enhanced credibility of their decisions in the eyes of stakeholders;
• More knowledge of the social feasibility of a decision, thus avoiding resistance from unhappy local groups, bad image for planners, useless mitigating measures, and simply missing the bigger picture;
• Improved economic efficiency because potential environmental stumbling blocks for economic development are better understood;
• Bringing promising alternatives into focus;
• A better understanding of the cumulative impact of a series of smaller projects, thus preventing costly and unnecessary mistakes; and
• Better insight in the trade-offs between environmental, economic, and social issues, enhancing the chance of finding win–win options.

---

practice have been published in Sadler and Verheem (1999), Therivel and Partidário (1996), and Dalal-Clayton and Sadler (2005), but apart from these, analyses of SEA practice are fragmented and often limited to a small selection of case studies. As a result, it is not very clear whether and under which conditions SEA can deliver what it promises. Many of the studies that have looked at how SEA is working on the ground have focussed on the procedural and legal requirements of country or regional SEA systems. Such studies (see, for example, Chaker *et al.*, 2006) tend to show the wide variety of SEA systems that have been put in place. They might, for instance, observe that SEA is generally initiated by the proponent of the planning process, that review of the SEA is delegated to an independent body in some countries, or that participation requirements are very limited in some places and more comprehensive in others.

Other investigations have analysed SEA processes as they unfolded in cases studies. Depending on the amount and range of cases selected such analyses can manage to paint quite an informative overview of SEA practice in a specific jurisdiction. Retief (2007), for example, examined SEA practice in South Africa by comparing a selection of cases against a set of SEA process principles. He noted that practitioners in those cases did not sufficiently appreciate the dynamics of real decision making, and applied SEA too rigidly. In an effort to produce 'independent' SEAs and deliver 'objective' results, the practitioners neglected to strive for full consultation and political buy-in. Such studies yield important lessons but give little insight into how effective SEA is in influencing decision making on plans, programmes and policies. There are few empirical studies that look at this aspect. In their stock-take of SEA practice Dalal-Clayton and Sadler (2005) describe cases where SEA made a difference to decision making but also conclude that the potential for SEA to lead to better plans is far greater than is utilised. Runhaar and Driessen (2007) inventoried a number of studies that have looked at whether changes had occurred in the proposed plans, programmes, or policies as result of the SEA. These reveal different levels of impact: some studies have found no discernible direct impact of the SEA on the plans, while in other sets of cases most decision processes had been influenced. They also studied the effect of SEA in their own selection of four cases. In this sample a modest impact was clear on the decisions made. SEA influenced, for example, the measures chosen in the plans and the rejection of a proposed plan alternative.

The influence of SEAs on planning in the UK has been studied quite extensively, and shows a more impressive result. Regular surveys of local authorities that undertake SEA for their plans show that changes have been made to just over 80 percent of the plans for which SEAs had been undertaken in recent times, up from the surveys that had been undertaken before (Therivel and Walsh, 2006). Aside from this direct impact, roughly three-quarters of those surveyed felt that SEA improved planners' awareness of sustainability generally and the plan's sustainability in particular. In about half of the cases plan making was considered to have become more transparent and to have improved planners' understanding of their plan. About half of the respondents also felt that the SEA was effective, given the time and cost needed (Therivel and Walsh, 2006).

State of the art: what is needed for effective SEA?

Despite the limited empirical evidence, there is a reasonable consensus amongst those working in the field about how SEA should be undertaken for it to be effective. In 2002 the International Association for Impact Assessment published the SEA performance criteria (see Box 5.3), which

---

Box 5.3. SEA performance criteria (IAIA, 2002)

A good-quality SEA process informs planners, decision makers, and the affected public on the sustainability of strategic decisions, facilitates the search for the best alternative, and ensures a democratic decision-making process. This enhances the credibility of decisions and leads to more cost-effective and time-effective EA at the project level. For this purpose, a good-quality SEA process:

Is integrated

- Ensures an appropriate environmental assessment of all strategic decisions relevant for the achievement of sustainable development.
- Addresses the interrelationships of biophysical, social, and economic aspects.
- Is tiered to policies in relevant sectors and (transboundary) regions and, where appropriate, to project EIA and decision making.

Is sustainability-led

- Facilitates identification of development options and alternative proposals that are more sustainable.[1]

---

[1] i.e. that contributes to the overall sustainable development strategy as laid down in Rio 1992 and defined in the specific policies or values of a country.

Is focussed

- Provides sufficient, reliable, and usable information for development planning and decision making.
- Concentrates on key issues of sustainable development.
- Is customised to the characteristics of the decision-making process.
- Is cost-effective and time-effective.

Is accountable

- Is the responsibility of the leading agencies for the strategic decision to be taken.
- Is carried out with professionalism, rigor, fairness, impartiality, and balance.
- Is subject to independent checks and verification.
- Documents and justifies how sustainability issues were taken into account in decision making.

Is participative

- Informs and involves interested and affected public and government bodies throughout the decision-making process.
- Explicitly addresses their inputs and concerns in documentation and decision making.
- Has clear, easily understood information requirements and ensures sufficient access to all relevant information.

Is iterative

- Ensures availability of the assessment results early enough to influence the decision-making process and inspire future planning.
- Provides sufficient information on the actual impacts of implementing a strategic decision, to judge whether this decision should be amended and to provide a basis for future decisions.

were the result of a wide debate amongst association members. These are still considered a benchmark for SEA practice, although it is important to note that thinking, and publication, on effectiveness criteria has generally been dominated by authors from a limited range of countries, and is mostly based on experiences in Europe (Fischer and Gazzola, 2006). The studies into SEA effectiveness have brought certain performance criteria more directly into focus. For example, different effectiveness studies show the importance of flexible SEA that follows the decision-making process, as well as the value of stakeholder participation. Others emphasise the importance of political will to use the results of the assessment, which can follow from regulatory requirements as well as good SEA public relations (Hildén et al., 2004). Contextual factors, such as the degree to which the different interests involved coincide with SEA recommendations, are also particularly relevant (Runhaar and Driessen, 2007). Dalal-Clayton and Sadler (2005) have emphasised the need to strengthen SEA systems by building institutional arrangements that provide for quality control. By and large those that have looked into effectiveness predominantly stress that the need to continue reviewing effectiveness systematically to better understand what works and what does not.

New thinking and challenges in EIA and SEA

Drawing the fields of SEA and EIA back together we would like to highlight three current trends in environmental assessment thinking and practice: (i) increased attention to the assessment context, (ii) integration of effects for sustainability assessment, and (iii) tailoring the assessment to the decision process. These are developments that show where environmental assessment is evolving in response to practical and theoretical challenges.

Becoming attuned to the environmental assessment context
This trend has been discernible in environmental assessment for a longer period. Practitioners and theorists alike have been realising that assessment which is viewed as a stand-alone product is not very effective. To improve the effectiveness of assessment, the links to the environmental management and planning system of which it is part needs to be strengthened. For EIA this has led to increased attention to the functioning of the whole system within which EIA operates – both the compliance system and the enabling context. EIA follow-up, a notoriously weak aspect of EIA practice, is getting more attention, especially in the developing countries. There is a growing awareness that adequate monitoring, inspection and enforcement is necessary for making EIA effective.

The same can be applied to SEA thinking. Recently, a wholly context-oriented approach to assessment has been coming out of experiences with policy level SEA. The World Bank has championed an institution-centred approach to SEA, which is also recognised in the OECD/DAC SEA guidance (OECD, 2006a; World Bank, 2006). The premise behind this type of SEA is that at a more strategic planning level, the planned actions are more abstract, and the direct relationships between these actions and concrete impacts is more difficult to identify and describe. Rather than attempt to assess, and avoid, negative impact, SEA at this level should assess the institutional context within which the policy is developed and implemented. The central question for such an SEA then becomes: how well equipped is the institutional capacity to manage environmental impact and to take advantage of environmental opportunities (OECD, 2006a:51)? Institutional capacity is interpreted broadly here, to mean not only government, but also public and private sector institutions. Of course, this shift in focus has major consequences for the SEA scope and approach to be taken. Although it is likely to still be necessary to determine broad categories of impacts, institutional analysis and strengthening will be central to this type of SEA. The World Bank is currently piloting this approach in a number of countries. The effectiveness of the institution centred approach in the pilots will be evaluated by the Netherlands Commission for Environmental Assessment and the University of Gothenburg.

The impact assessment framework described in Chapter 4 accommodates this increased attention to institutional context of EIA and SEA. It deliberately distinguishes between the biophysical subsystem and the social subsystem where ecosystem services provide the link between these two subsystems, while the resources management subsystem is the institutional context in which management decisions have to be taken on demand for and supply of ecosystem services.

From environmental assessment to sustainability assessment
More comprehensively described elsewhere, recent reviews of practices in SEA and related approaches show there is an emerging spectrum or 'continuum' of interpretation of the scope of assessment (Dalal-Clayton and Sadler 2005). At one end of the continuum, the focus is mainly environmental (what we might call 'conventional' SEA). It is characterised by the goal of mainstreaming and upstreaming environmental considerations into strategic decision making at the earliest stages of planning processes to ensure they are fully included and appropriately addressed. The 2001 SEA Directive of the European Union is an example of this approach. At the other end of the continuum is a more holistic and comprehensive approach which aims to assess environmental, social, and economic concerns in a more integrated manner and involves possible

Table 5.3 *Simplified characterisation of the continuum of impact assessment scope.*

Scope → Time/space horizons ↓	Environmental (ecological) aspects. . .	. . . & social aspects	. . . & economic aspects
Here and now. . .	Conventional EIA	Integrated EIA	
. . . & elsewhere . . . & later	Conventional SEA	Sustainability assessment SEA	

trade-offs between these considerations in strategic decision making at the earliest stages of planning processes. To a lesser degree the same development can be seen in EIA as well, where there is a broadening in types of effects being considered.

The ends of the continuum are characterised by Table 5.3, not just in terms of the dimensions that are considered, but also in terms of the time and space horizon that an assessment can cover. The upper left-hand corner represents conventional EIA, which concentrates on immediate environmental effects within a short time horizon. Integrated EIA covers the entire upper row in the table. Extending the time and spatial horizon is especially important in SEA. The strategic nature of the decisions that SEA supports necessitates a broader view in order to fully grasp the consequences and alternative options (Partidário 2007). For a national energy policy, for example, the SEA will need to explore the global implications as well as the national ones. Full sustainability assessment SEA, then, would cover the entire range: addressing environmental, social, and economic effects, those that are immediate, as well as those that manifest over an extended time horizon, and both at the directly impacted area, as well as beyond.

Of course, full incorporation of biodiversity in environmental assessment will in many cases require longer geographical and time horizons. Biodiversity very often depends on geographically different areas (e.g. migratory birds or fish) while biodiversity is important for the maintenance of life support systems for future generations to have at least the same quality of living as present generations have, while it also holds unknown potential for the future. The emergence of ecosystem services as a means to link biodiversity to stakeholders, conceptually captured by the impact assessment framework, requires a more-integrated approach including biophysical as well as social and economic aspects. Chapter 9 provides convincing examples.

There are clear advantages to a comprehensive and integrated assessment. For the decision makers, and indeed all involved stakeholders, the assessment information is conveniently brought together. This can concentrate the debate there where the real trade-offs are to be made between beneficial and detrimental effects. However, such assessment is methodologically more complex and requires intense cooperation among assessment professionals across disciplinary and organisational boundaries. This is not easy to achieve, especially in settings where such cooperation is not common place (Hilden, Rydevik and Bjarnadottir, 2007). It could also be argued (e.g. in Morrison-Saunders and Fischer, 2006) that full integration of economic and social effects into SEA en EIA is less effective in advocating for the biophysical environment. In an aggregate assessment the environmental effects may get snowed under the economic and social ones, which tend to be of more interest to decision makers anyway. Integrated assessment can furthermore obscure trade-offs that are being made against the environment. On the other hand, when biodiversity is translated into ecosystem services and is assigned its real value, in social or economic terms, the case for biodiversity can become stronger and decision making can be significantly influenced (see Chapter 9).

Tailoring the assessment to the decision process
This broadening of scope and application has occurred in conjunction with a change in attitude towards information and decision making in and outside the field of environmental assessment. It has become more common to doubt the supposed infallibility of scientific information, and the degree to which it is value-free and objective. Against this backdrop, and in concurrence with an increased appreciation of the complex sociopolitical context of the decisions for which assessment reports are produced, EIA and SEA are cast less as a technical information tool and more often as a process by which people communicate and integrate a variety of concerns in order to determine the future of the environment and themselves.

The shift away from a more technocratic approach to EIA and SEA has made room for a stronger decision-process orientation. At the time that EIA was developed, rational theories of decision making and planning were leading. Such theories suggest that sound decision making requires objective effects information that is the result from the application of scientifically proven techniques for information gathering. Good information, by those standards, will then lead to good decisions by means of a rational decision process. Since then, this way of thinking is gradually being overtaken by a more complex and realistic way of looking at decision making. There is now more recognition that effects information is often value-laden and uncertain, and that decision processes can be messy and unpredictable. The nature of decision making is also examined by the impact assessment community (Nitz and Brown 2001; Deelstra *et al.*, 2003; Leknes 2001; Weston 2000; Kørnøv and Thissen 2000; Nilsson and Dalkmann, 2001; Pischke and Cashmore 2006). Such examination continually stresses the importance of making the right kind of effects information available, to the right people, at the right time. In other words, environmental assessment should be adaptive to the decision process it is serving and provide relevant inputs when decision windows occur.

Partidário (2007) suggests that for SEA this requires a reorientation on the nature of strategic planning. To effectively support planning, she argues, SEA needs to be focussed on problem definition and solution finding in planning more directly. The SEA process should start when the plan objectives are defined and, in fact, more actively help to define these. Furthermore the assessment should evaluate how well strategic options meet the objective, as well as exploring their consequences. The SEA process itself needs to be centred around key activities that operate throughout the planning process, rather than follow a set sequence of steps. Partidario even suggests adopting a different lexicon for SEA, to differentiate it from more rigid EIA thinking, and to smooth communication with planners.

Application of this idea of decision-making–tailored SEA has presented a formidable challenge in practice. First, for obvious reasons adaptive assessment is difficult, because it relies on the decision process being both knowable and known, when it may in actual fact be fluid, and hidden. Second, such adaptive assessment requires flexibility in the assessment process, which can conflict with the procedural requirements that governments tend to set to ensure that at the very least a minimum quality of assessment takes place. Despite such difficulties, it has also led to some of the more innovative approaches that have emerged recently. Box 5.4, for example, shows the SEA process that was designed for the Free Trade Negotiations between the EU and Central America. This SEA process design tries to ensure that relevant assessment information is available for each stage of the negotiations.

The decision orientation in SEA and EIA also emphasises the importance of communication. A closer look at the decision processes that assessment reports feed into immediately makes clear the importance of presenting the information in an understandable and accessible way.

Box 5.4.   SEA for the EU–Central America Free Trade Agreement

In 2007, a tailor-made SEA approach was developed for the Free Trade Agreement (FTA) between the EU and Central America. Free Trade negotiations are characterised by changeable agendas and multiple decision-making moments. This puts high demands on the SEA. It has to be able to respond to negotiation processes that have their own (partly unpredictable) dynamics, as well as a confidential character. In addition, the impacts of trade agreements, while potentially extensive, can be difficult to accurately predict.

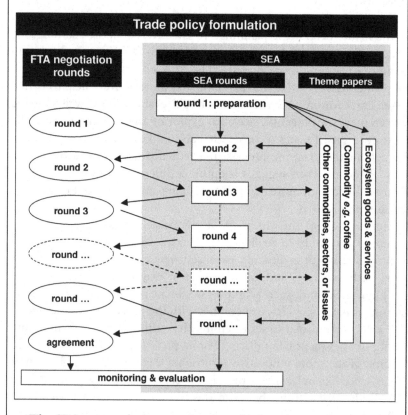

The SEA approach that was developed is based around an advisory group which undertakes both the stakeholder debate as well as impact analysis. During the negotiation process this group can channel information 'up', that is, provide advice to negotiators, as well as 'down', that is, disseminate information to relevant stakeholders. Naturally, membership needs to be broad; experts, government officials, representatives of the negotiation teams, nongovernmental organisations (NGOs), private sector, indigenous people, and so forth.

The SEA process itself has been separated into SEA rounds to match each FTA negotiation round. During each SEA round the advisory group scopes the upcoming negotiation topics, develops scenarios for different negotiation outcomes, consults with stakeholders, and makes recommendations for decision making, such as flanking measures. To be able to move quickly, the advisory group will commission 'theme papers' on a selected number of key issues at the outset. The theme papers will cover the main commodities to be addressed in the FTA. The papers serve as background document for the assessment rounds. Also, the advisory group will establish liaison officers, who can get rapid feedback from key stakeholder groups. The figure above depicts this experimental SEA approach. It is currently being trailed, and, if effective, will provide a good example of a decision sensitive SEA application.

Technical complexity in information, not only disenfranchises the public participants in decision making (Dryzek 1993), a point made more often in the environmental assessment literature, but it can also be difficult to understand for decision makers, who do not necessarily have technical expertise either.

Another insight that comes from decision analysis is the need to translate effects into clear impacts on the decision makers' priorities and interests. In this respect, the translation of biodiversity, too often considered a difficult topic by decision makers, into values for stakeholders helps to translate impacts into terms that decision makers can understand and relate to. Furthermore, the use of ecosystem services works to integrate different effect categories across the three pillars of sustainability. This ensures that the trade-offs remain transparent, and environmental effects stay visible. The impact assessment framework makes an effort to separate technocratic knowledge from the value based weighing of relevance of effects. Values only exist if they are recognised as such by stakeholders. How serious an effect is deemed to be depends on the value placed on the affected ecosystem service. This is the type of information that decision makers understand.

6 · Biodiversity in environmental impact assessment

Asha Rajvanshi, Vinod B. Mathur,
and Roel Slootweg

Introduction

As the importance of biodiversity for human well-being is increasingly
brought to the forefront by scientific endeavours in the fields of natural
sciences, resource economics, and conservation ecology, the need for
better representation of biodiversity in the processes that steer development decisions has clearly emerged. Furthermore, outspoken stakeholders actively participating in impact assessment processes bring the
importance of biodiversity to the forefront, contributing to the need for
better representation of biodiversity. Environmental Impact Assessment
(EIA) is one of few internationally recognised and often legally embedded
instruments to predict the potential consequences of human activities.
In recent years the impact assessment community realised that it was not
doing well in defining the impact on biological diversity. Implicitly biodiversity was always considered to be part of impact assessment, but the
tools to explicitly define biodiversity impacts were not enough. Box 6.1
provides some messages on biodiversity emerging from EIA practice.

   This chapter provides extensive first-hand background documentation
on the EIA guidelines adopted by the Convention on Biological Diversity (CBD, 2006) and further elaborates on the concepts behind the
guidelines. Reference is made to Chapter 2 elaborating CBD objectives
and explaining recent scientific developments in biodiversity science,
Chapter 4 providing the conceptual framework underlying the EIA as
well as SEA chapters, and Chapter 5 providing a general overview of
recent thinking on EIA. In this chapter the screening and scoping stages
of EIA are emphasised for two reasons. First, the need for an impact
assessment study when a proposed activity has potential significant effects
on biodiversity has to be defined by good screening criteria and procedures; second, the impact assessment study has to be carried out in such a
manner that all relevant issues are properly dealt with, providing sufficient

Box 6.1. Four messages on the relation EIA – biodiversity

(i) Svarpliene (2002) describes a road EIA in Lithuania in which the category of biodiversity-related impacts received highest attention and mitigation measures were predominantly concerned with biodiversity-related impacts in naturally flooded grassland and forest ecosystems (both ecosystems meet the European Union criteria for the Natura 2000 network). Message: EIA is used effectively as an instrument to bring biodiversity values into decision making on infrastructure projects.

(ii) Dahmer and Felley (2000) describe a case from Taiwan where growing concerns about a proposed steel mill project in a coastal wintering site of Black-faced spoonbills, a critically endangered water-bird of East Asia, has created an effective movement of NGOs, birdwatchers, and academics pressuring to declare a nature reserve at the proposed site and proposing alternative sources of local income. A first EIA study was rejected as it failed to address the importance of the site as wintering ground. A following EIA study was approved without public review, which is considered a blow to the credibility of the EIA system. Message: public review is a powerful tool in transparent decision making; absence of review jeopardises the process and neglects general opinion in society.

(iii) The Council of Andean Ministers of Foreign Affairs (Consejo Andino de Ministros DeRelaciones Exteriores, 2002) declared in its Regional Biodiversity Strategy for the Andean Countries that EIA and SEA are necessary tools for the assessment of potential impacts of regional transboundary mega projects. It emphasise the need to harmonise efforts, as EIA systems between countries differ. Especially the drafting of clear screening criteria is deemed necessary. Message: harmonisation in impact assessment for shared cross-border ecosystems is needed.

(iv) In a preliminary environmental evaluation of a landing site for offshore oil exploitation in Central Africa, an unknown area of marine corals was discovered. An investigation into alternative landing sites for the pipeline was recommended (Collins, personal communication). *Message*: EIA contributes to our knowledge of biological diversity and can help in early avoidance of negative impacts.

(*Source*: Slootweg, 2003)

information for decision makers to come to a well-founded decision. Because scoping determines the quality of the terms of reference of the impact study, good scoping procedures and guidance on the scoping process are of fundamental importance.

Nevertheless, flaws may appear during the following steps in the process. Therefore, further guidance on next steps in the project-level EIA process is provided, based on practical experiences. One of the recommendations from a workshop on the first version of the CBD Guidelines (Georgi and Peters, 2003) was that

the guidelines should not take over the function of general EIA-guidelines, but rather focus on aspects relevant to biological diversity. General provisions in terms of procedures should be kept as short as possible whilst remaining on a conceptual level and should be limited to the components essential for conducting sound EIA. This is the only way to ensure that the guidelines remain compatible to already existing planning systems and EIA procedures on national level.

This recommendation is taken seriously; as a consequence this chapter will follow a generally accepted series of procedural steps but will only provide the minimally required information on the procedural aspects of EIA.

Treatment of biodiversity in EIA: the experience so far

Although impact assessment processes and applications are currently in place in many countries, biodiversity is still considered implicitly in impact assessment. To some extent this may be explained by the historical timing of the pairing of EIA and biodiversity. The concept of biodiversity and its related Conventions are relative newcomers in the environmental scene. The CBD was only created in 1992 at the Rio Conference. EIA, on the other hand, traces its legacy back at least three decades to the passage of the National Environmental Protection Act in the United States in 1968. Experiences from around the world suggest that in the absence of having the appropriate tools to explicitly consider biodiversity in impact assessment, the outcome of assessments have generally been deficient in incorporating biodiversity in development decisions.

In 1995, the World Bank's East Asia Environment Unit undertook a review of the biodiversity components of several EIAs of projects in the infrastructure and forestry sectors to assess whether biodiversity was being appropriately studied and whether the information was geared to help the decision-making process (unpublished). The study found that in most biodiversity studies, quality of biodiversity information was generally weak, methodologies were poorly presented, the natural variability (at the gene, species, and ecosystem levels) was not accounted for, and the mitigation plans lacked provisions for biodiversity restoration opportunities. Quality reviews in the South Africa (LeMaitre and Gelderblom, 1998), United Kingdom (Byron *et al.*, 2000; Gray and Edwards-Jones, 1999; Thompson *et al.*, 1997; Treweek *et al.*, 1993), United States (Atkinson *et al.*, 2000), Australia (Warken and Buckley, 1998), Israel (Mandelik *et al.*, 2002, 2005a, 2005b), India (Rajvanshi, 2005), Japan (Tanaka, 2001) Sweden (de Jong *et al.*, 2004), Finland (Söderman, 2005), and for linear infrastructure (Geneletti, 2006) have all reflected on shortcomings in treatment of biodiversity in EAs.

Barriers to the incorporation of biodiversity in impact assessment include:

- a low priority for biodiversity and a limited capacity to carry out the assessments (Treweek, 1999, 2001);
- a lack of formalised procedures and inconsistency in methodologies (Thompson *et al.*, 1997; Geneletti, 2002);
- a lack of full treatment of biodiversity, combining the knowledge on the affected components, that is, composition, structure or function (Noss, 1990; LeMaitre and Gelderblom, 1998; Slootweg and Kolhoff, 2003; Slootweg, 2005); in general, there is a lack of attention to ecological processes (Pritchard, 2005);
- the concentration of assessments only around protected species and habitats and not including assessment of impacts on ecosystems, let alone ecosystem services (Gontier *et al.*, 2005) (see Knegtering, 2005, for an example of a totally species-oriented focus);
- geographically poorly defined study areas or a priori delimitation of study area not taking into account areas of impact (Geneletti, 2006);
- a lack of formal requirements for postproject monitoring (Treweek, 1996);
- a limited attention for positive planning for biodiversity;
- an incompatibility of timelines for EIA with seasonality for biodiversity surveys (Rajvanshi, 2005);
- poorly drawn connections between baseline studies and impact predictions resulting in superficial and inaccurate impact predictions based on data-deficient baselines (Söderman, 2005);

- a reliance on imprecise estimates of species distributions as provided by patchily distributed point data from field surveys or biological collections (Balfors and Mörtberg, 2002; Gontier *et al.*, 2005); and
- biodiversity assessments confined to local scales which do not allow prediction and assessment of effects of habitat loss and fragmentation on the landscape level or the consideration of scales of ecological processes (Treweek *et al.*, 1998 and Lehmann *et al.*, 2002).

The factor leading to the neglect of biodiversity in EIAs that is unique to most developing countries is the priority given to promoting development in key sectors to overcome poverty and improve economic well-being. Consequently, projects that are considered to be of national, political, and strategic importance, often override consideration of potential negative impacts on biodiversity. Furthermore, when locations of such priority projects tend to overlap with ecologically important areas, biodiversity issues are consciously underplayed in EIA reports to prevent these issues from becoming barriers to development (Rajvanshi *et al.*, 2007). On the other hand, the emergence of ecological economics in the mid-1980s emphasised the importance of protection of 'natural capital' as a precondition for societal development (Costanza and Daly, 1992; Jansson *et al.*, 1994) and lately, the Millennium Ecosystem Assessment (2005) brought the realization that the use of nature by humans is not simply a tool for economic development and that biodiversity and related ecosystem services are fundamental to human survival (i.e. physical, social, cultural, and spiritual) (see Millennium Ecosystem Assessment 2005a). These experiences from across the world are pointers to the growing need for the development of new and innovative EIA methods and tools for mainstreaming biodiversity and the enabling legal and policy support for their implementation.

Existing guidance for inclusion of biodiversity in EIA

In addition to the existing policies and regulations that already exist in most countries, a number of directives from convention bodies facilitate the integration of biodiversity issues in impact assessment through provisions of specific EIA-related obligations (Table 6.1).

Specific guidance has been produced over the last two decades. For instance, the Council on Environmental Quality (CEQ, 1993) and EPA (1999) in the United States, the Canadian Environmental Assessment Agency (CEAA, 1996, and the World Bank (1997, 2000) offer guidance

Table 6.1 *Provisions for biodiversity-inclusive assessments under various EIA-related obligations.*

CBD	'Each Contracting Party, as far as possible and as appropriate, shall (CBD, 1992: Article 14):
	Introduce appropriate procedures requiring environmental impact assessment of its proposed projects that are likely to have significant adverse effects on biological diversity with a view to avoiding or minimizing such effects and, where appropriate, allow for public participation in such procedure.
	(b) Introduce appropriate arrangements to ensure that the environmental consequences of its programmes and policies that are likely to have significant adverse impacts on biological diversity are duly taken into account.' (CBD, Article 14.1).
Ramsar	'Each Contracting Party shall arrange to be informed at the earliest possible time if the ecological character of any wetland in its territory and included in the List has changed, is changing or is likely to change as a result of technological developments, pollution or other human interference. Information on such changes shall be passed without delay to the organisation or government responsible for the continuing bureau duties specified in Article 8.' (Ramsar Convention, Article 3.2) (Ramsar Convention, 2007). Contracting Parties to reinforce and strengthen their efforts to ensure that any projects, plans, programmes and policies with the potential to alter the ecological character of wetlands in the Ramsar List, or impact negatively on other wetlands within their territories, are subjected to rigorous impact assessment procedures and to formalise such procedures under policy, legal, institutional and organizational arrangements;' (Resolution VII.16, para. 10)
Convention on Migratory Species	'Parties that are Range States of a migratory species listed in Appendix I shall endeavour: [. . .] to prevent, remove, compensate for or minimise, as appropriate, the adverse effects of activities or obstacles that seriously impede or prevent the migration of the species;' (CMS, article III, para. 4(b))
	– 'to the extent feasible and appropriate, to prevent, reduce or control factors that are endangering or are likely to further endanger the species, including strictly controlling the introduction of, or controlling or eliminating, already introduced exotic species.' (CMS, article III, para. 4(c))

on improving practices for ecological considerations and analysis of the impacts on biodiversity within EIA. Currently, biodiversity assessment within the Swedish International Development Cooperation Agency (SIDA) occurs as part of their Environmental Impact Assessment (EIA) process. EIA guidelines of SIDA (1998) contain a series of checklists that include questions on biodiversity. It is worth noting that SIDA has had the foresight to ask questions relating to impacts on genetic diversity – a rare feature in other donors' EIA guidelines or screening checklists. As part of SIDA's EIA, consideration is also given to how the proposed activities relate to the partner country's environmental legislation and its responsibilities under the biodiversity-related conventions and agreements. The RSPB (1996) has produced Good Practice Guides for prospective developers, which advocate a stepped approach, to first avoid impacts; to mitigate any residual impacts which cannot be avoided; to compensate for any losses (as a last resort); and always to seek opportunities to enhance the existing natural assets. The UK Institute of Ecology and Environmental Management published guidelines (IEEM, 2006) for Ecological Impact Assessment (EcIA). These guidelines, which are intended for ecologists, developers, planners, local and national planning authorities, environmental managers, statutory organisations, nongovernmental organisations (NGOs), and local groups, provide a recommended procedure for the ecological component of Environmental Impact Assessment. They set new standards for the assessment of the ecological impacts of projects and plans, so as to improve the consideration of the needs of biodiversity and thereby reduce the impacts of any development. The publication of *Biodiversity and Environmental Assessment Toolkit* by the World Bank (2000) suggested a more formalised framework for promoting biodiversity in the impact assessment practice. The review of experience and methods of integrating biodiversity in national EIA process supported by the Biodiversity Planning Support Programme (Treweek, 2001) was perhaps the first serious effort that highlighted the need to better integrate biodiversity in existing EIA procedures. The review focussed on the relevance of mainstreaming biodiversity in impact assessment and provided guidance on which levels and what elements of biodiversity need to be considered in each of the stages in the EIA process.

The efforts listed above did not address biodiversity in a consistent manner, following the definitions and the objectives for biodiversity management provided by the Convention on Biological Diversity (see Chapter 2). A first effort in this direction was done by the International Association for Impact Assessment (IAIA) by providing guidelines for the integration of biodiversity in EIA, adopted by the parties of the CBD in 2002 (Decision VI/7A in: CBD, 2002). Guiding principles to promote 'biodiversity-inclusive' impact assessment, including Environmental Impact Assessment (EIA) for projects, and Strategic Environmental Assessment (SEA) for policies, plans, and programmes were subsequently developed by International Association for Impact Assessment (IAIA, 2005). The Netherlands Commission for Environmental Assessment, in collaboration with IAIA, prepared Voluntary Guidelines on Biodiversity-Inclusive Impact Assessment, in response to Decision VIII/28 of the Conference of the Parties to the Convention on Biological Diversity, to fill remaining caveats in the earlier guidelines (CBD, 2006). This is the most recent in terms of a generic guidance that is available for the EIA community; this chapter is based on the same document but provides more scientific background and case material. The CBD guidelines have been adapted in the form of country-specific toolkits (Sawsan *et al.*, 2005) and best practice guidance in regional contexts (Rajvanshi *et al.*, 2007; Brownlie *et al.*, 2006).

Several sectoral initiatives have also seen the light, indicating that some segments of the private sector are similarly concerned about the way in which biodiversity has been treated (or not) in EIA.

(1) **The oil and gas sector.** This sector has, for obvious reasons, been a frontrunner in paying attention to the environmental effects of its activities. As early as 1986 the Oil Industry International Exploration and Production Forum produced a publication on EIA (E&P Forum, 1986). Biodiversity was specifically addressed in a joint publication with IUCN on oil and gas exploration in the Arctic (IUCN and E&P Forum, 1993), where EIA featured prominently as a tool to address the environmental challenges posed by work in the arctic region. The Energy and Biodiversity Initiative, a joint initiative by five major oil and gas companies and five conservation organisations provides a suite of guides, discussion papers, and resources for those interested to know more about how to integrate biodiversity into the oil and gas industry (Energy and Biodiversity Initiative, 2004). The guidance documents prepared by EBI (2003a, 2003b, 2003c) provide 'how to' approaches for integration of biodiversity considerations into upstream oil and gas development and are very useful for conservation organisations, governments, communities, and others with an interest in ensuring the effective integration of biodiversity considerations into oil and gas exploration and development.

It also refers to the underlying conceptual framework (Slootweg and Kolhoff, 2003) described in Chapter 4, although it does not use the framework in the practical manner for which it has been designed. Nevertheless, it is the first example of translation of the CBD decision on biodiversity in impact assessment into a suite of sector-oriented documents. Some of the guidance is unique in offering a 'menu' of sound biodiversity conservation practices from which the most appropriate measures that fit the operational and geographic setting can be chosen. Similarly, the International Petroleum Industry Environmental Conservation Association (IPIECA) comprising of oil and gas companies and associations from around the world that was founded in 1974 offer valuable guidance for integrating biodiversity conservation with existing company activities and processes throughout the oil and gas project life cycle. Guidance is specifically available for developing Biodiversity Action Plans (BAPs) corresponding with requirements of different sites and projects and for setting out key biodiversity questions that should be considered by both business managers and practitioners throughout all of the stages of the oil and gas project (IPIECA and OGP, 2005, 2006).

(2) **The transport sector.** An example of practical best practice guidance for planning roads through sensitive habitats and wildlife areas in South Asia is provided by Rajvanshi *et al.* (2001). The guide defines a basic step-by-step EIA process in order to provide a realistic backdrop to the wildlife–road transportation relationship to help practitioners identify wildlife-related concerns and incorporate wildlife and wildlife habitat conservation principles into road and rail planning. Another example on how to deal with biodiversity in road schemes is provided by Byron (2000). The guidelines provide an in-depth reference to existing regulations in the UK and to the way in which potential impacts on biodiversity by road schemes can be identified, assessed, and mitigated. Although this 'good practice guide' is focussed on the UK, the step-by-step approach could act as a reference to how such guidance documents can be structured.

(3) **The mining sector.** A similar long-awaited initiative comes from the mining sector with good practice guidance for mining and biodiversity. Sweeting and Clark (2000) made the initial attempts to review both the potential negative effects of large-scale metal mining on sensitive environments and cultures, and a range of technologies, practices, and strategic approaches for both minimizing

negative impacts and increasing the positive contribution of mineral development to conservation and community development. Although this guide is not meant to be a definitive guide to responsible mining, it does offer an important starting point for discussion and action on how all stakeholders can work towards 'lightening the load' of mining on sensitive ecosystems and cultures throughout the world. Subsequent guidance available in different documents (IUCN and ICMM, 2004; Rio Tinto, 2004; ICMM, 2006) show how good practice, collaboration, and innovative thinking can advance biodiversity conservation worldwide while ensuring that the minerals and products that society needs are produced responsibly. The varied guidance sources also aim to provide the mining industry with the steps required to improve biodiversity management throughout the mining cycle and demonstrate through case studies how management tools, rehabilitation, and restoration processes, together with improved scientific knowledge can help conserve biodiversity.

The recent leading practice handbook (Commonwealth of Australia, 2007) complements other publications, in addition to providing information specific to biodiversity management in the Australian context. This handbook outlines the key principles and procedures now recognised as leading practice for assessing biodiversity values, namely: identifying any primary, secondary, or cumulative impacts on biodiversity values, minimizing and managing these impacts, restoring conservation values, and managing conservation values on a sustainable basis.

(4) **The tourism sector.** A sector with biodiversity issues high on the agenda but with very little effective use of EIA is the tourism sector. The CBD (2001) has drafted international guidelines for activities related to sustainable tourism development in vulnerable terrestrial, marine, and coastal ecosystems and habitats of major importance for biological diversity and protected areas, including fragile riparian and mountain ecosystems. The document contains a section on impact assessment providing 20 categories of potential biophysical impacts, 8 categories of social and cultural impacts, and a number of potential benefits of tourism. The guidelines recommend that, as a minimum, impact assessment should address the impacts, effects, and information that are required to be covered in the notification process. Furthermore, impact assessment should be objective and transparent and based on recognised standards. It should also include assessment of cultural sustainability.

UNEP (2001) commissioned a series of thematic studies, each focussed on one aspect of sectoral integration. One of these thematic studies is Integration of Biodiversity into the National Tourism Sector. This Report provides global best practice guidelines for integrating biodiversity conservation planning into the tourism sector. These guidelines have emerged as a result of a careful analysis of the 12 country case studies. The specific objectives of these guidelines is to identify resources and clarify concepts for biodiversity planners on the subject of sustainable tourism and to provide practical tools for tourism planners and developers so that their activity will have a more positive interaction with biodiversity conservation planning. This document also builds on UNEP's Principles for Implementation of Sustainable Tourism, in an attempt to complement and reach further in attaining a symbiotic relationship between biodiversity conservation planning and tourism. The problem with the tourism sector is that EIA is not used very often because tourism activities usually are diverse, isolated, and relatively small interventions. Of course, Strategic Environmental Assessment (SEA) of higher level policies, plans, and programmes for the tourism sector provides the best solution to address the cumulative effects of multiple small-scale tourism activities. SEA will be treated in more detail in Chapter 7.

Sectoral initiatives discussed in this chapter have helped to a good extent in redrawing the caricatures of 'conservation versus development' by demonstrating that biodiversity-inclusive assessments can help achieve 'conservation and development'. Our main criticism of the guidance that is available in the sectoral guidelines is the narrow focus of biodiversity, which is predominantly addressed from a nature conservation position. Ecosystem services and human interests in biodiversity conservation have received little attention. There has also been an inherent failure in recognizing the links between biodiversity and human well-being which limit the consideration of benefits of biodiversity conservation from social and economic perspectives. In general, linkages to the objectives of the biodiversity convention are weak, contributing to the inconsistent and incomplete treatment of biodiversity.

Mainstreaming biodiversity in EIA: a stepwise explanation

Key challenges to address conservation and sustainable use of biodiversity through impact assessment are that impact assessment must give wider recognition to the ecosystem services concept as a means to translate biodiversity into societal values (ecological, social, and economic), and to not only look at potential negative impacts on biodiversity but to pay more attention to positive benefits for biodiversity. The latter is often referred to as a positive planning approach (IAIA, 2005). As new challenges emerge, new solutions are needed. The EIA process (Figure 6.1) needs to better mainstream biodiversity in impact assessment to create more opportunities for win–win situation that would not only prevent biodiversity losses but can also promote 'conservation through development'. A new approach should be based on the interpretation of biodiversity provided by the CBD, further enhanced by the Millennium Ecosystem Assessment (2003), which was extensively explained in Chapter 2 and conceptually developed in the impact assessment framework in Chapter 4. This new approach must essentially be able to

(1) recognise the links between human well being and ecosystem services in the assessment of impacts;
(2) present the ecological rationale behind the impact mechanisms through which ecosystem services are affected, by including both biodiversity pattern (composition and structure) and key ecological processes; and
(3) incorporate valuation of ecosystem services for deriving economic, social, and ecological values of biodiversity. (See Box 6.2 summarizing the economic measures for biodiversity management defined by the CBD).

The rest of this chapter is structured according to the generally acknowledged sequence of steps in EIA presented in Figure 6.1, based on the UNEP EIA Handbook (UNEP, 2002). The steps are discussed in the context of the impact assessment framework from Chapter 4; 'entry points' for mainstreaming biodiversity in impact assessment will be identified.

Screening

Screening is the step to determine which proposals should be subject to impact assessment, to exclude those unlikely to have harmful environmental impacts and to indicate the level of environmental appraisal required. The outcome of the screening process is a screening decision. Based on an overview of case evidence, Slootweg (2003) recommended that rules for screening need biodiversity-specific criteria in order to

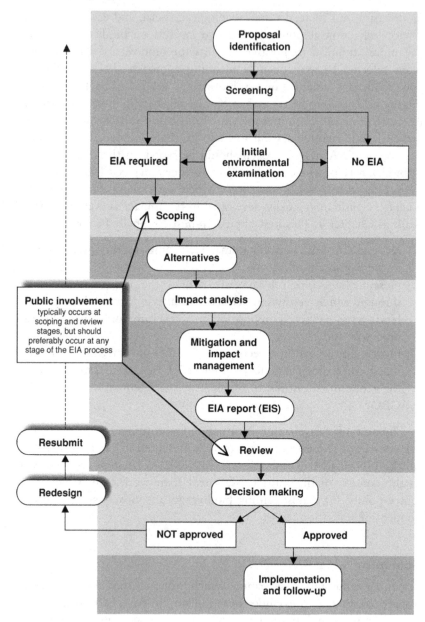

Figure 6.1 Generalised EIA process flowchart (adapted from UNEP, 2002).

Box 6.2. Economic valuation of biodiversity in the CBD

The valuation of biodiversity gathers momentum, exemplified by the Articles of the Convention on Biological Diversity that provide insight in the importance of economic measures linked to biodiversity.

Relevant CBD article:	6	7	8	9	10	11	12	14	15	16	20	21
Economic assessment	•							•				
Economic incentives	•		•		•	•		•		•	•	•
Financial resources			•	•		•		•		•	•	•
Economic valuation		•										

(*Source*: Emerton and Muramira, 1999)

provide enough 'legal clout' against developers that want to proceed with their projects. Because legal requirements for EIA may not guarantee that biodiversity will be taken into account, consideration should be given to incorporating biodiversity criteria into existing, or the development of new, screening criteria (see Box 6.3). Important information for developing screening criteria can be found in National Biodiversity Strategies and Action Plans (NBSAPs) or equivalent documents. These strategies provide detailed information on conservation priorities and on types and conservation status of ecosystems. Furthermore, they describe trends and threats at ecosystem as well as species level and provide an overview of planned conservation activities. Even although this chapter aims to provide guidance, countries have to design there own biodiversity-inclusive screening mechanism that fit the individual country characteristics.

The screening criteria for biodiversity must set out circumstances in which EIA is justified on the basis of biodiversity considerations. The screening criteria for biodiversity can be evolved based on the following:

- Legal triggers including existing and future legislation for biodiversity conservation.
- Maps indicating areas important for conserving biodiversity, protecting ecosystem components and maintaining important ecosystem services.
- Categories of activities that are likely to be the drivers of change in biodiversity (e.g. harvest or removal of species; habitat diversion, fragmentation and isolation; external inputs such as emissions,

---

Box 6.3. Screening in practice

In a case study from Israel, Mandelik *et al.* (2002, 2005a, 2005b) evaluated 52 environmental impact statements. Probably the most fundamental problem to address biodiversity in EIA in Israel is the lack of direct and clear legal requirements for ecological attention in impact assessment. Because of this, any elaborated ecological guidelines are vulnerable to objections by the developers. Recently, however, there has been progress in this respect as there is an initiative now of drafting ecological guidelines and staff has been assigned to this (Mandelik, personal communication).

A good example of how biodiversity is represented in an existing national EIA system is provided by Hong Kong (Hong Kong Environmental Protection Department, 1997). The 'Technical Memorandum on Environmental Impact Assessment Process' is an example of integrating biodiversity considerations into a national EIA system (Dahmer, personal communication). It was established before the development of CBD guidelines on EIA. One omission in the light of the CBD objectives may be that the guidelines are predominantly focussed on the conservation of biodiversity and less so on the sustainable use.

The Republic of Yemen (2006) has recently proposed a new directive on EIA of Dam Projects. Article 6 on screening provides criteria related to activities as well as geographical location of the activities. Article 6.1 provides a list of activities for which EIA shall be applied; Article 6.2 distinguishes between different classes of EIA, depending on type and size of project (comprehensive EIA, general EIA, or site-related EIA); Article 6.3 states that all projects, independent of their type and size, which are located in protected areas are subject to comprehensive EIA. This is the first example of screening criteria following the 2006 CBD guidelines, further explained in this chapter.

---

effluents, radiations, or noise; introduction of alien, invasive or genetically modified organisms; or change in ecosystem composition, structure or key processes).

- Thresholds referring to size of the intervention area and/or magnitude, duration and frequency of the activity that have a bearing on the significance of the impacts.

Pertinent questions which need to be answered from an ecological perspective at the screening stage in an EIA must take into account all three objectives of the Convention on Biodiversity (conservation of biological diversity, sustainable use of its components, and the fair and equitable sharing of the benefits arising out of the utilization of genetic resources) and be applied to all the three levels of diversity. Fundamental questions which need to be answered in an EIA study include:

(a) *Conservation*: would the intended activity affect the biophysical environment directly or indirectly in such a manner or cause such biological changes that it will increase risks of extinction of genotypes, cultivars, varieties, populations of species, or the chance of loss of habitats or ecosystems?

(b) *Sustainable use*: would the intended activity surpass the maximum sustainable yield, the carrying capacity of a habitat/ecosystem or the maximum allowable disturbance level of a resource, population, or ecosystem, taking into account the full spectrum of values of that resource, population, or ecosystem?

(c) *Equitable sharing*: would the intended activity result in changes to the access to, and/or rights over biological resources?

To facilitate the development of screening criteria, the questions above have been reformulated for the three levels of diversity (Table 6.2). In the original version of the table, appearing in the CBD guidelines (CBD, 2006), the objective of equitable sharing was omitted, as this was not considered to be of relevance to environmental assessment but moreover a statement based on political choice. In hindsight, taking into account the rapid shift towards more integrated assessment of environmental, social and economic impacts, this was not the right thing to do. The emphasis we nowadays place on the valuation of ecosystem services for human well-being puts sharing of benefits on the agenda of decision makers when considering the consequences of large interventions. Therefore, we have put the third objective of the convention back in the table again. This last objective was triggered by the fact that large multinational pharmaceutical companies tapped the knowledge of traditional healers on the use of medicinal herbs, often leading to significant profits for these companies, while the providers of the knowledge, often living in great poverty, did not receive anything for their contribution. Since the establishment of the biodiversity convention many examples of benefit sharing have become available, recognising the rights of local communities to share in benefits obtained using their knowledge but

Table 6.2 *Questions pertinent to screening of biodiversity impacts.*

Level of diversity	Conservation of biodiversity	Sustainable use of biodiversity	Equitable sharing of benefits
Genetic diversity	Would the intended activity result in extinction of a population of a localised endemic species of scientific, ecological, or cultural value?	Does the intended activity cause a local loss of varieties/cultivars/breeds of cultivated plants and/or domesticated animals and their relatives, genes or genomes of social, scientific and economic importance?	Would the intended activity lead to a diminution in the share that vulnerable groups in society accrue from benefits arising from the use of genetic resources?
Species diversity	Would the intended activity cause a direct or indirect loss of a population of a species?	Would the intended activity affect sustainable use of a population of a species?	Would the intended activity diminish the access of vulnerable groups in society to the (sustainable) use of a population of a species?
Ecosystem diversity	Would the intended activity lead to loss of (an) ecosystem(s) or impair ecosystem services that create challenges for conservation?	Does the intended activity affect the sustainable human exploitation of (an) ecosystem(s) or land use type(s) in such manner that the exploitation becomes destructive or nonsustainable (i.e. the loss of ecosystem services)?	Does the intended activity diminish the access of vulnerable groups in society to the sustainable human exploitation of (an) ecosystem(s) or land use type(s).

also recognising the rights of countries to share in benefits arising from the commercial and other utilization of genetic resources from these countries (CBD, 2000b, 2004b). Because local knowledge and exploitation of ecosystem services goes beyond the boundary of genetic diversity we have decided to apply the third objective to all three levels of diversity.

The scale at which ecosystems are defined depends on the definition of criteria in a country, and should take into account the principles of the ecosystem approach. Similarly, the level at which 'population' is to be defined depends on the screening criteria used by a country. For example, the conservation status of species can be assessed within the boundaries of a country (for legal protection), or can be assessed globally (IUCN Red List of Threatened Species (www.iucnredlist.org/); see also Meynell, 2005). The definition of vulnerable groups in society definitely is a political choice. Vulnerable people are generally described as being marginalised, because of least access to economic resources, to decision making, or to social services.

A screening decision defines the appropriate level of assessment. The result of a screening decision can be that

- The proposed project is 'fatally flawed' in that it would be inconsistent with international or national conventions, policies or laws. It is advisable not to pursue the proposed project. Should the proponent wish to proceed at his/her risk, an EIA would be required;
- An EIA is required (often referred to as Category A projects);
- A limited environmental study is sufficient because only limited environmental impacts are expected; the screening decision is based on a set of criteria with quantitative benchmarks or threshold values (often referred to as Category B projects);
- There is still uncertainty whether an EIA is required and an initial environmental examination has to be conducted to determine whether a project requires EIA or not; or
- The project does not require an EIA.

A suggested approach to the development of biodiversity-inclusive screening criteria, includes the following steps: (i) design a biodiversity screening map indicating areas in which EIA is required; (ii) define activities for which EIA is required; (iii) define threshold values to distinguish between full, limited/undecided or no EIA (see Appendix 6A for a generic set of screening criteria and Box 6.4 for an example). The suggested approach takes account of biodiversity values (including valued ecosystem services) and activities that might act as drivers of change of

---

Box 6.4. Example of biodiversity-sensitive screening criteria for linear infrastructure combining spatial and activity-related information, including threshold criteria

- Category A: always EIA

  – All linear infrastructure above a minimum threshold (e.g. 5 km length) in all areas.

- Category B: limited EIA/preliminary study needed

  – Linear infrastructure below threshold in areas providing key ecosystem services indicated on the biodiversity screening/ecosystem services map.

- Category C: no EIA required

  – Linear infrastructure below threshold, outside indicated areas.

---

biodiversity. If possible, biodiversity-inclusive screening criteria should be integrated with the development (or revision) of a national biodiversity strategy and action plan. This process can generate valuable information such as a national spatial biodiversity assessment, including conservation priorities and targets, which can guide the further development of EIA screening criteria.

Step 1

According to the principles of the ecosystem approach (CBD, 2000a, 2004a), a *biodiversity screening map* is designed, indicating important ecosystem services (replacing the contested and unclear concept of sensitive areas). The map is based on expert judgement and has to be formally approved (see Geneletti 2008 for an advanced GIS-based approach). Suggested categories of geographically defined areas, related to important ecosystem services, are:

- Areas with important regulating services in terms of maintaining biodiversity;
- *Protected areas*: depending on the legal provisions in a country these may be defined as areas in which no human intervention is allowed, or as areas where impact assessment at an appropriate level of detail is always

required; the stated reasons for the site's designation (its conservation objectives) and any management plan that may exist can give hints to the factors an assessment should address (Pritchard, 2005);

- Areas containing *threatened ecosystems outside formally protected areas*, where certain classes of activities (see Step 2) would always require an impact assessment at an appropriate level of detail;
- Areas identified as being important for the *maintenance of key ecological or evolutionary processes*, where certain classes of activities (see Step 2) would always require an impact assessment at an appropriate level of detail;
- Areas known to be *habitat for threatened species*, which would always require an impact assessment at an appropriate level of detail;
- Areas with *important regulating services for maintaining natural processes with regard to soil, water, or air*, where impact assessment at an appropriate level of detail is always required. Examples can be wetlands, highly erodable or mobile soils protected by vegetation (e.g. steep slopes, dune fields), forested areas, coastal, or offshore buffer areas, and so forth;
- Areas with *important provisioning services*, where impact assessment at an appropriate level of detail is always required. Examples can be extractive reserves, lands, and waters traditionally occupied or used by indigenous and local communities, fish breeding grounds, and so forth;
- Areas with *important cultural services*, where impact assessment at an appropriate level of detail is always required. Examples can be scenic landscapes, heritage sites, sacred sites, and so forth;
- Areas with *other relevant ecosystem services* (such as flood storage areas, groundwater recharge areas, catchment areas, areas with valued landscape quality, etc.); the need for impact assessment and/or the level of assessment is to be determined (depending on the screening system in place); and
- All other areas: no impact assessment required from a biodiversity perspective (an EIA may still be required for other reasons).

Step 2

Define activities for which impact assessment may be required from a biodiversity perspective. The activities are characterised by the following direct drivers of change:

- Change of land-use or land cover, and underground extraction: above a defined area affected, EIA always required, regardless of the location of the activity — define thresholds for level of assessment in terms of surface (or underground) area affected;

- Change in the use of marine and/or coastal ecosystems, and extraction of seabed resources: above a defined area affected, EIA always required, regardless of the location of the activity — define thresholds for level of assessment in terms of surface (or underground) area affected;
- Fragmentation, usually related to linear infrastructure. Above a defined length, EIA always required, regardless of the location of the activity — define thresholds for level of assessment in terms of the length of the proposed infrastructural works;
- Emissions, effluents or other chemical, thermal, radiation, or noise emissions — relate level of assessment to the ecosystem services map; and
- Introduction or removal of species, changes to ecosystem composition, ecosystem structure, or key ecosystem processes responsible for the maintenance of ecosystems and ecosystem services — relate level of assessment to the ecosystem services map.

It should be noted that these criteria only relate to biodiversity and serve as an add-on in situations where biodiversity has not been fully covered by the existing screening criteria.

Step 3

Determining norms or threshold values for screening is partly a technical process and partly a political process — the outcome of which may vary between countries and ecosystems. The technical process should at least provide a description of:

(a) *Categories of activities* that create direct drivers of change (extraction, harvest or removal of species, change in land-use or cover, fragmentation and isolation, external inputs such as emissions, effluents, or other chemical, radiation, thermal or noise emissions, introduction of invasive alien species or genetically modified organisms, or change in ecosystem composition, structure or key processes), taking into account characteristics (such as type or nature of activity, magnitude, extent/location, timing, duration, reversibility/irreversibility, irreplaceability, likelihood, and significance) and the possibility of interaction with other activities or impacts;

(b) *Where and when*: the area of influence of these direct drivers of change can be modelled or predicted; the timing and duration of influence can be similarly defined;

Box 6.5. Scoping in practice

In Israel, Mandelik *et al.* (2002, 2005) provide a unique and detailed study of 52 environmental statements produced over a six-year period. The main conclusions of the analysis were that, with respect to biological diversity:

- Lack of quantitative data, meaningful analyses, and broad perspective were apparent throughout the EISs reviewed.
- Most EISs presented baseline information and potential ecological impacts but failed to give any quantified data or predict indirect, cumulative effects.
- Many EISs failed to perform field surveys, and their qualitative nature hampered meaningful impact prediction.
- Most EISs mentioned the need for mitigation measures but provided no description of these measures or their likely success.

The reason for the apparent lack of attention for biological diversity could largely be traced back to a lack of ecological requirements in the guidelines (or term of reference) for the EIA study. There was a generally high correlation between the (ecological) quality of the EIS and their corresponding guidelines. Suggested underlying reasons are the lack of ecological experts directly involved in the scoping phase of the EIA process. This obviously bears on both the awareness and the ecological capacity of the EIA staff.

(c) *Map of valued ecosystem services* (including maintenance of biodiversity itself) on the basis of which decision makers can define levels of protection or conservation measures for each defined area. This map is the experts' input into the definition of categories on the biodiversity screening map referred to above under Step 1.

Scoping

Scoping is designed to focus the impact assessment study on relevant issues (see Box 6.5 for an example on the relevance of the scoping stage). It is used to derive terms of reference (sometimes referred to as guidelines) for environmental impact assessment. Scoping also enables the competent authority (i) to guide study teams on significant issues and alternatives to be assessed, clarify how they should be examined (methods of prediction and analysis, depth of analysis), and according to which guidelines and criteria; (ii) to provide an opportunity for stakeholders to have their interests taken into account in the environmental impact assessment; and (iii) to ensure that the resulting environmental impact statement is useful to the decision maker and is understandable to the public. During the scoping phase, promising alternatives can be identified for in-depth consideration during the EIA study (CBD, 2006).

Scoping is thus an important stage in EIA that helps in setting temporal and spatial boundaries for the assessment, identifying the key impacts and controlling the quality of the EIA. Scoping should be seen as a flexible, adaptive, and iterative process, usually based on preliminary consultations, literature searches, site visits, and reconnaissance surveys. There is generally no set process for conducting scoping. Scoping workshops have sometimes been considered to provide all major stakeholders with an opportunity to discuss a project and reach consensus on the scope of the assessment. This approach can significantly reduce consultation time and avoid delays caused by stakeholders requesting additional survey or other work at a later stage. Scoping should ensure that the EIA team is able to perform 'good focussing' that helps to 'count the best and leave the rest' (See Box 6.6). Depending upon the legal requirements and country-level guidance that steer the national EIA process, the results of the scoping are in most countries publicly shared through a formal report or letter or as 'scoping opinion'. In some countries, scoping guidelines

---

Box 6.6. Good scoping practices (Swanson, 1999)

- Make early site visits in order to ensure that matters related to important biodiversity and ecosystem values and conservation sites are identified at an early stage.
- Establish appropriate consultation arrangements with interested parties including the competent authority.
- Conduct the scoping exercise in a systematic manner using scoping checklists and matrices and producing a Scoping Report where appropriate.
- Develop a consensus on baseline survey requirements, prediction methods, and evaluation criteria with appropriate bodies, including planners and decision makers.
- Review the costs and benefits of development choice alternative options including the option.

Box 6.7. Examples of gender/specific impacts of development projects

The Pak Mun Dam destroyed the Mun River's natural fisheries, which prevented the seasonal reproductive and feeding migrations of fish species between the Mun and Mekong rivers. Women were usually involved in the processing and marketing of the catch in fishing communities. As the means of livelihood got affected, communities were faced with shrinking economic opportunities. This has resulted in male migration, leaving women to face an increasingly uncertain economic future (OED, 1998).

The EIA of West Frontier Province Road Development Sector of Pakistan was conducted by ADB. The study of the Peshawar–Torkham subregional connectivity project highlighted that the section of the expressway from 22 km to 27 km (following the existing highway) will pass above several villages. The local people in this area were concerned about the privacy of their women and families. It was apprehended that the road users will be able to see down into the houses and this may be interpreted as an invasion of privacy. Planting or roadside barriers were recommended in order to shield the view of the villages from passing vehicles and additionally serve as sound barriers (ADB, 2006).

---

are statutory requirements for decisions on structure and contents of the report.

As gender, class, caste, ethnicity, and age are integral to understanding the social relations and decision-making processes concerning access to, and use and management of biodiversity resources, a categorical inclusion of women and indigenous communities is highly desirable in the scoping teams for reviewing impacts especially on biodiversity (for a good example of collaboration with indigenous First Nations in Canada, see Sherrington, 2005). The feminisation of agriculture, marketing of nonwood produce and handicrafts made from biodiversity resources and involvement of women in collection of resources, such as fish, water, and fuel wood, have a direct link with development-induced changes in biodiversity. The Participation of women in scoping can be particularly useful in visualisation of several impacts that need to be reviewed in their context (Box 6.7). Similarly, the need to integrate traditional knowledge in EIA to capture impacts on traditional resource utilization practices has been well exemplified. In Uganda, for example, pressure from communities regarding environmental problems associated with the poor location of Kalangala Oil palm project in an ecologically sensitive ecosystem within a gazetted forest reserve led planning authorities to take action against developers (Ecaat, 2004).

Suggested approach for incorporating biodiversity in scoping

Scoping often is an unstructured process. Based on a list of potential impacts provided by experts, a proponent, competent authorities, stakeholders, and the general audience open a 'bidding process' culminating into a list of issues deemed important to be studied further. In order to provide a more structured way to derive a list of potential impacts, a sequence of questions has been developed, based on the impact assessment framework presented in Chapter 4. It provides examples of the kind of information that should be requested for in the terms of reference of an impact study. It should be noted that this list of steps represents an iterative process. Scoping and the actual impact study can be considered to be two formally embedded rounds of iteration; however, during the scoping process as well as during the study further iterative rounds may be needed, for example, when alternatives to the proposed project design have to be defined and assessed. (The approach has been field trialled in several World Bank funded irrigation and drainage projects; see for example, Abdel-Dayem (2004) and Slootweg *et al.*, 2007). Figure 6.2 illustrates the links between the impact assessment framework and the iterative, stepwise list of questions:

(a) Describe the type of project, and define each project activity in terms of its nature, magnitude, location, timing, duration, and frequency.
(b) Describe expected biophysical changes (in soil, water, air, flora, fauna) resulting from proposed activities or induced by any socioeconomic changes caused by the activity.
(c) Determine the spatial and temporal scale of influence of each biophysical change.
(d) Describe ecosystems and land-use types lying within the range of influence of each biophysical change.
(e) Determine, for each of these ecosystems or land-use types, if biophysical changes are likely to have impacts on biodiversity in terms of composition, structure, and key processes (see Chapter 2 for an elaborate description of these aspects of biodiversity). Guidance for determining levels of acceptable change to biodiversity need to be

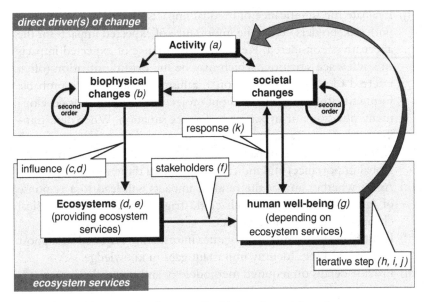

Figure 6.2 Impact assessment framework with iterative scoping steps.

developed at country level. Highlight any potentially irreversible impacts and any irreplaceable loss; take into account potential cumulative effects of different biophysical changes working on the same ecosystem.

(f) Identify, in consultation with stakeholders, the current and potential ecosystem services provided by the affected ecosystems or land-use types. Determine the values these services represent for society, in social, economic, and ecological terms. Give an indication of the main beneficiaries and those adversely affected from an ecosystem services perspective. Take into account the vulnerability of stakeholders.

(g) Determine which of these services will be significantly affected (positive or negative) by the proposed project.

(h) Define possible alternatives, including 'no net biodiversity loss' or 'biodiversity restoration' alternatives. Alternatives include location alternatives, scale alternatives, siting or layout alternatives, and/or technology alternatives; such alternatives may not be readily identifiable at the outset of the impact study, and one would need to go through the impact study to determine such alternatives.

(i) If impact cannot be avoided by alternative solutions, define possible measures to minimise or compensate for significant damage to, or loss of, biodiversity and/or ecosystem services; define possibilities to enhance biodiversity. Make reference to any legal requirements.

(j) Evaluate the significance of residual impacts, that is, in consultation with stakeholders define the importance of expected impacts for the alternatives considered. Relate the importance of expected impacts to a reference situation, which may be the existing situation (often referred to as baseline situation[1]), a historical situation, a probable future situation (e.g. the 'without project' or 'autonomous development' situation), or an external reference situation. When determining importance (weight), consider geographic importance of each residual impact (e.g. impact of local/regional/national/continental/global importance) and indicate its temporal dimension

(k) Assess whether any of the residual impacts will lead to a response of (groups in) society, which could trigger social and biophysical changes leading to more impacts.

(l) Identify necessary surveys to gather information required to support decision making. Identify important gaps in knowledge.

(m) Provide details on required methodology and timescale.

One should bear in mind that not implementing a project may in some cases also have adverse effects on biodiversity. In rare cases the adverse effects may be more significant than the impacts of a proposed activity (e.g. projects counteracting degradation processes). The opposite is equally true; it is important to acknowledge that compensation does not mean never having to say no to development proposals (Quigley and Harper, 2006a, 2006b; see Box 6.8).

Alternatives and mitigation

The scoping stage is fundamentally important for the identification of alternatives that avoid, mitigate and/or enhance proposed activities. Yet, during the EIA study further refined or even new ways to mitigate

---

[1] In many impact assessment frameworks a baseline description is asked for, even before the nature and severity of impacts is established. In our view a relevant baseline description can only be produced when the overview of expected impacts is presented through an analysis as described in this section (qualitative in scoping, as far as possible quantitative during study). How can one know what should be described? Conditions are dynamic, implying that present and expected future developments if the proposed project is not implemented (autonomous development) may also need to be included. Furthermore, a baseline in the sense of present situation, or autonomous development is not always the preferred reference situation, as described under item 'j'. As the term 'baseline' is not used unambiguously, we try to avoid the use of the term.

Box 6.8.   Does habitat compensation work?

Quigley and Harper (2006a, 2006b) report on habitat compensation in Canada as required by the Fisheries Act. A 'harmful alteration, disruption, or destruction to fish habitat' (HADD) cannot occur unless authorised with legally binding compensatory habitat to offset the HADD. His study demonstrated that fish habitat compensation, is at best, slowing the rate of habitat loss in Canada. Essentially, adaptive management has not been occurring because follow-up monitoring (by proponents and Fisheries and Oceans Canada, DFO) and independent, quantitative evaluations are rarely completed. Compensation habitat is almost always smaller than authorised and the habitat impacts are often larger than authorised. The ability to replicate ecosystem function is limited and both improvements in compensation science and institutional approaches are necessary. Recommendations to improve success include larger compensation ratios, creation and documentation of the functionality of compensation habitats prior/concurrent to HADDs, maintenance programs, increased monitoring and enforcement, and attention to limiting factors on a watershed basis among others. It is important to acknowledge that compensation does not mean never having to say no to development proposals. It is not possible to compensate for all habitats. Failure to acknowledge the limitations of compensatory science raises the disturbing proposition that Canada's efforts to conserve fish habitat will not be achieving the goal of 'no net loss'.

---

impacts can be proposed. Therefore we put a separate chapter on alternatives and mitigation in between the scoping and study stages; it is at the heart of good environmental impact assessment and is fundamentally important to scoping as well as the actual study. The purpose of mitigation in EIA is to look for ways to achieve the project objectives while avoiding negative impacts, or reducing them to acceptable levels. The purpose of enhancement is to look for ways of optimising biodiversity benefits. Both mitigation and enhancement of impacts should strive to ensure that the public or individuals do not bear costs, which are greater than the benefits that accrue to them. Remedial action can take several forms, that is, avoidance (or prevention), mitigation (including restoration and rehabilitation of sites), and compensation (often associated with residual impacts after prevention and mitigation).

With the 1997 EIA law in Japan, 'ecosystem' assessment became a requirement in addition to traditional 'flora and fauna' assessment, and mitigation provisions, such as 'avoid-minimise-compensate for', were regulated. Consequently 'ecological mitigation', especially compensatory mitigation, became the most controversial provision in EIAs. Because clear guidelines on ecological mitigation had not been published yet, some ecological restoration projects have been proposed as remissions of ecologically unsound development projects without reconsideration of original project plans. In response Tanaka (2000, 2001, 2002) has proposed elements of such guidelines. 'Avoidance' means to evade impacts by avoiding a whole or parts of original development plan by providing alternatives. 'Minimization' means to minimise impacts by the proposed development plan. 'Compensation' means to compensate impacts by restoring/creating ecosystems similar to the ones affected. According to Tanaka this sequencing is considered too general for EIA practitioners who need to propose substantial mitigation measures to their project proponents. For ecological mitigation measures, assessment aspects of both 'quality' and 'quantity' are indispensable. The aspect of 'quantity' is further divided into 'space' and 'time.' The following sequencing is proposed by Tanaka:

(1) Do we really need the proposed project? (We may be able to avoid the whole project.)
(2) Do we really need the project here/now? (We may be able to avoid the site/time of the original proposal.)
(3) Do we really need proposed project as a whole? (We may be able to avoid some parts of the original proposal.)
(4) Can we minimise the size/duration of the original proposal?
(5) When we cannot avoid/minimise the impacts of the original proposal, we must compensate for the remaining impacts (i.e. loss of ecosystems/habitats of the site).

First, compensatory measures, such as the restoration or creation of ecosystems can thus only be considered after considering 'avoid' and 'minimise' mitigation measures. Second, compensatory measures must always be proposed when ecological impacts cannot be avoided. Ecologically speaking, the following solutions are desirable:

• Quality: restore/create similar type of habitats, not different.
• Location: restore/create habitats within the development site.

- Size: restore/create similar (or larger) sized habitats.
- Time: restore/create before/at the same time of the development project.

One should acknowledge that compensation will not always be possible: there are cases where it is appropriate to reject a development proposal on grounds of irreversible damage to, or irreplaceable loss of, biodiversity (see Box 6.8. and Chapter 10 on biodiversity offsets for more information on compensation).

In his proposed methodology, Tanaka presents concepts very similar to the European Union Habitat Directive (European Council, 1992). This framework directive is presently being implemented at country level by the members of the European Union. From the perspective of biological diversity the mandatory sequencing of avoidance–minimization–compensation, where a next level can only be chosen if an earlier level is proven to be unfeasible, must be prepared to avoid 'excuse' type compensation. The EU habitat directive provides an elaborate procedure that can serve as an example for countries. The EU defines mitigation in Directive 85/337/EC as 'measures envisaged in order to avoid, reduce and, if possible remedy significant adverse effects' (European Council, 1985). Treweek (1999) defined mitigation as 'any deliberate action that is taken to alleviate adverse effects, whether by controlling the sources of impacts or the exposure of ecological receptors to them'. A particularly useful and influential definition of mitigation in the context of designated European Wildlife Sites was provided by the European Commission's guidance note on Article 6 of the Habitats Directive (European Commission, 2000), defining mitigation as 'measures aimed at minimizing or even negating the negative impact of a plan or project, during or after its completion'.

Compensation is distinct from 'mitigation' in the sense that it involves undertaking measures to replace lost or adversely impacted environmental values that should be equal to existing environmental values. Cowell (2000) defined environmental compensation as 'the provision of positive environmental measures to correct, balance or otherwise atone for the loss of environmental resources'. Kuiper (1997) talked about compensation in terms of 'the creation of new values, which are equal to the lost values'. In the United States, for the purposes of the Clean Water Act, under which wetland permits are issued, mitigation is defined as 'sequentially avoiding and minimizing impacts and compensating for remaining unavoidable impacts'. This sequential approach is also favoured by Canada.

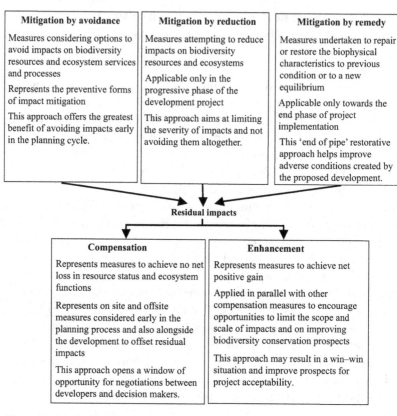

Figure 6.3 Approaches for mitigation of impacts.

The circumstances for application of different approaches of mitigation of impacts on biodiversity and their relative merits and outcomes are presented in Figure 6.3. See Appendix 6B in this chapter which provides an overview of various approaches with examples.

Several factors determine the reliability, practicality, and successful implementation of mitigation measures. In this context, Tomlinson (1997) warned that 'promises' and commitments to mitigation made in EIAs may not be delivered unless built in the consent procedures. Mitigation measures must therefore be translated into action in the correct way and at the right time if they are to be successful. Good Environmental Management Plans should expand on the mitigation measures described in the EA. Inclusion of technical details, justification for measures proposed, financial allocations, and schedules for implementation become essential requisites to increase the likelihood of implementation of mitigation measures.

Impact analyses

The impact assessment process comprises of a sequence of activities that are part of an iterative process. As new information is gathered through studies and stakeholder engagement, new issues emerge, as do different project alternatives, which all feed back into the assessment process, the project design and decision-making processes. The main tasks of impact analysis and assessment are:

- Refinement of the understanding of the nature of the potential impacts identified during scoping and described in the terms of reference. This includes the identification of indirect and cumulative impacts, and of the likely causes of the impacts. Identification and description of relevant criteria for decision making is an essential element.
- Evaluation of impacts; comparison of the alternatives; review and redesign of alternatives; consideration of mitigation measures; planning of impact management.
- Reporting of study results in an environmental impact statement.

The main steps in the actual impact analysis are: (i) data collection on the existing conditions, (ii) getting an overview of applicable regulations and standards, (iii) prediction of impacts and benefits, (iv) evaluation to establish the 'significance' of the predicted impacts, and (v) to determine if all, some, or none of the predicted impacts can be mitigated. The starting point for an assessment can be an existing National Biodiversity Strategy and Action Plan, a document asked for by the CBD. Such NBSAPs provide the policy with respect to biodiversity conservation at all three levels and usually contain relevant baseline data. The sequence of analytical steps provided in the scoping subsection can be followed iteratively to quantify to the extent possible the expected impacts. Apart from quantifying the expected impacts it is similarly important to provide a description of the autonomous development, or the 'no project' situation, for those ecosystem services that are expected to be affected by the proposed project. Impacts resulting from the autonomous development and the project interventions, including alternatives, can be compared.

A number of practical lessons with respect to the study process have emerged including that the assessment should:

- Allow for enough survey time to take seasonal features into account, where confidence levels in predicting the significance of impacts are low without such survey.

- Focus on ecosystem processes and ecosystem services, which are critical to the integrity of ecosystems and to human well-being. Explain the main risks and opportunities for biodiversity.
- Apply the ecosystem approach and actively seek information from relevant stakeholders and indigenous and local communities. Address any request from stakeholders for further information and/or investigation adequately. This does not necessarily imply that all requests need to be honoured; however, clear reasons should be provided where requests are not honoured.
- Consider the full range of factors affecting biodiversity. These include direct drivers of change associated with a proposal (e.g. land conversion, vegetation removal, emissions, disturbance, introduction of invasive alien species or genetically modified organisms, etc.) and, to the extent possible, indirect drivers of change, including demographic, economic, sociopolitical, cultural, and technological processes or interventions.
- Evaluate impacts of alternatives with reference to the autonomous development (no project situation). Compare against legal standards, thresholds, targets and/or objectives for biodiversity. Use national biodiversity strategies and action plans and other relevant documents for information and objectives. The vision, objectives and targets for the conservation and sustainable use of biodiversity contained in local plans, policies, and strategies, as well as levels of public concern about, dependence on, or interest in, biodiversity provide useful indicators of acceptable change.
- Take account of cumulative threats and impacts resulting either from repeated impacts of projects of the same or different nature over space and time, and/or from proposed plans, programmes, or policies.
- Recognise that biodiversity is influenced by cultural, social, economic, and biophysical factors. Cooperation between different specialists in the team is thus essential, as is the integration of findings, which have bearing on biodiversity.
- Provide insight into cause–effect chains. Also explain why certain chains do not need to be studied.
- Explain the expected consequences of the loss of biodiversity associated with the proposal, including the costs of replacing ecosystem services if they will be adversely affected by a proposal.
- Indicate the legal provisions that guide decision making. List all types of potential impacts identified during screening and scoping and described in the terms of reference and identify applicable legal provisions. Ensure that relevant impacts to which no legal provision applies are taken into account during decision making.

In practice it is not common that all three levels of diversity are treated in impact assessment; most attention goes to legally protected species and their habitats. Yet, there are some biological and/or societal reasons why certain issues merit special attention without necessarily being covered by regulations. Conducting genetic studies per se for determining impacts of projects at the *genetic level* is both extremely difficult and not usually feasible within the time frame in which EIA studies are generally conducted. One aspect which may merit special attention at genetic level is the risk of genetic erosion, specially for (i) highly threatened or legally protected species in the wild; (ii) varieties/cultivars/breeds of cultivated plants and domesticated animals and their relatives; (iii) species which are limited in numbers and/or have highly separated populations (rhinoceros, tigers, etc.); or (iv) ecosystems that may become isolated and thus obstruct gene flow (this applies to many species that depend on construction of so-called ecoducts across major lines of infrastructure). Another issue at genetic level is the introduction of living modified organisms that can possibly transfer transgenes to endemic plant or animal species.

Species receiving attention in EIA represent animal or plant species that have restricted distribution ranges, occupy specialised habitats, are endemic to an area, are locally distinct subspecies, are already vulnerable on account of existing threats to its habitats; have small isolated populations, or are rare or uncommon, either internationally, nationally, or locally. The species selection can be guided by a list of nationally protected species under country law as these would represent species that command highest conservation priority at the local level. Additionally, the policy documents, such as Biodiversity Action Plans, should be useful in prioritizing species recommended for conservation action. For a globally and regionally important species, IUCN Red List serves is a good guide to species selection. (See Box 6.9.)

Ecosystems include habitats that are critical for survival of rare and endangered species; perform critical functions such as routes for migration, dispersal, and genetic exchange of wild species; or serve as buffer areas of designated conservation units such as national parks and habitats suitable for reintroduction of species in alternative habitats. Assessment of magnitude and nature of impacts on habitats provide adequate guidance for determining the significance of impacts to suggest a 'no go' option or alternatively, suggest appropriate mitigation strategies for timely action for conservation. Ecosystems and the plants and animals within them provide humans with *ecosystem services* that would be very difficult to duplicate. For example, coral reef and associated mangrove forests play a critical but often undervalued role for coastal residents throughout the

188 · Asha Rajvanshi *et al.*

Box 6.9. Example of species-specific issues in impact assessment

The Indian wild buffalo is an endangered species listed in the Red Data Book (IUCN, 1994). It is found in four relict populations in the Bastar district of Chattisgarh state in India. One of the populations is found in Bhairamgarh Sanctuary. The ecological impact assessment of the Bodhghat hydropower project proposed in Bastar identified that the flooding of the river bed grasslands during the water release at the peaking hours (8 p.m. and 11 p.m.) would be one of the direct impacts of the project on wild buffaloes in Bhairamgarh Sanctuary. It was feared that the loss of foraging habitat of wild buffaloes in summer, when such riverbed grasslands offer critical food resources, would cause significant impacts on wild buffalo populations. These threats to species conservation posed by the project proposal led to the rejection of the project even after substantial progress was made in the construction activities (Rajvanshi, 2002).

---

world. Coastal protection from waves and storm surge, food security, recreation, and tourism are but a few of the ecosystem services provided by coral reef ecosystems. While many of these services are performed seemingly for 'free', they are worthy for assessment and monetary valuation. Environmental impact assessment requires consideration of how biodiversity can be sustained as the basis for provision of ecosystem services and the support of livelihoods. The fundamental concept is to understand the relationship between the biophysical status of ecosystems (and biodiversity) in terms of provisioning of ecosystem services and the links to livelihood of people. Valuation of the changes to the provision of ecosystem services under alternative project scenarios becomes relevant to assess changes in costs and benefits from the alternative project scenarios for different groups, and highlights the incremental cost or benefit of changing the biophysical status of a particular ecosystem (Box 6.10).

After a detailed analysis of the impacts in terms of their nature, magnitude, extent, and duration, a judgement of their significance is required; in other words: are the impacts acceptable to stakeholders and society as a whole, do they require mitigation and/or compensation, or are they simply unacceptable? Biodiversity–inclusive evaluation involves:

Box 6.10. Examples of critical ecosystem services (Chivian, 2002)

For centuries, the oyster population of the Chesapeake Bay was capable of filtering a volume of water equal to the complete volume of the Bay in a three-day period. Pollution, habitat destruction, overharvesting, and other pressures have dramatically reduced the oyster population, greatly diminishing this critical filtering service. With the diminished oyster population the filtering now takes a year, and the waters of the Bay are poorer in oxygen and generally more polluted.

The oil palm was introduced into Malaysia from the forests of Cameroon in West Africa in 1917, but the weevil that pollinates the African oil palm was not introduced at the same time. For decades, the palm growers of Malaysia relied upon expensive, labor-intensive hand pollination. In 1980, the weevil was imported to Malaysia, boosting fruit yield in the palms 40–60 percent, and generating savings in labor cost of US$140 million per year.

- Identifying impact that, by its magnitude, duration or intensity alters an important aspect of biodiversity.
- Assessing sensitivity of the ecological features to provide a benchmark against which changes can be evaluated to determine the vulnerability of species or ecosystem characteristics.
- Determining the overall significance of the anticipated impacts including the economic, social, and ecological values affected positively or negatively.
- Recommend what impacts essentially need to be managed through impact reduction measures.

Approaches for evaluation of project impacts are presented in Box 6.11. Guidance is available to assist practitioners in assigning significance of ecological impacts (Canter and Canty, 1993; Hilden, 1995; Canter, 1996). Decision makers in developing countries tend to focus on economic growth and poverty reduction imperatives. In this sense, when links between project objectives are juxtaposed with Millennium Development Goals and poverty reduction, decision makers take greater notice of impact evaluation. This has been the case in the integration of environment in most developing countries' poverty reduction strategies (Rajvanshi et al., 2007).

Box 6.11. Approaches for impact evaluation

- Opinions of qualified decision makers in municipalities or ministerial departments.
- Opinion of specialists (environmentalists, ecologists, economists, hydrologists, engineers, social scientists, and urban planners).
- Past experience of evaluating similar projects.
- Public opinion (Public hearing reports are mandatory requirements in many countries in the region and these are helpful in evaluating the significance of project-related impacts).
- Compatibility of the proposed project with the Government's development policy, in general.
- Link between project objectives and Millennium Development Goals (biodiversity conservation; livelihood security and eradication of poverty).

Reporting

The purpose of an EIA report is not to reach a decision but to present the consequences of the proposed project for (i) the proponent to plan, design, and implement the proposal in a way that eliminates or minimises the negative effect on the biophysical and socioeconomic environments and maximises the benefits to all parties in the most cost-effective manner; (ii) the Government or responsible authority to decide whether a proposal should be approved and the terms and conditions that should be applied; and (iii) the public to understand the proposal and its impacts on the community and environment, and provide an opportunity for comments on the proposed action for consideration by decision makers. Some adverse impacts may be wide ranging and have effects beyond the limits of particular habitats/ecosystems or national boundaries. Therefore, environmental management plans and strategies contained in the environmental impact statement should consider regional and transboundary impacts, taking into account the ecosystem approach. The inclusion of a nontechnical summary of the EIA, understandable to the interested general audience, is strongly recommended.

Good practice guidelines in impact evaluation advise practitioners to:

- Give a clear and transparent summary of the positive and negative impacts of project on ecosystem services, where possible in quantitative terms.
- Indicate the benefits and costs of the project to society.

Box 6.12. Accreditation schemes for EIA professionals

In India the National Registration Board for Personnel and Training (NRBPT), a constituent of Quality Council of India, has launched the scheme for registration of EIA consultant organisations which has been duly recognised by Ministry of Environment and Forests, Government of India (at http://moef.nic.in). The Mozambican EIA legislation has provisions for the accreditation of EIA professionals. Similarly, South Africa and Uganda have put in place regulations for the certification and registration of EIA practitioners. The Uganda regulations of 2003 set minimum standards and criteria for qualification as an EIA practitioner. The Regulations also establish an independent Committee of Environmental Practitioners whose roles are, among others, to regulate the certification, registration, practice, and conduct of all environmental impact assessors and environmental auditors (SAIEA, 2003; Ecaat, 2004).

- State clearly any critical assumptions and uncertainties.
- Show clearly any spatial distributional effects.
- Indicate the criteria adopted for evaluation of impacts on biodiversity components.
- Present the argument for the decision choice by highlighting the significance of impacts and the variations in impacts associated with different alternatives where applicable.
- Highlight trade-offs.

Several factors can influence the quality of biodiversity-related information in an EIA report. Some of the more obvious of these include limitations with respect to quality and adequacy of biodiversity data, limited capacity of consultants to give a fair treatment to biodiversity in the assessment of project-related impacts and the constraints of costs to undertake long-term and focussed studies for biodiversity assessment. Some countries are improving the EIA practices and reporting skills of the practitioners through targeted capacity building initiatives and by promoting reforms that are aimed at introducing accreditation systems and registration schemes for ensuring checks on competence of individual practitioners and EIA consulting firms in relevant functional areas or disciplines (see Box 6.12).

Experts preparing an EIA must appreciate that the EIA report will be read by a wide range of people and that the subject matter may appear technically complex for some readers. Senior administrators and planners responsible for making decisions may not understand the importance of technical arguments unless they are presented carefully and clearly. In order to do so, the environmental impact statement usually consists of (i) a technical report with annexes; (ii) an environmental management plan, providing detailed information on how measures to avoid, mitigate, or compensate expected impacts are to be implemented, managed, and monitored; and (iii) a nontechnical summary.

Review

The purpose of the review of the environmental impact statement is to ensure that the information for decision makers is sufficient, focussed on the key issues, and is scientifically and technically accurate. In addition, the review should evaluate whether:

- The likely impacts would be acceptable from an environmental viewpoint.
- The design complies with relevant standards and policies, or standards of good practice where official standards do not exist.
- All of the relevant impacts, including indirect and cumulative impacts, of a proposed activity have been identified and adequately addressed in the EIA. To this end, specialists should be called upon for the review and information on official standards and/or standards for good practice to be compiled and disseminated.
- The concerns and comments of all stakeholders are adequately considered and included in the final report presented to decision makers. The process establishes local ownership of the proposal and promotes a better understanding of relevant issues and concerns.

Review should also guarantee that the information provided in the environmental impact statement is sufficient for a decision maker to determine whether the project is compliant with or contradictory to the objectives of the Convention on Biological Diversity. The effectiveness of the review process depends on the quality of the terms of reference defining the issues to be included in the study. Scoping and review are therefore complementary stages. Reviewers should as far as possible be independent and different from the persons/organisations who prepare the environmental impact statement.

Generic frameworks for review are provided by Lee and Colley (1992), European Commission (1994) and VROM (1994). This can be

Table 6.3 *Evaluation of EIA reports.*

Quality grade	Quality remark	Explanatory note on quality grade/remark
A	Excellent	The work has generally been well-performed with no important omissions of biodiversity-related issues.
B	Good	Task performed satisfactorily and is complete with only minor omissions/inadequacies.
C	Satisfactory	Task is satisfactory despite some omissions or inadequacies.
D	Weak	Indicates that parts are well attempted but, on the whole, are just unsatisfactory because of omissions or inadequacies.
E	Poor	Task is not satisfactory, revealing significant omissions or inadequacies.
F	No opinion	The work is insufficient to base judgment.

appropriately adapted for promoting 'biodiversity driven' review of EIA reports. Good practices demand that the review should be guided by mainstreaming criteria for biodiversity. The advantage of criteria-based review is two-fold – it can facilitate the process of review even in absence of regulatory guidelines and it can also ensure that the results comply with the requirement of integrated assessment incorporating ecological-economic issues. The final evaluation of the report presented as a 'report card' can be extremely helpful in making recommendations for facilitating the decision making. A grading system presented in Table 6.3 in line with several grading systems that are already in place (Lee and Colley, 1992; UNEP, 2002) can be adopted for evaluation of EIA reports for adequacy and completeness of information on biodiversity.

Decision making

Formal and informal decision making takes place throughout the process of environmental impact assessment. Depending upon the EIA arrangements that are in place, these interim decisions are made by different players, for example, because the screening decisions and drafting of terms of references after scoping are usually the task of responsible authorities. During EIA preparation, the proponent in consultation with experts often makes choices between alternatives and mitigation measures. The final decision between refusal and authorization of the project is normally a political decision, often taken by the national government, planning authority, or other equivalent body. In some EIA systems a formal clearance is needed from the environmental authority, based on the EIA report. In other systems the EIA report is accepted by the environmental authority as good information for decision making; final decision making can overrule the outcome of the EIA report. In such cases, the transparency of the EIA procedure requires the decision making authority to provide good arguments for such a decision.

It is observed that impact studies are directed by legal obligations, while the aim of impact assessment is the provision of information for good decision making. By leapfrogging from expected impact to legal requirement, one runs the risk of losing relevant information on those biodiversity issues that cannot be caught under the legal umbrella but which may represent valued elements from a biological or from a social perspective. The impact assessment study should provide insight in the cause–effect chains and come to conclusions regarding impacts on biological diversity (Brouwer and Van den Tempel, 2006). The next step is to indicate the legal issues that may create the boundary conditions for decision making (Slootweg, 2005). A workshop on the CBD guidelines concluded that a strict separation between data-collecting and impact analysis on the one hand and evaluation of the predicted environmental impacts on the other hand is essential. Besides, how the results of the evaluation become part of the decision-making process requires separate attention. Generally, a strict separation of the impact analysis, support of the decision, and decision making is of great importance (Georgi and Peters, 2003).

Decision makers should take the following into consideration:

- The systematic assessment of the societal values of biodiversity, in ecological, economic, and social terms.
- Consideration of the uncertainty factors affecting the impact evaluation. The consideration of uncertainty factors is generally missing in the forecast of impacts, and in particular of impacts concerning ecological aspects (Geneletti, 2002). For example, Culhane et al. (1987) conclude from their analysis of United States' EISs that nearly two-thirds of the impact forecasts 'fall into a grey area between accuracy and clear inaccuracy' The dependence on uncertainty analyses is likely to increase when different alternatives have to be weighed and the decision makers decision need to benefit from indications about the stability of the resulting ranking and the degree of confidence with which an alternative appears to perform better than another one.
- The precautionary approach should be applied in decision making in cases of scientific uncertainty about risk of significant harm to

biodiversity. As scientific certainty improves, decisions can be modified accordingly.

- Finally, the decisions should preferably not put conservation goals against development goals that may stir irresolvable controversies. The rationale for encouraging the principles of sustainable development should be underlined.

The following rules and conventions for decision making adopted by leading EIA systems (Wood, 1995) would also be applicable to decision making in the context of biodiversity:

- No decision will be taken until the EIA report has been received and considered.
- The findings of the EIA report and review are a major determinant of approval and condition setting.
- Public comment on the EIA report is taken into account in decision making.
- Approvals can be refused or withheld, conditions imposed, or modifications demanded at the final decision stage.
- The decision is made by a body other than the proponent.
- Reasons for the decision and the conditions attached are made transparent.
- There is a public right of appeal against the decision (where procedures have not been followed or they have been applied unfairly).

Follow-up

EIA does not stop with the production of a report and a decision on the proposed project. Activities that have to make sure the recommendations from EIS or EMP are implemented are commonly grouped under the heading of 'EIA follow-up'. The main components and tools of EIA implementation and follow up include:

- surveillance and supervision – to oversee adherence to and implementation of the terms and conditions of project approval;
- effects or impact monitoring – to measure the environmental changes that can be attributed to project construction and/or operation and check the effectiveness of mitigation measures;
- compliance monitoring – to ensure that applicable regulatory standards and requirements are being met, for example, for storing mine overburden and it stabilization;

- environmental auditing – to verify the implementation of terms and conditions, the accuracy of the EIA predictions, the effectiveness of mitigation measures, and the compliance with regulatory requirements and standards;
- ex post evaluation – to review the effectiveness and performance of the EIA process as applied to a specific project; and
- postproject analysis – to evaluate the overall results of project development and to draw lessons for the future.

The follow-up phase of environmental impact assessment process (EIA) is generally considered as a major shortcoming in many jurisdictions (Arts, 1998 and Arts *et al.*, 2001). Experience suggests that follow-ups have particularly paid little attention to the impacts on biodiversity. This is more so because traditionally the balance of effort applied to EIA has been skewed towards project preparation rather than on monitoring the 'actual' environmental impacts of projects and effectiveness of mitigation plans. This trend has transformed the basic character of EIA making it a predictive tool that has become linear rather than iterative in nature. Factors that have been recognised to hinder the effective implementation of the Environmental Management Plans (EMPs) include insufficient commitment on the part of project executors, absence of a framework for implementation of an EMP for the proposed project and weak internal and external protocols for supervision of implementation (Kamppinen and Walls, 1999). A review of 30 projects with EIA approvals executed by Shell Petroleum Development Company–Eastern Division (SPDC-E), Nigeria, between 1997 and 2002 revealed that only three implemented the EMP as stipulated in EIA approvals obtained for the projects. Further investigation showed that the practice among project teams was to 'forget' the EIA report on the shelf as soon as the approval for project development was obtained (Dada and Okubokimi, 2004). Without appropriate implementation and follow-up to decision making, EIA becomes a paper exercise to secure an approval, rather than a practical exercise to achieve environmental, economic. and biodiversity benefits. It therefore becomes even more important to mainstream biodiversity in the implementation and follow-up stages of impact assessment (see Box 6.13).

For biodiversity, continuous monitoring may not be required. It is necessary to monitor impacts on biodiversity at relevant stages throughout the life of a project. By systematically comparing and assessing changes to biodiversity against baseline data, developers can evaluate the accuracy of the impact predictions. This is particularly important where uncertainty exists (e.g. in the prediction of impacts and availability of baseline

---

Box 6.13. Monitoring and follow-up for biodiversity

Great crested newts (*Triturus cristatus*) are protected under European and UK legislation, but are frequently the subject of conflict between development and conservation in England. When this occurs, the developer is legally obliged to undertake in situ translocations of newts to areas within or adjacent to the development site. Where postproject monitoring was not undertaken, the number of newts translocated per project declined and was related to the total area of habitat destroyed and work effort. About 27 percent of the great crested newt terrestrial habitat was destroyed during the developments along with about half of all ponds. Where follow-up monitoring of translocations was conducted, there was evidence of breeding at most sites one-year postdevelopment, but it is unclear whether these populations were sustainable in the long term (Edgar *et al.*, 2005).

Based on experiences with large infrastructure projects in the Netherlands, Van Den Tempel and Brouwer (2005), however, state that monitoring and evaluation focussed on counting of species and measuring of surface areas only does not provide sufficient information; understanding and monitoring the mechanisms behind these changes leads to better understanding of the effects of the intervention and the actual results of mitigation and/or compensation.

---

data). Monitoring allows developers to check effectiveness of mitigation measures to take any subsequent actions necessary to ameliorate problems not identified earlier in the assessment stage and to undertake audit and evaluation to strengthen future EIA applications. Monitoring also allows postdevelopment problems to be identified and rectified. It is critically important that the monitoring program is well-structured and includes monitoring at each of the project stages.

Monitoring for biodiversity changes usually has to be tailored to particular circumstances. The use of indicators is common to all, typically used to summarise trends in particular habitats or species, and acting as warning lights of adverse as well as positive trends. Good practices dictate that standard techniques/methods of data collection and quality control mechanisms should be used so that the data can be used for comparative purposes, both over time for the project at hand and with other projects elsewhere as appropriate. Although the purpose of EIA implementation and follow-up is to ensure that the conditions attached to project approval are carried out and function effectively, and to gain information that can be used to improve future EIA practice, by itself, this process cannot turn around an environmentally unsound project.

Monitoring should focus on those components of biodiversity most likely to change as a result of the project. The use of indicator organisms or ecosystems that are most sensitive to the predicted impacts is thus appropriate to provide the earliest possible indication of undesirable change. Because monitoring often has to consider natural fluxes as well as human-induced effects, complementary indicators may be appropriate in monitoring. Indicators should be specific, measurable, achievable, relevant, and timely. Where possible, the choice of indicators should be aligned with existing indicator processes. The results of monitoring should provide information for periodic review and alteration of environmental management plans and for optimising environmental protection through good, adaptive management at all stages of the project. In accordance with 'best practice', the biodiversity data collected for assessment and any subsequent monitoring should be made publicly available to provide opportunities to link into the national planning and nature conservation management processes. Provision should be made for regular auditing in order to verify the proponent's compliance with the EMP, and to assess the need for adaptation of the EMP (usually including the proponent's license). An environmental audit should be encouraged as an independent examination and assessment of a project's (past) performance.

Appendix 6A: Indicative set of screening criteria to be further elaborated at national level[2]

Category A:[2] Environmental impact assessment mandatory for:

• Activities in protected areas (define type and level of protection);
• Activities in threatened ecosystems outside protected areas;
• Activities in ecological corridors identified as being important for ecological or evolutionary processes;
• Activities in areas known to provide important ecosystem services;
• Activities in areas known to be habitat for threatened species;
• Extractive activities or activities leading to a change of land-use occupying or directly influencing an area of at minimum a certain threshold size (land or water, above or underground – threshold to be defined);

---

[2] Note: These criteria only pertain to biodiversity and should therefore be applied as an add-on to existing screening criteria.

- Creation of linear infrastructure that leads to fragmentation of habitats over a minimum length (threshold to be defined);
- Activities resulting in emissions, effluents, and/or other means of chemical, radiation, thermal or noise emissions in areas providing key ecosystem services (areas to be defined); and
- Activities leading to changes in ecosystem composition, ecosystem structure or key processes responsible for the maintenance of ecosystems and ecosystem services in areas providing key ecosystem services (areas to be defined).

Category B: The need for, or the level of environmental impact assessment is to be determined for:

- Activities resulting in emissions, effluents and/or other chemical, thermal, radiation or noise emissions in areas providing other relevant ecosystem services (areas to be defined);
- Activities leading to changes in ecosystem composition, ecosystem structure, or ecosystem functions responsible for the maintenance of ecosystems and ecosystem services in areas providing other relevant ecosystem services (areas to be defined); and
- Extractive activities, activities leading to a change of land-use or a change of use of inland water ecosystems or a change of use of marine and coastal ecosystems, and creation of linear infrastructure below the Category A threshold, in areas providing key and other relevant ecosystem services (areas to be defined).

Appendix 6B: Mitigation hierarchy and illustrative examples of various approaches and options

Approaches and options	Illustrative examples
AVOIDANCE Identification of the least damaging alternative	Planning the route of new linear projects through existing route corridors (e.g. for road, rail, pipeline, canal, and transmission line) to avoid impacts on sensitive environments, such as human settlements, biodiversity rich areas, habitats of endangered species, archeological and cultural sites within the route corridor of the proposed projects.

(*cont.*)

Appendix 6B (*cont.*)

Approaches and options	Illustrative examples
Sensitive design plan for avoiding impacts	The application of 'nature engineering' concepts in the designing of ecosensitive structures. Construction of culverts, underpasses, and bridges to avoid obstruction of animal movement across home ranges and landscape is a common practice in planning of transportation infrastructure in many countries. The main reason for the declining population of swifts and sparrows in the Netherlands is the lack of nesting spaces due to spread of towns and modernization of roof designs. Lafarge has designed special bird tiles which contain a cavity to let birds build their nests. These tiles designed to allow nesting of swifts have helped stem the decline of birds in the Netherlands. (Lafarge, 2000).
Environmentally sustainable technology options	Sustainable technology options for controlling impacts and making good environmental choices during construction, postconstruction, and progressive phases of the project. For laying pipeline across major rivers in India, Horizontal Directional Drilling has been adopted as opposed to the open cut method to avoid impacts on several endangered species like the mugger crocodile and Gangetic dolphin. (WII, 1993).
Development restrictions in sensitive areas	Restriction on locating projects in sensitive areas. In many countries, siting ordinances and regulations govern the project location. In Hong Kong, for general multistorey industrial sites without chimneys, a buffer distance of at least 100 m from sensitive uses is normally required (Environmental Protection Department, 1997). The UK Planning Policy Statement 22 (Anonymous, 2004) for renewable energy stipulates that priority should be given to locate renewable energy projects in less sensitive parts of the countryside and coasts and that these should be designed to minimise adverse impact on landscape, wildlife, and amenity.
Avoidance of certain key areas by adopting the 'precautionary approach'	Application of the Precautionary Principle recognises the merit of delaying development consent until the best available information can be obtained through consultation with local stakeholders/experts and/or new information can be consolidated. (Cooney and Dickson, 2006). Exclusionary criteria for designation of 'no development' zones have provided controls in many countries based on legal and policy directives for safeguarding biodiversity resources of the country.

Appendix 6B (*cont.*)

Approaches and options	Illustrative examples
	A general consensus on the 'no go' zones have emerged based on various guidelines (WWF, 2002; EBI, 2004; IFC, 2004) that have been developed in the context of sector-specific developments around the world.
Suitable timing of activities	Timing of various activities planned under a project to avoid overlaps with key life cycle events (e.g. flowering and seeding, nesting or breeding seasons) has been recognised as a common and effective approach for avoiding impacts on protected species.

MINIMIZATION

Control measures to prevent pollution	Specific examples include installation of appropriately designed chimneys to regulate emissions; sound-proofing of buildings to reduce noise; treatment of effluents to reduce pollution load in wetlands; and arresting soil erosion to reduce loss of productive soils.
Minimization of physical disturbances	Responsible operations and adoption of good practices while undertaking activities involving physical alteration of land can bring about significant reduction in land degradation. Use of nonintrusive techniques, such as remote sensing and global positioning systems during exploration for oil and the use of lighter drilling rigs or helicopter-assisted drilling programs to transport the equipment into sensitive or rugged terrain (White *et al.*, 1996).
Creative land management	Creative land management, landscaping, and development of alternative land use to reduce physical impacts during construction/operation and improve post project aesthetics. The pit in Sanquelim mine in Goa, India has been managed as a pisiculture pond, and the fishery resources are being used by local communities. The mine overburden dumps are planted with native species of economic value. (Patil, personal communication, 12 February 2006).
Technological fixes	Transportation departments in many countries are incorporating innovative designs in the development of roadways to minimise barrier effects of roads and to enhance connectivity functions of passages for animals across highways. Several agencies including Florida Department of Transportation (FDOT), Colorado Department of Transportation (CDOT), and the public Works department in the Netherlands are working to develop strategies for reducing road-related impacts on wildlife in new transportation projects.

(*cont.*)

Appendix 6B (*cont.*)

Approaches and options	Illustrative examples

REMEDIATION

Onsite repair, reinstatement and restoration activities

Restoration, rehabilitation, and conservation efforts to restore the historic type of ecosystem. Application of such measures is common in mining projects.

Offsite restoration and conservation of biodiversity

Translocation of plants, animals, and habitats from the sites of proposed development to parts of their former range. The capture and translocation of dwarf chameleons was successfully carried from the proposed light industry park to adjacent Durban Metropolitan Open Space System (D'MOSS) area in South Africa. The developer of the industrial park, Cato Manor Development CMDA, also provided the funding support for maintenance of the habitat for chameleons in the release site. (Armstrong, 2004). For the crocodilian species to be impacted, by construction of Narmada Dam in India, captive rearing and release in other suitable rivers that have recorded distribution of mugger crocodile have been recommended. (WII, 1994). A protected species of pitcher plants (*Nepenthes mirabilis*) was successfully transplanted from the damaged sites of the North Lantau Expressway in Hong Kong (*Source*: Environmental Protection Department, 1997).

COMPENSATION

Onsite compensation of ecological values

Examples of this form of compensation include restoration of natural areas in an urban context where original ecological or hydrologic conditions cannot be restored or where an altered environment can no longer support any previously occurring type of regional ecosystem forest. Examples of compensation include artificially created lakes in mined out pits and managed on scientific principles as wetland ecosystems to serve as excellent replacement habitats for a wide variety of wetland birds.

Offsite compensation

Efforts of strengthening conservation of species threatened by a proposed development elsewhere through a third party where a developer purchases biodiversity credits or pays a third party to provide an offset ex ante. This also includes the costs to biodiversity dependent communities for the foregone uses of areas for hunting, cultivating and collecting forest product.

Appendix 6B (*cont.*)

Approaches and options	Illustrative examples
	BP has three petrochemical plants in Terengganu, Malaysia and there are significant oil and gas reserves off the east coast of the state. Terengganu is also home to about 70 percent of Malaysia's turtles and the sanctuary is an important nesting habitat for three species of marine turtles and the painted terrapin. In June 1999, BP Petronas Acetyls, a joint venture between BP and Petronas, partnered with the Malaysian Department of Fisheries and the World Wide Fund for Nature, Malaysia to create the Ma'Daerah Turtle Sanctuary in the state of Terengganu, Malaysia. It is the first turtle sanctuary to be funded by the private sector and the second largest sanctuary in Malaysia. (EBI, 2003a, 2003b). The 1985 EIS of the trans–Chugoku highway construction project overlooked the presence of an important wetlands area as there was no information on these wetlands, and field surveys were not required by the early guidelines. Local NGOs protested during construction, which led to the creation of a new wetland as a compensatory measure. This is the first case in Japan in which a proponent tried to implement compensatory measures based on biological diversity considerations (Tanaka, 2001).
In-kind compensation	A range of in-kind compensation measures involving use of trading instruments to offset impacts on biodiversity and ecosystems to assure the sustainability of development proposals are being promoted. Carbon trading and the wetland and conservation banking schemes, developed in the context of the Endangered Species Act and the Clean Water Act of US regulatory regimes, are perhaps the best examples of trading instruments. Estimates indicate that these trading schemes have created 72,000 hectares of wetland and endangered species habitat in more than 250 approved 'banks' selling habitat 'credits' in more than 45 states in United States (Wilkinson and Kennedy, 2002; Fox and Nino-Murcia, 2005).
Out-of-kind compensation	Approaches involving making direct monetary payments in the form of user fees, charges, taxes, and royalties for enhancing biodiversity conservation in designated PAs and on private lands.

(*cont.*)

Appendix 6B (*cont.*)

Approaches and options	Illustrative examples
ENHANCEMENT	
Options aimed at achieving net positive gain for biodiversity	Options aimed at providing new benefits for biodiversity through improved management, better conservation practices, and higher level of protection. Examples of enhancements include using sustainable drainage schemes so that drainage infrastructure also acts as biodiversity habitat; landscaping in additional areas so that planting within them forms a wildlife corridor and habitat link between areas of habitat adjacent to the site; and creating new protected areas for protection of endangered species.

7 · Biodiversity-inclusive Strategic Environmental Assessment

Roel Slootweg

Biodiversity in SEA: a new field of expertise

To facilitate consideration of cumulative impacts and the early consideration of environmental and social constraints in a planning process, there has been a growing demand for Strategic Environmental Assessment (SEA) and an increase in the number of countries introducing SEA legislation. It is generally agreed that effective safeguarding of biodiversity is only possible if ecological constraints and possibilities are identified early in the development planning cycle, well in advance of individual development proposals. The SEA tool has been developed for this purpose as it identifies impacts further 'upstream' in the planning process, including biodiversity impacts. A classical case of late realisation of biodiversity impacts is provided by the US$17.6 billion Korean High-Speed Railway Project, of which construction was delayed for five years because of unacceptable biodiversity impacts and related social uprising and court cases. Conservationists have called for legislators to update the EIA system to encourage biodiversity conservation during the early planning stages of development. Towards this end, the Korean Ministry of Environment introduced SEA, which requires environmental impact studies during the early stages of development planning, thus allowing them to be used as a decision-making tool (Sang, 2005).

In general, SEA enables consideration of the status of biodiversity over a longer time frame and for larger geographical areas. It offers solutions to some of the shortcomings commonly attributed to project-level Environmental Impact Assessment (EIA), including the difficulties inherent to considering cumulative or landscape-scale ecological effects (Treweek, 2001). Many threats to the long-term survival of biodiversity are individually insignificant but collectively serious. By definition, 'cumulative' environmental effects are not attributable to any one source of activity and cannot be regulated in isolation. Planning for new development must therefore take account of cumulative threats to biodiversity as well as those posed by individual proposals.

The methods by which biodiversity considerations would be incorporated into SEA have not been elaborated in much detail; there is little common agreement on how to address biodiversity in SEA (see Uprety, 2004, 2005, for a well-documented case from Nepal). In 15 country case studies, Treweek (2001) identified only 3 countries as having SEA procedures that pay some attention to biodiversity. However, ten countries indicated that impact assessment improves consideration of biodiversity in development planning. There obviously is scope and need for the improvement of methods to incorporate biodiversity in SEA. Treweek refers to a number of cases of SEAs to assess management plans for national parks. In such SEAs biodiversity is obviously taken into account. However, few references exist of biodiversity being fully integrated into an SEA process outside typical biodiversity-related cases, such as the national parks example. This is explained by the relatively recent introduction of the SEA instrument; there is a general lack of accessible SEA documentation as most material is written in local languages and SEA reports usually have no broad distribution in the scientific community. There is an obvious need for the scientific community to collect case material and analyse the way in which biodiversity is treated in these cases. The European SEA Directive came into force in 2006. This should result in a wealth of information becoming available in the coming years. Even the development of this Directive had an impact on the preparation of the Estonian Development Plan for 2003–2006, as outlined in Box 7.1.

In order to overcome the general lack of scientific information on how to treat biodiversity in SEA, the CBD has invited the impact assessment community through its professional association, the International Association for Impact Assessment (IAIA) (see: www.iaia.org) to collect case material through its worldwide membership. In a first compilation of available case material among members of IAIA, Slootweg (2003) drew a number of conclusions serving as a first step towards the development of an approach to integrate biodiversity in SEA:

- The need for SEA is widely recognised and important frameworks have been developed (such as the EU directive on SEA). Although biodiversity considerations do surface in many SEAs, the theme is not systematically and consistently addressed.
- Case material points towards the importance of participatory approaches, with the obvious objective of harmonising the interests

---

Box 7.1. The 2006 European SEA-directive will provide many examples

An interesting example of case material which can be collected throughout countries in Eastern Europe recently entered into the EU, is the SEA for the Estonian National Development Plan 2003–2006. It provides some insight in the dilemma faced by a country that has put its economic and social development into higher gear. At the moment of study the country was preparing to become a member of the EU.

Estonia features so many natural and semi-natural landscapes that it requires an enormous amount of work even to get a comprehensive picture of these landscapes. For example, little has yet been done to ensure protection of habitats of water biota. The natural springs that are unique in Europe are still waiting to be researched. On the other hand, a system of protected areas and other environmental limitations needs to be accomplished soon, in order to avoid unreasonable delay of economic development. (Regional Environmental Centre, 2002)

---

of all having a stake in biodiversity; more simply stated, a participatory approach is necessary to perform the balancing act between present day uses and the need to safeguard biodiversity for future generations and purposes (see Box 7.2). This points towards the relevance of the ecosystem approach for SEA, because it stresses the importance of stakeholder involvement (CBD, 2000a, 2004a).

- There hardly is experience on the assessment of key ecological processes, that is, those processes that are essential to maintain or restore biological diversity and its functioning. The present status of assessment is static and reflects impacts on groups of species. It can be argued that more focus on key ecological processes could avoid large data collection exercises and provide more relevant information for ecosystems that almost by definition are dynamic. Because SEA usually deals with less well-defined areas, compared to project level EIA, the notion of key ecological processes (see Chapter 2) may provide a better tool for impact assessment under conditions of higher uncertainty (see Box 7.3).
- The provision of multiple ecosystem services by one ecosystem is neglected. By definition an ecosystem provides multiple ecosystem

---

Box 7.2. Public participation and biodiversity in SEA

Two examples of the role of public participation in SEAs where biodiversity issues were at stake:

(i) In Estonia (Jalakas, 1998) a pilot project was initiated to conduct SEA during the development of comprehensive planning for a selected municipality. The aim of the pilot project was to use the experience for the development of an SEA methodology for the Estonian conditions. The pilot area was an island belonging to a protected area with recreational objectives, having a history of military use and consequent pollution problems. One of the most important and successful stages of the process was public involvement and participation. Timely and early informing of the public enabled to avoid the arising of conflicts, find new creative solutions, and receive information concerning the preferences of interested parties and inhabitants. The implemented pilot project proved well that the integration of SEA into the very process of development planning is the only way to reach a solution optimum from the viewpoint of both the natural environment and the society while using the minimum of resources.

(ii) Sheate (2003) and Sheate *et al.* (2008) describe a research into scenarios for reconciling biodiversity conservation with declining agriculture use, focussed on mountain areas in six study areas in Europe. A key component is the sustainability assessment of alternative scenarios both for agriculture and rural policy and for biodiversity management. A particular aspect of the approach is the engagement of stakeholder panels in each study area throughout the research, emphasising the participatory nature of the methodology.

---

services. However, responsibility for management of ecosystem services usually is divided over different sectoral institutions, separating, for example, nature conservation from economic activities, thus neglecting potentially viable alternatives in which biological diversity can be enhanced and used sustainably (see Box 7.4).

More in-depth case material has been specially developed for the preparation of the presently most visible effort to put biodiversity on the SEA agenda, that is, the CBD Decision VIII/28 on Voluntary Guidelines

Box 7.3. Key ecological processes in SEA

An example of how the concept of key ecological processes can be used in complex situations is provided by the review of an EIA for the Hidrovia canalisation project, a project aiming to enhance navigability of the Paraguay River, which runs through the world's largest freshwater wetland, the Pantanal. The independent review commission concluded that the EIA provided insufficient information on a key ecological process, essential for the maintenance of the biological diversity in the entire 140,000 km^2 area, that is, the expected changes in the hydrological (flooding) regime at subbasin level. So, even without having to go into any detail on the potential effects on species or ecosystems, extremely relevant questions related to biological diversity could be formulated (Commission for Environmental Impact Assessment, 1997). Knowing the incredible diversity of the area and its hydrological (and maybe even climatological) importance for the entire Paraguay–Parana basin, it was decided that the precautionary principles applied to this case: nothing should be approved until there is certainty about the impacts of dredging on the hydrology of the seven subbasins in the area.

Box 7.4. A sectoral approach cannot deal with multifunctionality of ecosystems

Analysis of two SEAs in the Netherlands shows the difficulty of dealing with multiple ecosystem services provided by the same ecosystem (author's personal observation). In the western lower part of the country there is a need for floodwater storage, a need for recreational space around highly urbanised areas, and a need to restore nature. Reasoning from an ecosystem services perspective and the need to restore key ecological processes in a delta, these three needs can be combined by optimising a combination of ecosystem services. Yet, a regional planning SEA dealt with these demands in a technocratic, sectorally divided manner, aimed at maximisation of each function in different locations. In a similar SEA study, the potential combination of water storage, nature development, and new housing developments is dealt with in a strictly separate way. Yet, many studies have indicated the feasibility of floating homes, floating offices, floating greenhouses, and even floating roads. A combination of floodwater storage (safety) in a dynamic, more naturally functioning delta system is proposed by environmental NGOs but is not taken up, because multiple ecosystem services cannot be addressed in an integrated manner by a sectorally divided public administration.

for Biodiversity in SEA (CBD, 2006). By its nature, a CBD decision is not a suitable vehicle to distribute any academic analysis or conceptual thinking. The text is very concise and predominantly of a procedural nature. The background document to the CBD decision (Slootweg *et al.*, 2006) provides more insight. Even although it has been thoroughly peer-reviewed by the SEA and biodiversity communities, it remains an informally published technical report. This chapter (and this book) is this first formal publication providing a comprehensive elaboration of the concepts and case evidence behind the CBD guidelines. This chapter contains many references to unpublished sources of information; yet these represent an important source of primary field data, which was indispensable to support the drafting of this chapter. Where possible the authors are identified and Internet references provided; all cases in preparation of the CBD Guidelines have been included in the convention's online database (www.biodiv.org/programmes/cross-cutting/impact/search.aspx) and in the SEA database of the Netherlands Commission for Environmental Assessment (www.eia.nl/ncea/database/index.htm). Parallel to the process of preparing the CBD guidelines, the Journal of Environmental Assessment Policy and Management prepared a special issue on biodiversity in SEA (Byron and Treweek, 2005), to our knowledge the first scientific journal to pay such special attention to the topic. Apart from a number of good case descriptions, this special edition also contains guidelines on the integration of biodiversity into SEA (Treweek *et al.*, 2005). The document is based on experiences in the UK and proposes an EIA-type of approach to SEA, thus limiting the scope of SEA and making it less suitable for nonindustrialised countries. However, it does provide valuable and practical guidance. In 2004 the Capacity Building for Biodiversity in Impact Assessment (CBBIA) programme started its activities under the umbrella of the International Association for Impact Assessment. A steady stream of relevant outputs can be expected from this source, one of the first being a practical biodiversity manual for EIA and SEA in Lebanon (Sawsan *et al.*, 2005).

The approach in this chapter

The approach to integrate biodiversity in SEA as described in this chapter does not follow a structured procedure, as is the case with EIA. The principal reason is that good practice SEA should ideally be fully integrated into a planning (or policy development) process. Since planning processes differ widely, there is no typical sequence of procedural steps in SEA. Moreover, there is no general agreement on what a typical SEA procedure might be. This chapter intends to provide guidance on how to integrate biodiversity issues into SEA, which in turn should be integrated into a planning process. Because the planning process may vary, the SEA is not described as a separate process but as an integral component of the applicable planning process. The SEA process needs to be structured according to the needs of a specific situation. SEA is not a mere expansion of an EIA and it does not usually follow the same stages (Partidario, 2007). This is the reason why, for example, the highly praised South African Guideline Document on SEA (CSIR, 2000) only recognises procedural principles, not procedural steps (see Box 7.5). The

---

Box 7.5. Substantive and procedural principles for SEA in South Africa

Substantive/content principles:
1. SEA is driven by the concept of sustainability.
2. SEA identifies the opportunities and constraints which the environment places on the development of plans and programmes.
3. SEA sets the criteria for levels of environmental quality or limits of acceptable change.
4. Procedural principles:
5. SEA is a flexible process which is adaptable to the planning and sectoral development cycle.
6. SEA is a strategic process which begins with the conceptualisation of the plan or programme.
7. SEA is part of a tiered approach to environmental assessment and management.
8. The scope of an SEA is defined within the wider context of environmental processes.
9. SEA is a participative process.
10. SEA is set within the context of alternative scenarios.
11. SEA includes the concepts of precaution and continuous improvement.

approach and language used in this chapter are therefore conceptual in nature, not procedural.

As explained in Chapter 2, the approach is fully consistent with the Ecosystem Approach (CBD, 2004a). It focusses on people–nature interactions and the role of stakeholders in identifying and valuing potential impacts on biodiversity. For the identification of stakeholders and the valuation of biodiversity, the concept of ecosystem services as elaborated by the Millennium Ecosystem Assessment (MA), explained in Chapter 4, provides a useful tool. It translates biodiversity into ecosystem services that represent (present and future) values for society, thus providing a mechanism to 'translate' the language of biodiversity specialists into language commonly understood by decision makers.

The way in which biodiversity is interpreted in this book has been described in detail in Chapter 2, and Chapter 4 provided the conceptual framework for impact assessment. In this SEA chapter both sources of information will be combined into one approach to address biodiversity in SEA. In order to do so we first need to expand the conceptual framework from Chapter 4. In SEA, biodiversity can best be defined in terms of the *ecosystem services* provided by biodiversity. These services represent ecological or scientific, social (including cultural), and economic values for society and can be linked to stakeholders. Stakeholders can represent biodiversity interests and can consequently be involved in an SEA process. Maintenance of biodiversity (or nature conservation) is an important ecosystem service for present and future generations but biodiversity provides many more ecosystem services. The Millennium Ecosystem Assessment (MA) (2003) provided us the term 'direct drivers of change', which are human interventions (activities) resulting in biophysical and social effects with known impacts on biodiversity and associated ecosystem services. Figure 7.1 positions the ecosystem services and drivers of change in the impact assessment framework, as explained in Chapter 4.

The MA also defined 'indirect drivers of change' as societal changes, which under certain conditions may influence direct drivers of change, ultimately leading to impacts on ecosystem services (Figure 7.1, top arrow with question mark). Indirect drivers of change may also influence human well-being, invoking new social effects leading to direct drivers change (Figure 7.1, bottom arrow with question mark). A direct driver unequivocally influences ecosystem processes and can therefore be identified and measured to differing degree of accuracy. An indirect driver operates more diffusely, often by alternating one or more direct

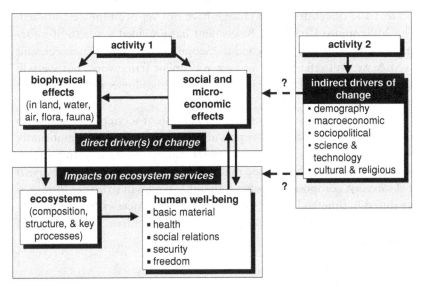

Figure 7.1 Modified impact assessment framework.

drivers; its influence is established by understanding its effect on a direct driver (Millennium Ecosystem Assessment, 2003). The indirect drivers of change are primarily:

• *demographic* (such as population size, age, and gender structure, and spatial distribution);
• *economic* (such as national and per capita income, macroeconomic policies, international trade, and capital flows);
• *sociopolitical* (such as democratisation, the roles of women, of civil society, and of the private sector, and international dispute mechanisms);
• *scientific and technological* (such as rates of investments in research and development and the rates of adoption of new technologies, including biotechnologies and information technologies); and
• *cultural and religious* (such as choices individuals make about what and how much to consume and what they value).

Actors can have influence on some drivers (endogenous driver), but others may be beyond the control of a particular actor or decision maker (exogenous drivers). The way in which indirect drivers of change influence direct drivers is complex, surrounded by uncertainties and subject to many ongoing research efforts. As the figure shows, the links between indirect and direct drivers of change have not yet been fully established. The Millennium Ecosystem Assessment has provided a wealth of information on this, going far beyond the scope of this book. Where relevant to SEA we will refer to the impact assessment framework, but we will not go into any detail with respect to the methodologies to establish such relationships.

To determine potential impacts on ecosystem services, one needs to assess whether the ecosystems providing these services are significantly impacted by the policies, plans, or programmes under study. Impacts on biodiversity can best be assessed in terms of changes in one or more *aspects of biodiversity*: composition (what is there), changes in structure (how is it organised in time and space), or changes in key processes (what physical, biological or human processes govern creation and/or maintenance of ecosystems).

Views on SEA and biodiversity

SEA is a rapidly evolving field with numerous definitions and interpretations in theory, in regulations, and in practice. SEA is required by legislation in many countries and carried out informally in others. Approaches exist that use some or all of the principles of SEA without using the term SEA to describe them. Stinchcombe and Gibson (2001) observed that

although methodological issues in SEA have received a great deal of attention a host of interrelated problems remain. SEA methodologies are not yet well developed, nor widely agreed upon by those involved. These difficulties are hardly surprising given the youth of SEA as a concept and practice; and the great diversity of applications in the broad range of policies, plans and programmes.

However, practices in SEA and related approaches show an emerging continuous spectrum of interpretation and application. At one end of the continuum, the focus is mainly on the biophysical environment. It is characterised by the goal of mainstreaming and upstreaming environmental considerations into strategic decision making at the earliest stages of planning processes to ensure they are fully included and appropriately addressed. The SEA Directive of the European Union (European Commission, 2001) and SEA Protocol to the Convention on Environmental Impact Assessment in a Transboundary Context (Kiev protocol, 2003) are examples of this approach. At the other end of the spectrum is an approach, which addresses the three pillars of sustainability and aims to assess environmental, social, and economic concerns in an integrated manner. Integrated assessment and sustainability appraisal are examples of this approach (OECD-DAC, 2006). Depending on the needs of SEA users and the different legal requirements, SEA can be applied in different ways along this spectrum using a variety of methodologies (see Chapter 5 for more details). Accordingly, SEA is referred to as 'a family of tools that identifies and addresses the environmental consequences and stakeholder concerns in the development of policies, plans, programmes and other high level initiatives'. (OECD-DAC, 2006). The present spectrum of SEAs, ranging from those with a focus on the biophysical environment, to broadly sustainability-oriented SEAs, results in different perspectives on biodiversity in SEA. Although the CBD Convention text is very clear on how biodiversity should be interpreted (see Chapter 2), day-to-day practice shows widely different interpretations.

Biodiversity conservation as nature conservation. SEA traditionally focusses on the biophysical environment. Other instruments are used to represent the economic and social interests of stakeholders. Biodiversity therefore tends to be considered from a nature conservation perspective in which protection, rather than sustainable or equitable use of biodiversity is highlighted. In this manner nature conservation becomes segregated from, and potentially conflicting with, economic and social development. The problem with the sectoral approach in conventional impact assessment is that responsibility for biodiversity is divided among a number of sectoral organisations. For example, the exploitation of fish or forest resources, agriculture, or water quality and quantity management, all have to do with (sustainable) use of biodiversity, but regulations and policies are defined by different entities that do not refer to their activities as sustainable use of biodiversity.

Biodiversity for social and economic well-being. In recent years, environmental assessment practices have been adopted in most developing countries. In these countries the biophysical environment, including biodiversity, is not only looked at from a nature conservation perspective, but moreover as the provider of livelihoods. Especially in rural areas the main objective of development is the social and economic improvement of the situation of poor communities. Both socioeconomic and biophysical environments are seen as complementary. Consequently, an integrated assessment approach has been developed in many of these countries. Biodiversity conservation and sustainable use are equally important issues in SEA; decision makers have to deal with the equitable sharing of benefits derived from biodiversity, including those derived from the utilisation of genetic resources, in societies characterised by unequal distribution of wealth. Such integrated approaches reflect a broad perspective on biodiversity in accordance with the CBD and the Millennium Development Goals. The CBD SEA guidelines explicitly state the intention to contribute to Goal 7 of the Millennium Development Goals, that is, to 'ensure environmental sustainability', and its Target 9 to 'integrate the principles of sustainable development into country policies and programs and reverse the loss of environmental resources'. Environmental sustainability in this context means 'using natural resources wisely and protecting the complex ecosystems on which our survival depend. Overcoming present environmental problems will require greater attention to the plight of the poor and an unprecedented level of global cooperation' (United Nations, 2005).

Merging perspectives. Both the integrated and sectorally divided approaches are converging as it is being realised that the environment, including its biodiversity components, provides goods and services that cannot be assigned to a sector (biodiversity provides multiple goods and services simultaneously) or a geographically defined area (goods and services are not limited to protected areas only). At the same time it is generally recognised that certain parts of the world are of such importance for the conservation of biodiversity, that these areas should be safeguarded for the future and require strict protective measures.

Why special attention to biodiversity in SEA?

Many nonbiodiversity experts in environmental assessment may view the extra attention to biodiversity as unnecessary. The presented description of biodiversity in Chapter 2 may be seen as an all-ecompassing concept. That is, it includes many aspects of environmental assessment that are already common practice without necesarily being described as biodiversity. Indeed, biodiversity is a broad concept and present-day SEA already deals effectively with many asepcts of biodiversity. However, as stated earlier, improvements and more methodological consistency with the internationally agreed principles are needed. Moreover, a number of practical problems are repeatedly observed. Barriers to effective incorporation of biodiversity in environmental assessment include (i) a low priority for biodiversity, (ii) a lack of awareness of biodiversity values and importance, (iii) a lack of capacity to carry out assessments, (iv) a lack of adequate data, and (v) a lack of guidance (Treweek, 2001). Apart from the reasons given above, there are also reasons particularly linked to SEA to have a specific biodiversity focus. These reasons are grouped in four categories, explained below.

Ecological reasons

Cumulative effects on biodiversity are best anticipated at a strategic level. In situations with a high risk of cummulative effects, biodiversity merits special attention (For example, various upstream and downstream, water-related activities in one river (sub-)basin will have cummulative effects on biodiversity in the entire basin. For a detailed overview see Abdel-Dayem *et al.*, 2004). By applying the principles of the ecosystem approach the cumulative effects of activities on biodiversity and its ecosystem services which support human well-being can be addressed. At the same time, it is appropriate to define levels of acceptable change or desired levels of environmental quality at the strategic (ecosystem or catchment) level.

Maintaining the genetic base of evolution for future opportunities. The conservation of biodiversity for future generations is one important aspect of sustainability. It seeks to maintain options for the wealth of yet unknown potential uses of biodiversity. Moreover, maintaining the capacity of biodiversity to adapt to changing environments (e.g. climate change) and to continue providing viable living space for people is critical to human survival. Any long-term sustainability assessment has to make provisions for safeguarding that capacity.

Time and space. From a biodiversity perspective spatial and temporal scales are of particular importance. In conventional SEA, the planning horizon is often linked to economic planning mechanisms with planning horizons of around 15 years. Assessing the impacts on biodiversity often requires a longer time horizon. Biophysical processes, such as soil formation, forest (re)growth, genetic erosion, and evolutionary processes, or the effects of climatic changes and sea level rise, operate on far longer time scales and are rarely taken into account in conventional SEAs. A longer time horizon is required to address the fundamental processes regulating the world's biological diversity. Similarly, flows of energy, water, and nutrients link the world's ecosystems. Effects in an area under assessment may have much wider biodiversity repercussions. The most visible example is the linkage of ecosystems on a global scale by migratory species; on a continental or regional scale ecosystems are linked by hydrological processes through river systems and underground aquifers; and on a local scale pollinators, on which important commercial species depend, may have specific habitat needs beyond the boundaries of an SEA. Biodiversity considerations may consequently require a geographical focus that exceeds, or at least differs from, the area for which an SEA is carried out.

Reasons linked to social and economic development

Opportunities and constraints versus cause–effect chains. Biodiversity underpins ecosystem services on which human well-being relies. Biodiversity thus represents a range of opportunities for, and constraints to, sustainable development. Recognition of these opportunities and constraints as the point of departure for informing the development of policies, plans, and programmes at a strategic level enables optimal outcomes for sustainable development (De Villiers, personal communication). The question at the SEA level is therefore 'how does the environment affect or determine development opportunities and constraints?' An interesting SEA approach based on the development potentials and constraints of the natural environment is the Strategic Environmental Analysis (SEAN). In a structured series of steps local stakeholders are invited to elaborate and discuss ideas about the development of their territory (www. seanplatform.org/). This approach contrasts with the largely reactive approach adopted in project EIA, where the key question being asked is 'what will the effect of this project be on the environment?'

Safeguarding livelihoods. The identification of stakeholders through recognition of ecosystem services can lead to a better understanding of how the livelihoods of people who depend on biodiversity will be affected. In many countries, especially in developing countries, a large proportion of rural society is directly dependent on biodiversity. As these groups may also belong to the poorer and less educated strata of society, they may go unnoticed as they are not always capable to participate meaningfully in an SEA process (see Box 7.6). (Of course, indirectly all of humanity depends on ecosystem services as these regulate the mechanisms that provide us with food, air, and water, protect us from cosmic radiation, and provide us with a suitable environment in which to live.)

Box 7.6. Stakeholders and participation

Environmental assessment is concerned with: (i) information for, (ii) participation, and (iii) transparency in decision making. Public involvement consequently is a prerequisite for effective environmental assessment and can take place at different levels: informing (one-way flow of information), consulting (two-way flow of information), or 'real' participation (shared analysis and assessment). In all stages of the process public participation is relevant. The legal requirements for and the level of participation differ among countries, but it is generally accepted that public consultation at the scoping and review stage are minimally required; participation during the assessment study is generally acknowledged to enhance the quality of the process.

With respect to biodiversity, four groupings of stakeholders can be distinguished:

1. Beneficiaries of the policy, plan or programme – target groups making use of or putting a value to known ecosystem services which are purposefully enhanced by the policy, plan, or programme;

2. Affected (groups of) people – that is, those people that experience, as a result of the policy, plan, or programme, intended or unintended changes in ecosystem services that they value in positive or negative manner;

3. General stakeholders:

   • National or local government bodies having a formal government responsibility with respect to the management of defined areas (town and country planning departments, etc.) or the management of ecosystem services (fisheries, forestry, water supply, coastal defence, etc.);

   • Organisations representing affected people (water boards, trade unions, consumer organisations, civil rights movements, ad hoc citizens committees, etc.);

   • Organisations representing (the intrinsic value of) biodiversity itself (nongovernmental nature conservation organisations, park management committees, scientific panels, etc.).

   • The general audience that wants to be informed on new developments in their direct or indirect environment (linked to transparency of democratic processes).

4. Stakeholders of future generations, who may rely on biodiversity around which we make decisions. Formal and informal

organisations are increasingly aware of their responsibility to take into account the interests of these 'absent stakeholders'. (Charl de Villiers is acknowledged for introducing the important idea of future generations being regarded as absent stakeholders.)

In general it can be observed that the role of institutionalised stakeholders becomes more important at higher strategic levels of assessment; at lower level the actual beneficiaries and affected people will become more important. There are a number of potential constraints to effective public participation. These include:

- Poverty: involvement means time spent away from income-producing tasks;
- Rural settings: increased distances make communication more difficult and expensive;
- Illiteracy: or lack of command of nonlocal languages, can inhibit representative involvement if print media are used;
- Local values/culture: behavioural norms or cultural practice can inhibit involvement of some groups, who may not feel free to disagree publicly with dominant groups (e.g. women versus men);
- Languages: in some areas a number of different languages or dialects may be spoken, making communication difficult;
- Legal systems: may be in conflict with traditional systems, and cause confusion about rights and responsibilities for resources;
- Interest groups: may have conflicting or divergent views, and vested interests;
- Confidentiality: can be important for the proponent, who may be against early involvement and consideration of alternatives.

Sound economic decision making. Ecosystem services such as erosion control, water retention and supply, coastal defence, and recreational potential can be valued in monetary terms, thus providing a figure on potential economic benefits and/or losses caused by the implementation of planned activities. Good representation of ecosystem services and potential changes in these services simply contributes to a sound economic analysis supporting any strategic decision.

Formal legal obligations

A reason to pay particular attention to biodiversity in SEA is a national, regional, or international legal obligation to do so. A number of legal obligations can be distinguished:

Protected areas and protected species: ecosystems, habitats, and species can have a form of legal protection, ranging from strictly protected to restrictions on certain activities.

Valued ecosystem services can be subject to some form of legal regulation triggering the need for environment assessment. Examples are fisheries and forestry activities, coastal protection (by dunes or forested wetlands), water infiltration areas for public water supply, recreational areas, landscape parks, and so forth. (See Box 7.7 on ecosystem services in their regulatory context). Lands and waters traditionally occupied or used by indigenous and local communities represent a special case of ecosystem services.

Acts of Parliament. For example, the Nature Conservation (Scotland) Act 2004 places, in its first clause, a duty on every public body and

---

Box 7.7. Ecosystem services in their regulatory context

SEA provides information on policies, plans, and programmes for decision makers, including their consistency with the regulatory context. It is important to realise that ecosystem services often have formal recognition by some form of legal protection. Legislation often has a geographical basis (e.g. protected areas) but this is not necessarily always the case (species protection is not always limited to demarcated areas). Of course, the legal context in any country or region is different and needs to be treated as such.

Some examples of ecosystem services linked to formal regulations:

Ecosystem service: preservation of biodiversity:

- Nationally protected areas/habitats, protected species;
- International status: Ramsar convention, UNESCO Man and Biosphere, World Heritage Sites;
- Subject to national policies such as the UK Biodiversity Action Plans (BAP), or regional regulations such as the European Natura 2000 Network;
- Marine Environmental High Risk Areas (sensitive areas prone to oil pollution from shipping);
- Sites identified and designated under international agreements (e.g. OSPAR Marine Protected Areas);
- Sites hosting species listed under the Convention on the Conservation of Migratory Species of Wild Animals or the Convention on International Trade in Endangered Species of Wild Flora and Fauna;

- Sites hosting species listed under the Bern Convention (Annexes 1 and 2 of the Convention on the Conservation of European Wildlife and Natural Habitats, 1979).

Ecosystem service: provision of livelihood to people:

- Extractive reserves (forests, marine, agriculture)
- Areas of indigenous interest
- Touristic (underwater) parks (service: maintaining biodiversity to enhance tourism)

Ecosystem service: preservation of human cultural history / religious sites:

- Landscape parks
- Sacred sites, groves
- Archaeological parks

Other ecosystem services, in some countries formally recognised:

- Flood storage areas (service: flood protection or water storage)
- Water infiltration areas (service: public water supply)
- Areas sensitive to erosion (service: vegetation preventing erosion)
- Coastal defences (dunes, mangroves) (service: protecting coastal hinterlands)
- Urban or periurban parks (service: recreational facilities to urban inhabitants)
- Ecosystem functioning (soil biodiversity, pollination, pest control)

office holder in exercising any functions to further the conservation of biodiversity. The Act brings the Convention on Biological Diversity into Scottish law. Within the UK this is certainly the strongest measure to conserve biodiversity, since in other countries within the UK public bodies are only obliged 'to have regard to the conservation of biodiversity' (M. Usher, personal communication).

International treaties, conventions and agreements such as the World Heritage Convention, Ramsar Convention, the UNESCO Man and Biosphere Programme, or the Regional Seas agreements. By becoming a Party to these agreements, countries agree to certain obligation to manage these areas according to internationally agreed principles.

Practical reason – Facilitation of stakeholder identification

The concept of biodiversity-derived ecosystem services provides a useful tool to identify potentially affected groups of people. Ecosystems are multifunctional and provide multiple services. By applying the ecosystem approach and focussing on ecosystem services in describing biodiversity, directly and indirectly affected stakeholders can be identified and, as appropriate, invited to participate in the SEA process. Public participation is generally regarded as a prerequisite for effective environmental assessment as it (i) fosters justice, equality and collaboration (ii) informs and educates stakeholders on planned interventions and its consequences, (iii) gathers relevant data and information on the social and biophysical environment, (iv) seeks input to enhance positive outcomes and ways to reduce or mitigate negative impacts, (v) contributes to better analysis and more creative development, and (vi) contributes to mutual learning and the improvement of environmental assessment practice (André et al., 2006).

Apart from the need to integrate biodiversity in existing SEA procedures, the reasoning can also be the other way around. By promoting the use of SEA, biodiversity stands a better chance of being recognised as an area of concern in the formulation of policies, plans, and programmes. Compared to EIA, SEA is applied much earlier in the preparation of human interventions and provides more room to proactively formulate alternative options and start timely data collection efforts. Treweek (2001) describes a number of situations in which SEA should be considered as a potential tool for incorporation of biodiversity in policies, plans, or programmes:

• Comprehensive biodiversity monitoring has not been instituted, and consequently biodiversity data are largely lacking. The SEA process can be used to obtain such data.
• Ecosystem behaviour is poorly understood, so longer lead-times are required to collect reliable baseline information. Similarly, unstable or fluctuating ecosystems require more baseline data for predictions to be reliable.
• Important biodiversity resources are limited and fragmented, or threatened throughout their range, justifying the need for investigation through SEA into the consequences of any plan potentially affecting these resources.
• Avoidance or mitigation options are limited or replacement options are all long-term, giving very little room to find sustainable options.

For example, mining activities cannot be relocated to alternative sites; old growth forests take generations to establish.

- Multiple threats to biodiversity exist which can only be assessed at strategic level through SEA. For example, urban expansion, including housing schemes, industrial estates, traffic corridors, and, for example, a need for water supply, sanitation, and flood protection. At the project level each of these activities might have moderate impacts, but the combined impacts can be overwhelming.
- Many stakeholders depend on local use of biodiversity, which needs to be sustained. Examples are diverse but could include coastal communities depending on low-technology fisheries, migratory pastoralist depending on uplands in rainy season and wetlands in dry season, indigenous people depending on multiple forest resources, and so forth.

Identifying potential impacts on biodiversity

Biodiversity 'triggers' for SEA

Impact assessment by definition has to deal with uncertainty; in the case of SEA the level of uncertainty is even higher as compared to EIA. Furthermore, the tiered nature of SEA creates a variety of contexts in which SEA is applied. To be able to make a judgement if a policy, plan, or programme has potential impact on biodiversity, two types of information are of overriding importance:

(i) affected area of land or water: if the plan affects a geographically defined area it is possible to define the ecosystems or types of land use in the area, identify ecosystem services linked to these ecosystems or land use types, and identify related stakeholders;

(ii) type of planned activities: if the plan provides information on planned activities it is possible to identify activities that can act as direct or indirect drivers of change in biodiversity.

Based on these two information elements three conditions are defined that 'trigger' the need for special attention to biodiversity. When any one or a combination of these conditions below applies to a policy, plan, or programme, special attention to biodiversity is required in the SEA of this policy, plan, or programme.

(1) *Important ecosystem services*. When a geographically defined area, affected by a policy, plan, or programme, is known to provide one

or more important ecosystem services, these services and their stakeholders should be taken into account in an SEA. Geographical delineation of an area provides the most important biodiversity information, as it is possible to identify the ecosystems and land use practices in the area, and identify ecosystem services provided by these ecosystems or land use types. For each ecosystem service, stakeholder(s) can be identified who preferably are invited to participate in the SEA process. Area-related policies and legislation can be taken into account (e.g. those referred to in Box 7.7 earlier).

(2) *Interventions acting as direct drivers of change.* If a proposed intervention leads to biophysical changes, directly or through social changes, with known impact on biodiversity, these interventions can be considered as drivers of change in biodiversity and special attention needs to be given to biodiversity (see Box 7.8 for an overview of these social and biophysical changes). It depends on characteristics of society and the environment to know whether these changes will indeed lead to impacts on biodiversity and ecosystem services. Knowledge of affected stakeholders and the affected area is needed. If the intervention area of the policy, plan, or programme has not yet been geographically defined, such as the case of a sector policy, the SEA can only define impacts on biodiversity in conditional terms: impacts are expected to occur in case the policy, plan or programme will affect certain types of ecosystems, or when the plan affects certain stakeholders. Usually at a lower level of elaboration of such policy or plan into an implementation programme it is possible to identify the actual impacts when area information becomes available. If the intervention area is known it is possible to link drivers of change to ecosystems, ecosystem services, and their stakeholders.

(3) *Interventions acting as indirect drivers of change.* When a policy, plan or programme leads to activities acting as indirect driver of change (e.g. for a trade policy, a poverty-reduction strategy, or a tax measure), it becomes more complex to identify potential impacts on ecosystem services. Linkages between indirect and direct drivers of change are difficult to establish. In broad terms, biodiversity attention is needed in SEA when the policy, plan or programme is expected to significantly affect the way in which a society (i) consumes products derived from living organisms, or products that depend on ecosystem services for their production, (ii) occupies areas of land and water; or (iii) exploits its natural resources and ecosystem services.

Box 7.8. Direct drivers of change

Direct drivers of change are human interventions (activities) resulting in biophysical effects, directly or through social effects, with known impacts on biodiversity and associated ecosystem services. Interventions characterised by one or a combination of the following biophysical and/or social effects can be considered drivers of change in biodiversity:

• Land conversion: the existing habitat is completely removed and replaced by some other form of land use or cover. This is the most important cause of loss of biodiversity and ecosystem services.

• Fragmentation by linear infrastructure: roads, railways, canals, dikes, powerlines, and so forth affect ecosystem structure by cutting habitats into smaller parts, leading to isolation of populations. A similar effect is created by isolation through surrounding land conversion. Fragmentation is a serious reason for concern in areas where natural habitats are already fragmented.

• Extraction of living organisms is usually selective since only few species are of value, and leads to changes in species composition of ecosystems, potentially upsetting the entire system. Forestry and fisheries are common examples.

• Extraction of minerals, ores, and water can significantly disturb the area where such extractions take place, often with significant downstream and/or cumulative effects.

• Wastes (emissions, effluents, solid waste), or other chemical, thermal, radiation, or noise inputs: human activities can result in liquid, solid, or gaseous wastes affecting air, water, or land quality. Point sources (chimneys, drains, underground injections) as well as diffuse emission (agriculture, traffic) have a wide area of impact as the pollutants are carried away by wind, water, or percolation. The range of potential impacts on biodiversity is very broad.

• Disturbance of ecosystem composition, structure, or key processes: Chapter 2 provides an overview of how human activities can affect these aspects of biodiversity.

Some social effects are known to lead to one of the above-mentioned biophysical effects (nonexhaustive, based on van Schooten *et al.* (2003)):

• Population changes due to permanent (settlement/resettlement), temporary (temporary workers), seasonal in-migration (tourism) or opportunistic in-migration (job-seekers) usually lead to land occupancy (land conversion), pollution and disturbance, harvest of living organisms, and introduction of nonnative species (especially in relatively undisturbed areas).

- Conversion or diversification of economic activities: especially in economic sectors related to land and water, diversification will lead to intensified land use and water use, including the use of pesticides and fertilisers, increased extraction of water, introduction of new crop varieties (and the consequent loss of traditional varieties). Change from subsistence farming to cash crops is an example. Changes to traditional rights or access to biodiversity goods and services falls within this category. Uncertainty or inconsistencies regarding ownership and tenure facilitate unsustainable land use and conversion.
- Conversion or diversification of land use: for example, the enhancement of extensive cattle raising includes conversion of natural grassland to managed pastures, application of fertilisers, genetic change of livestock, and increased grazing density. Changes to the status, use, or management of protected areas are another example.
- Enhanced transport infrastructure and services, and/or enhanced (rural) accessibility; opening up of rural areas will create an influx of people into formerly inaccessible areas.
- Marginalisation and exclusion of (groups of) rural people: landless rural poor are forced to put marginal lands into economic use for short term benefit. Such areas may include erosion of sensitive soils, where the protective service provided by natural vegetation is destroyed by unsustainable farming practices. Deforestation and land degradation are a result of such practices, created by nonequitable sharing of benefits derived from natural resources.

The three boxes in the conceptual framework of Figure 7.1 thus position the biodiversity 'triggers': affected ecosystem services (1), and activities acting as direct (2) or indirect drivers of change (3) in ecosystem services. If any of these triggers result from a plan, the SEA should include biodiversity as a special area of attention.

Table 7.1 below provides a summary overview of the conditions under which a strategic environmental assessment should pay particular

Table 7.1 *Summary overview of when and how to address biodiversity in SEA.*

Biodiversity triggers	When is biodiversity attention needed	How to address biodiversity issues
Trigger 1. Plan area known to provide important ecosystem services	*Does the policy, plan, or programme influence: important* ecosystem services, both protected (formal) or nonprotected (stakeholder values) Areas with legal and/or international status; Important biodiversity to be maintained for future generations?	*Area focus.* Systematic conservation planning for nonprotected biodiversity. Ecosystem services mapping. Link ecosystem services to stakeholders. Invite stakeholders for consultation.
Trigger 2. Policy, plan, or programme creates direct drivers of change	*Does the policy, plan or programme lead to: biophysical effects* known to significantly affect biodiversity (e.g. land conversion, fragmentation, emissions, introductions, extraction) *nonbiophysical effects* with known biophysical consequences (e.g. relocation migration of people, migrant labour, change in land use practices, enhanced accessibility, marginalisation)?	*Focus on direct drivers of change and potentially affected ecosystem.* Identify drivers of change (i.e. biophysical changes known to affect biodiversity). Within administrative boundaries to which the policy, plan, or programme applies, identify ecosystems sensitive to expected biophysical changes.
Combined triggers 1 and 2. Interventions with known direct drivers of change affecting area with known ecosystem services	Combination of 1 and 2 above	*Knowledge of intervention and area of influence allows prediction of impacts on composition or structure of biodiversity or on key processes maintaining biodiversity.* Focus on direct drivers of change (i.e. biophysical changes known to affect biodiversity). Define spatial and temporal influence. Identify ecosystems within range of influence. Define impacts of drivers of change on composition, structure, or key processes. Describe affected ecosystems services and link services to stakeholders. Invite stakeholders into SEA process. Take into account the absent (future) stakeholders.
Trigger 3. Policy, plan, or programme creates indirect drivers of change, but without direct biophysical consequences	*Are indirect drivers of change affecting the way in which a society:* produces or consumes goods, occupies land and water, or exploits ecosystem services?	*More research and case material needed.* MA methodology potentially valuable to identify linkages between indirect and direct drivers of change.

attention to biodiversity issues and how they should be addressed. Each row of the table will be subsequently treated in the following sections.

Trigger 1: The area influenced by the policy, plan, or programme provides important ecosystem services.

SEA can roughly be divided into two broad approaches: the reactive cause–effect chain approach where the intervention is known and the cause–effect chains from activity to impact are fairly clear (comparable to EIA), and the 'bottom up' opportunities and constraints of the natural environment approach, where the environment effectively shapes the policy, programme or plan. The latter is most often used in land use planning/spatial planning where interventions are potentially wide-ranging and the objective is to tailor land uses to be most suited to the natural environment. Trigger 1 can be recognised in such policies, plans, or programmes with a focus on a defined geographical area, without necessarily having precisely defined activities. Because the area is geographically defined, the biodiversity in the area can be described in terms of ecosystem services providing goods and services for the development and/or well-being of people and society. The opportunities and constraints of the area 'guide' the planning of regional development. Box 7.9 provides two examples of such approaches.

The procedure to deal with biodiversity in this situation is as follows:

• Identify ecosystems and land use types in the area to which the policy, plan or programme applies (human land use can be considered as an attempt by humankind to maximise one or few specific ecosystem services, for example, productivity in agriculture, often at the cost of other services). Identify and map ecosystem services provided by these ecosystems or land use types;
• Identify which groups in society have a stake in each ecosystem service; invite such stakeholders to participate in the SEA process. Identification and valuation of ecosystem services is an iterative process initiated by experts (ecologists, natural resources specialists) but with stakeholders playing an equally important role. The frequency of reliance on ecosystem goods or services should not necessarily be used as an indication or measure of their value because ecosystem services on which local communities rely even on an occasional basis can be critical to the resilience and survival of these communities during surprise or extreme natural conditions (e.g. coastal mangroves or dunes provide protection against infrequent tsunami or storm surges);

Box 7.9. Examples of SEA in geographically defined areas

Two case studies provice examples of how biodiversity information from geographically defined areas provides relevant information for effective SEA.

In the first case, a SEA has been carried out for the planning of open space in UMhlathuze, a rapidly developing and urbanising municipality in South Africa. River catchments provided an effective environmental entity for assessing synergistic impacts of urban development. A catchment is a functional unit, because it constrains key energy and material flows; it also provides an easy unit of comparison. A strategic catchments assessment had to provide criteria for measures of protection and planning of development in nondeveloped lands. It accounted for the balance between supply of environmental goods and services provided by the natural environment and the demand for these goods and services by people. By using a pressure, state, response indicator model it was possible to make a status quo report of each catchment, indicating required management actions where needed. It furthermore calculated the economic benefits provided by 'free' ecosystem services at R 1.7 billion annually (about €200 million at the 2004 exchange rate). Important benefits included water supply and regulation, flood and drought management, nutrient cycling, and waste management. Monetisation of ecosystem services made decision makers react much more openly to the need for conservation measures, even when reputed for not listening to biodiversity arguments. In this case biodiversity is not necessarily being rare or endangered. The case provides evidence of the economic and social sense it makes to maintain biodiversity for the services it provides. It shows a good example of mapping and monetisation of ecosystem services in a known geographical area as an input for informed decision making on priorities for interventions. It strongly emphasises the value of the concept of ecosystem services as a means to translate biodiversity information into the language of decision makers. (Case from Van Der Wateren et al., 2004.)

The second case provides a mechanism to focus on the need to conserve unique and important biodiversity in a situation of overwhelming presence of nonprotected biodiversity, without jeopardising the need of the country to develop. Since 2000, municipalities in South African have to prepare Spatial Development Frameworks and carry out associated SEAs. In two regions systematic biodiversity planning was applied to support this process in an attempt to improve effective consideration of biodiversity in Environmental Assessment. Most biodiversity in South Africa, including priority areas for conservation, does not fall within existing protected areas. Changing land use patterns have a major impact on biodiversity. Under such conditions sound SEA in land use planning is critical to decision making. Systematic biodiversity planning aims at conserving a representative sample of species/habitats and key ecological and evolutionary processes. The focus on priority areas allows for recognition of competing land uses and development needs. It sets target for conservation and defines limits of acceptable change within which human impacts have to be kept. Although driven by conservation objectives, the process is very similar to SEA and outputs are easily integrated in the SEA process. (Case from Brownlie *et al.*, 2005b.).

The two South African cases provide an excellent example of how to deal with both conservation of irreplaceable but nonprotected biodiversity and with sustainable use (and conservation) of biodiversity derived ecosystem services.

• For absent stakeholders (future generations), identify important protected and nonprotected biodiversity which is representative of species, habitats and/or key ecological and evolutionary processes (e.g. by applying systematic conservation planning or similar approaches; see Box 7.9);
• Ecosystem services identified by experts but without actual stakeholders may represent an unexploited opportunity for social, economic, or ecological development. Similarly, ecosystem services with conflicting stakeholders may indicate overexploitation of this service representing a problem that needs to be addressed.

Trigger 2: The policy, plan, or programme is concerned with interventions producing direct drivers of change

An approach based on analysis of cause and effect chains fits best for SEAs where the intervention is known. As explained above, interventions resulting from a policy, plan, or programme can directly, or through socioeconomic effects, lead to biophysical effects that affect biodiversity and ecosystem services derived from biodiversity. When activities are defined by the plan but the locations of these activities have not been clearly defined, impacts on ecosystems services can only be defined in terms of *potential impacts*. For example, one can state that the drivers of change resulting from this plan are known to have particular strong effects on certain types of ecosystems. This situation often occurs in sectoral policies, such as policies or plans on energy, public water supply, or transport. Box 7.10 provides two examples.

Box 7.10. Example of SEA of defined activities

Two cases illustrate that even without a concrete geographical focus, ways exist to describe impacts on biodiversity in general terms, design mitigation measures, and provide guidance for the further study at lower level of assessment.

The Netherlands National Policy on Water Supply is a sectoral policy without predefined locations of interventions. The SEA of this plan focussed on the most important biophysical effect of water extraction, that is, a change in the hydrology of underground aquifers and surface waters. A major issue at the national scale is the desiccation of various types of landscapes, predominantly wetlands converted over centuries into agricultural land use types, rich in biodiversity and highly valued for characteristic 'Dutch' landscape features. Quantitative information on potential impacts of water extraction was deemed necessary. The national scale of the study forced the study team to focus on simple vegetation indicators for hydrological changes. Combination of potential hydrological changes (modelled) with nationally available vegetation data provided a computational model which served the purpose of national decision making. Further elaboration of the policy into concrete plans and programmes requires further site-specific field observations to quantify potential impacts. The national Policy SEA identified potentially sensitive areas that require special attention. (Case: van Schooten, 2004a.)

In Bolivia, an SEA for a 600-km road corridor had to deal with an area of potential influence twice the surface of the Netherlands (see the example above). The SEA followed a broad, integrated approach, including social and economic processes. The relatively pristine and untouched character of the area made such an approach essential in order to capture all relevant impacts on biodiversity. The SEA identified social and economic changes as the main drivers of change associated to the road scheme. Economic development, creation of employment, and immigration from the Andean highlands were considered main threats to biodiversity and ecosystem services as these would lead to increased land conversion. The extent of potential influence of the road is immense. Road trajectories as well as area of direct and indirect influence were not clearly defined. Therefore, the identification of affected ecosystems was impossible. Instead, an inventory of major types of ecosystems in the entire region was made, processes of key importance for the maintenance of these systems were identified, and potential impacts induced by road development were identified. A hierarchy was designed, assigning types of ecosystem into categories with differing levels of protection. An extensive mitigation programme accompanies the road scheme, including assistance to management of national parks in the region and social support programmes. (Case: Consorcio Prime Engenharia, 2004.)

The procedure to deal with biodiversity in this situation is as follows:

• Identify drivers of change, that is, activities leading to biophysical effects known to affect biodiversity (see Box 7.8 earlier);
• A plan usually applies to an area within administrative boundaries, such as a province, state, or an entire country. Define the administrative boundaries to which the plan applies.
• Identify and, if possible, map the major ecosystems within the area.
• Determine the sensitivity of each of these ecosystems to the drivers of change resulting from the plan.
• Develop a mechanism to avoid, mitigate, or compensate potential negative impacts to ecosystems which are most sensitive to the drivers of change (and which may provide important ecosystem services). Identification of alternative locations with least damage to biodiversity provides good opportunities in this approach; these can be taken into account at lower planning levels when activities and locations become more precise.

Triggers 1 and 2 combined: activities and intervention area both defined

For many SEAs, often at lower level of planning, activities as well as the location of activities are more precisely defined. In such cases special attention to biodiversity is triggered when the planned activities are known drivers of change in combination with a planning area known to provide important ecosystems services. This combined knowledge allows relatively detailed assessment of potential impacts by defining changes in composition or structure of ecosystems, or changes in key processes maintaining ecosystems and associated ecosystem services. This combination of triggers is often associated with SEAs carried out for programmes. In their implementation these SEAs resemble complex, large-scale EIAs. Examples are detailed spatial plans, programme level location, and routing alternatives or technology alternatives.

The procedure to deal with biodiversity in this situation is a combination of the procedures for Triggers 1 and 2, but the combination allows for greater detail in defining expected impacts:

- Identify direct drivers of change and define their spatial and temporal range of influence;
- Identify ecosystems lying within this range of influence (in some cases species or genetic level information may be needed);
- Describe effects of identified drivers of change on identified ecosystems in terms of changes in composition or structure of biodiversity or changes in key processes responsible for the creation or maintenance of biodiversity;
- If a driver of change significantly affects either composition, or structure, or a key process, there is a very high probability that ecosystem services provided by the ecosystem will be significantly affected;
- Identify stakeholders of these ecosystem services and invite them to participate in the process. Take into account the absent (future) stakeholders.

Most available case evidence comes from this type of EIA-like SEA. Strong opinions exist on whether these cases should be considered SEAs at all, or whether they should be considered large-scale EIA. Partidário (2007) states that 'while it is recognised that most practitioners and authorities around the world have adhered to the idea and name of SEA, practice still lacks the innovative methodological capacity of SEA as a strategic tool'. At the IAIA 2007 conference the same author even stated that those EIA-based SEAs should simply be named EIA. The procedural steps in the SEA cases used in this chapter indeed very much resemble EIA. This discussion is largely beyond the scope of this chapter, as the available material provides interesting learning on how to deal with biodiversity. It provides lessons on common SEA practice, and the results are relevant for day-to-day SEA. The examples below are structured around some of the important concepts that we consistently use in our approach.

Direct drivers of change. A case from Sweden takes biophysical effects resulting from urban development (the driver of change) as the basis for identifying indicators to measure change in biodiversity. The case focusses on biodiversity conservation as important ecosystem service. The case has similarities to the systematic biodiversity planning case from South Africa; it shows that the concept can also be used for urban planning in a different setting. As a result nonprotected biodiversity is taken into account.

• Urban planning of the area surrounding Stockholm (Sweden) requires strategic decision making on the model of urban expansion in a biodiversity rich environment. A biodiversity analysis at ecosystem level is carried out to support the SEA process. The analysis results in (i) operational targets for biodiversity, translating biodiversity policies into concrete objectives for the region; (ii) distinctive indicators for habitat change; (iii) reliable prediction methods; and (iv) sensible scenarios for future urban growth as a base for comparison. The indicators were linked to the major biophysical effects resulting from urban development affecting biodiversity: habitat loss, isolation/fragmentation, and disturbances. (Case: Balfors et al., 2005.)

Similarly biophysical effects were used as indicators to model the impacts of major interventions in river hydrology (the driver of change) in the Netherlands. The case further illustrates the concept of ecosystem services and shows that ecosystem level information provides sufficient information for decision making.

• An SEA for a river management project along the Meuse River in the Netherlands had to study potential combinations of seemingly contradictory ecosystem services: flood control, shipping, and nature restoration. The main objective was to reduce river peak flows as a safety measure. The SEA took a historical perspective and portrayed major services of the ecosystems throughout the ages – biodiversity has been managed and exploited to such an extent that the resulting ecosystems depend on human management to maintain their appreciated features. Based on this information four alternatives were developed. Water depth, flood duration and groundwater level were considered key biophysical effects affecting biodiversity. These were modelled in a computational model and linked to the requirements of different 'ecotypes' (small-scale managed ecosystems). It provided sufficient information to compare alternatives, although further field observation are required for detailed intervention planning. (Case: Schooten, 2004b.)

The availability of biodiversity inventory data greatly enhances SEA studies by allowing computational models to link computed biophysical effects to indicator species or ecosystems. As the distribution of these indicator species is known, effects of the interventions can be estimated at a level of detail sufficient for strategic decision making.

Aspects of biodiversity: Impacts on biodiversity can best be described in terms of changes in composition (what is there), changes in structure (how is it organised in time and space), or changes in key processes (what physical, biological, or human processes govern creation and maintenance of ecosystems).

A case from Nepal shows that prior knowledge on how a biophysical effect influences a specific aspect of biodiversity provides a means to focus an SEA study. In this case forestry (intervention) leads to selective removal of trees (biophysical effect), affecting species composition.

• Plan level SEAs were carried out in Nepal to assess the environmental impacts of district forestry plans. Forestry practices were considered to impact on biodiversity by changing the species composition of forests; this consequently was the focus of the study. The SEA resulted in recommendations on how to include conservation principles in forestry activities. (Case: Uprety, 2004.)

From India two examples were provided where the need for an SEA was triggered by protected species, but where the SEA study focussed on ecosystem and foodweb structure to provide relevant and sufficient information.

• SEA was used in India as a diagnostic tool to assess siting alternatives for a nuclear power facility. The facility was partially projected on one of India's prominent tiger reserves. The facility also affected traditional land use practices. Regulations limited the study area to a 25-km radius. Within this radius protected areas and ecologically sensitive areas were defined. The study focussed on contiguity of habitats for endangered species (such as tiger, leopard, Indian wolf, and others) and the area needed for predators to have sufficient stock of prey animals.

In other words, the study focussed on ecosystem structure: the spatial structure of habitat and foodweb structure. (Case: Rajvanshi and Matur, 2004a.)

- An SEA approach was followed in India to review an EIA of a planned dam and irrigation scheme which resulted in deadlock. The deadlock resulted from a lack of attention to wildlife migration routes (including tigers). The SEA aimed at enhancement of conservation planning and mediation to steer environmental decision making. Again vital habitat links (corridors) and foodweb structure were the focus of study. The creation of a new reservoir provided important new habitats; the design of a canal created fragmentation of major habitats. Redesign of a new migration corridor upstream of the canal mitigated this problem, and the SEA resulted in renewed decision making. (Case: Rajvanshi and Matur, 2004b.)

Changes in key processes as a means to identify impact on ecosystem services appear in a number of cases throughout the text.

Ecosystem services. Translating biodiversity into ecosystem services is an effective means to make biodiversity tangible in impact assessment. Services represent ecological, social, and economic values for society that can be linked to stakeholders. Stakeholders can speak on behalf of biodiversity and can consequently be involved in an SEA process. Maintenance of biodiversity (or nature conservation) is an important ecosystem service.

A case from Uzbekistan provides an early example of how ecosystem services provided by biological diversity (representing direct use values for the population living in the area as well as nonuse, future, and external use values), are taken as a point of departure in a planning process aimed at the restoration of the Amu Darya Delta for the benefit of present and future generations. It also provides a good example of how biodiversity can be enhanced to contribute to human well-being.

- Within the framework of the World Bank Co-ordinated Aral Sea Programme, a consortium of expatriate consultants and local Uzbek institutes developed a coherent strategy for the restoration of the Amu Darya Delta, taking the ecosystem services of a dynamic seminatural wetland system as the point of departure, and using participative evaluation techniques as a means to structure the decision making process on a future development strategy for the delta. A pilot project has successfully been implemented recently, with astonishingly rapid results as regards to the return of productivity of fish resources and the sheer numbers of animals present in the restored wetlands. The case is a clear illustration of the effectiveness of participatory evaluation of wetland ecosystem services at strategic level; the use of a services–values matrix appeared to be an effective communication tool in visualising the multiple services of a wetland system, their societal values, and their stakeholders, and for comparison of alternative strategies. Use of local expertise was key to the identification and quantification of wetland services. (Case: Euroconsult and the Wetland Group, 1996; Schutter, 2002; Dukhovny and Schutter, 2003.)

A case from the UK shows that by taking an ecosystem services approach with active involvement of stakeholders, an important contribution to the definition of viable SEA alternatives was made.

• The availability of Biodiversity Action Plans (BAPs) and Species Action Plans (SAPs) provided biodiversity objectives for an SEA on a local flood management strategy in the UK. Within the wetland ecosystem, priority habitats and priority species have been defined in the BAP. Furthermore, ecosystem services were considered an important economic asset of the region, with biodiversity based tourism as most important sector. Opportunities to use wetlands for flood attenuation provided additional important benefits. Flood management was considered to be a key driver of change, as flooding is a key ecological process in wetlands. The study area was defined on the basis of likely limits of impacts. For the assessment it was considered appropriate to identify risks and the main ecological processes likely to affect outcomes for biodiversity in relation to objectives for the area. Public participation was action–oriented and focussed on identifying preferred changes to achieve outcomes compatible with stakeholder interests; local knowledge was an important source of information. Biodiversity specialists were able to provide effective flood control alternatives that were also beneficial for biodiversity (making use of ecosystem services). (Case: Treweek, 2004.)

A case from the Wadden Sea in the Netherlands shows that natural ecosystems provide multiple services. Exploitation of one service leads to potential impacts on others when key ecosystem processes are affected. Stakeholder involvement reoriented the SEA study to be more focussed on these key processes, instead of looking at the exploited ecosystem service only.

• The Netherlands national policy on large scale extraction of shells in marine environment required an SEA. Shell mining also takes place in protected areas, representing important international ecosystem services for the maintenance of pathways of migratory birds and breeding grounds of North Sea fish, tourism, and so forth. Focus of the permitting procedure was on whether the ecosystem service was not overexploited; in other words the natural regeneration of shell deposits was studied in relation to exploitation pressure. However, the mining process itself also influences key ecological processes essential to other ecosystem services. Bottom morphology and related bottom life were consequently included in the SEA study. Stakeholder contributions highlighted the lack of knowledge on the function of shells and shell banks in the ecosystem. As a result more alternatives were included in the study. The study concluded that natural regrowth fully compensates mining; it was concluded however that ecological processes should define mining conditions. Potential mining locations were ranked according to these conditions. In small parts of the area the precautionary principle was applied because too little was known of the function of shell banks and mining was prohibited. An interesting equity discussion erupted. Shell mining was a monopolised business; the SEA process triggered a discussion on public tender procedures for other interested operators. This request was granted. (Case: van Schooten, 2004c.)

A case from the Scheldt River in Belgium shows that restoration and conservation of biodiversity was sought after as a means to optimise other ecosystem services provided by the river, representing social and economic values, in this case safety from flooding and navigability and accessibility of the Antwerp port.

• The Sigma plan intends to guarantee safety against inundations in the valley of the Scheldt River and its tributaries. The study area incorporates over 250 km of river valley. Most of it is subjected to twice-daily tides and much of the valley would be inundated every day were it not for the presence of dikes. The freshwater tidal areas are unique to North-western Europe. Construction of dikes resulted in considerable loss of the original biodiversity and its flood retention capacity as an ecosystem service. Partial restoration of this biodiversity and its associated flood retention function is still feasible. Nature conservation was an important element in the SEA. However, nature conservation is not seen as an end in itself, but as a way to obtain a 'solid and robust' ecological system in the estuary, capable of supporting intense shipping activities (accessibility of Antwerp port). Other ecosystem services addressed by the SEA study are pollution breakdown and recreation. (Case: Dijk, 2005.)

The presented cases are a selective sample of good practice cases. In reality, many facets of biodiversity go unnoticed in SEA. Even with this selective, biodiversity-friendly sample of cases, it has become clear that the concept of ecosystem services does not yet receive wide recognition. As stated earlier, many of the ecosystem services are considered to be the responsibility of a sector department (fisheries, irrigation department, public works department, etc.) that has no obvious linkage with biodiversity issues and usually does not consider its activities in an integrated, cross-sectoral manner. This explains that many ecosystem services go unnoticed, thus losing an opportunity to describe the actual values of biodiversity. (An irrigation department will not automatically see the downstream fisheries impacts of its measures; a public works department considers flood storage by wetlands as suboptimal and designs flood storage basins; a forestry department is not inclined to change forestry practices and reduce revenues in order to enhance tourism or leisure activities; etc.).

Levels of biodiversity. Three levels are distinguished (genetic, species, ecosystems) but in general, the ecosystem level is the most suitable level to address biodiversity in SEA, as most cases above have shown. Even in cases where the trigger to start an SEA was at species level (protected tigers in India), the studies focussed on ecosystem structure. Similarly, the Nepal case focusses on species composition only and does not go into further detail of individual species. In other studies individual species only serve the purpose of being an indicator for changes in key ecosystem processes. The large extent of study areas, the limited resources available for SEA, and a lesser level of detail required for strategic decision making explain this focus on more generic biodiversity issues and a 'loss' of focus on species level information. In a study of five SEAs linked to spatial plans in the Netherlands Kolhoff and Slootweg (2005) draw a similar conclusion that 'impacts in biodiversity are considered at ecosystem level'.

However, situations exist with a need to address lower levels. A case from UK shows that for local-level plans it may be needed and possible that the SEA looks at species level information. The limited extent of the study area and the presence of many protected species in nonprotected areas required detailed analysis of these species. Yet, the study focussed on indicator species for each biophysical effect in order to reduce data collection effort.

- In the UK a Local Transport Plan requires an SEA. In an area renown for its biodiversity, the SEA focussed on species and their habitats. Roads are considered to cause a number biophysical effects: barrier effects (e.g. cutting of routes to foraging areas of bats), road mortality, emission into air and water, hydrological changes, and the fragmentation of habitats. For each effect a 'focal species' was used as an indicator. Many protected species rely on unprotected countryside and species-level attention. Furthermore, the study included alternatives that would minimise impacts on priority habitats as listed in the Biodiversity Action Plan. (Case: Burrows, 2004.)

Legal protection – a word of caution. A case from the Netherlands shows the far-reaching influence of a formal system of protected areas and a policy for the enhancement of this system. It forces spatial planners to take biodiversity into account, and it defines the setting for SEA of such plans. Similarly formal policies trigger biodiversity attention within SEA through Biodiversity Action Plans in the UK. Analysis of four spatial planning SEAs at national, provincial, and municipal level in the Netherlands revealed the overwhelming importance of the National Ecological Network (NEN, predecessor to and part of the European Nature 2000 network of protected areas). The NEN is intended to create a continuous network of protected areas; the area has been formally defined, but in broad terms. All spatial plans coinciding with the NEN have to include nature restoration measures in order to comply with the NEN policy and SEAs strictly assess proposed alternatives on this aspect. The focus consequently is on ecosystems; species level diversity does not play a role as the NEN includes species-related protected areas (EU birds and habitat directives). Further biodiversity attention is focussed on restoration of key hydrological processes in existing protected areas. Because most activities focus on enhancing the quality of existing nature and increasing the surface area of protected areas, nonprotected biodiversity is lost out of sight. The down-side of the strong Netherlands policy on the National Ecological Network is that nonprotected biodiversity and ecosystem services other than maintenance of biodiversity get out of focus in spatial planning ànd similarly so in the SEAs of such plans (Kolhoff and Slootweg, 2005). SEA is supposed to picture the impacts of plans on protected and nonprotected biodiversity. The built-in argument is that if biodiversity is not protected it probably is not worth taking into account and it consequently does not appear in the SEA. The uMhlathuze strategic catchments assessment in South Africa (Van Der Wateren *et al.*, 2004) has already shown that nonprotected and nonthreatened biodiversity still represents highly valued ecosystem services.

Public participation. An important observation from a number of cases above is that public participation may lead to a broader perspective of biodiversity resulting in formulation of different alternatives. The UK flood management case (Treweek, 2004) and the Dutch shell mining case (Schooten, 2004b) both show that public participation resulted in enhanced studies, including a significant contribution of viable alternatives. Public participation may also be the key to biodiversity-inclusive SEA in cases where attention to biodiversity is not triggered by objectives of the study or by formal regulations.

Trigger 3: Interventions as indirect drivers of change

Plans leading to interventions that indirectly influence biodiversity, so-called indirect drivers of change, are diverse. As stated earlier, anything ranging from changes in human population density, to changes in consumption pattern, or to changing human behaviour, technology, or tax measures can act as an indirect driver of change in biodiversity. In recent years there has been an increased attention towards the development of instruments to assess the environmental consequences of such changes, with varied levels of success. There is a rapid proliferation of instruments such as sustainability assessment, integrated impact assessment, integrated impact analysis, sustainability appraisal all aiming to deal with the triple bottom line of sustainability for varying purposes. Dalal-Clayton and Sadler (2005) in their recent overview of international SEA experiences provide a list of 15 SEA-type of approaches. Obviously, the world has not fully come to terms with the concept of sustainability and is searching for ways to predict the impacts of proposed plans in terms of sustainability. In this world of methodological approaches, biodiversity simply is one of the many issues in need of attention. Consequently, there is very little biodiversity-specific guidance for plans triggering biodiversity attention through indirect drivers of change. Probably the Millennium Ecosystem Assessment provides the best available information. The MA methodology is potentially valuable to identify linkages between indirect and direct drivers of change. The scenarios working group of the MA considered the possible evolution of ecosystem services during the twenty-first century by developing four global scenarios exploring plausible future changes in drivers, ecosystems, ecosystem services, and human well-being. The reports on global and subglobal assessments (still to be published as we are writing this book) may also provide suitable material. However, from a practical SEA point of view, these approaches are methodologically complex and require amounts of data that may go beyond the possibilities of 'ordinary' SEA.

One field of rapidly expanding knowledge with biodiversity specific information is the impact assessment of measures related to international trade agreements. The EU applies sustainability impact assessments to its trade agreements. The approach is to project effects of trade measures on consumer and producer behaviour, and hence on production systems. Baseline conditions, trends, and characteristics of the production and socioeconomic systems determine whether indirect consequences will actually affect biodiversity. Impacts on biodiversity are described in very broad terms, mainly as changes in quantity (surface area) and quality of biodiversity (species richness). Grouping of countries with relatively similar characteristics provides some further detail (George and Kirkpatrick, 2003). In each group of countries a case-study country is studied more in-depth. The difficulty in the identification of biodiversity-related impact lies in the definition of impact mechanism. The EU sustainability impact assessment of WTO trade agreements on agriculture and forest products has been used as a case example (George, 2004). This SEA works with a combination of economic modelling studies, empirical evidence from literature, case-study analysis, and causal chain analysis. Impacts are described only in terms of change in quality and quantity. By addressing specific sectors in economy it was possible to broadly define the ecosystems under pressure, such as forests in the forestry sector, without any specific indication of the location of these ecosystems. The available case study, however, predicted that the major impacts on forests (and other relatively untouched ecosystems) can be expected from trade liberalisation in agriculture. The need for agricultural land is a much stronger driving force leading to forest conversion than the forestry sector itself.

A study carried out within the framework of the Convention on Biological Diversity synthesised eight existing approaches and assessment frameworks (CBD, 2004d). All frameworks offer entry points to address impacts on biodiversity in the assessment process, and many offer some additional guidance on what effects to expect in particular sectors and what indicators to use. However, most organisations and states questioned showed considerable dissatisfaction with the state of affairs of integrating biodiversity concerns into trade-related assessments. A number of deficits were identified, and corresponding needs for further research stressed. According to the EC, the different cause–effect chains of trade liberalisation on agricultural biodiversity have not been adequately analysed, and only limited policy responses have been proposed. The need to develop comprehensive and more practical indicator sets for biodiversity was repeatedly stressed. For instance, although progress has been made in developing aggregate quantitative indicators capable of showing changes in air and water pollution, progress in indicators capable of showing changes in biodiversity, forest cover, habitats, and ecosystems remain less-developed and certainly less-quantitative than pollution-related indicators. On the one hand, there is a need expressed for aggregated biodiversity indicators that can be integrated into formal economic models in order to allow for insights into the overall impacts of economic changes on biodiversity. On the other hand, in order to provide meaningful advice to policy makers on where corrective policy action may be needed, it was said that indicators should also be able to point to the spatial distribution of specific impacts. As was pointed out, understanding that production patterns may change due to trade liberalisation is of only limited value in terms of trying to determine where on the landscape the change would actually occur and, in a next step, of trying to estimate how biodiversity would be impacted. Of course, extensive gaps in environmental data exist within countries, while analysis attempting to examine cross-border issues often runs into pronounced problems regarding the comparability of environmental data. None of the methodologies seems to endorse specific formal tools or models for conducting the actual assessment. The methodologies either point to a plethora of possible quantitative as well as qualitative tools or keep silent on the issue.

Concluding it can be observed that there is an obvious need for further conceptual as well as practical development of this assessment tool to better include biodiversity. A good example is provided by the work of Kessler et al., (2007) who have developed a transparent methodology to quantify impacts on biodiversity resulting from changes in land use. The problem here is that the quantification is based on the relative loss of species diversity (Natural Capital Index); it does not provide any information on how ecosystem services are affected. It goes beyond the objectives and capacities of the authors of this book to go into any further detail of this expanding field of expertise. For further information we refer to the CBD (2004d) document as a good starting point and to the mayor players in this field of assessments:

International organisations:

- Integrated Assessment Methodology of Trade-Related Policies of the Economics and Trade Branch of the United Nations Environment Programme (UNEP framework); UNEP's has supported several rounds of Country Projects to analyse environmental, economic, and social effects of trade liberalization and trade-related policies (www.unep. ch/etb/areas/IntTraRelPol.php).
- Environmental and Trade Reviews of the Organization for Economic Cooperation and Development (www.oecd.org/document/8/0,2340, en_2649_34183_36629256_1_1_1_1,00.html).

Regional economic organisations:

- The European Commission (2006) introducing the sustainability impact assessment (ec.europa.eu/trade/issues/global/sia/index_en. htm).
- The Assessment Framework of the North American Commission for Environmental Cooperation (www.cec.org/programs_projects/trade_ environ_econ/index.cfm?varlan=english).

States:

- The Canadian National Framework for Conducting Environmental Assessments of Trade Negotiations (www.international.gc. ca/tna-nac/env/env-en.asp).
- The U.S. Guidelines for Environmental Review of Trade Agreements (www.ustr.gov/Trade_Sectors/Environment/Guidelines_ for_Environmental_Reviews/Section_Index.html).

Dealing with uncertainty

The reasons to start an SEA process may not be linked to biodiversity, but biodiversity often is influenced by policies, plans, and programmes subject to SEA. Ample reasons have been provided to identify biodiversity issues during the scoping phase, but also to focus the study on relevant issues as to avoid unnecessary detailed studies. Two important elements have been introduced which facilitate the identification of potential impacts on biodiversity: (i) delineation of geographical boundaries and (ii) identification of drivers of change. Three triggers where described which stress the need for specific attention to biodiversity. Environmental assessment by definition has to deal with uncertainty; in the case of SEA the level of uncertainty is even higher as compared to EIA. Furthermore, the tiered nature of SEA creates a variety of contexts in which SEA is applied. The kind of biodiversity information that can be obtained in situations where information is usually incomplete has been elaborated in this chapter. In addition to this, two more elements are important in defining the way how biodiversity can be assessed, that is, *extent of the study* in relation to required *level of detail*.

The required level of detail in a study depends on a variety of factors, such as the spatial and temporal scale of the study, the number of relevant issues to be studied, the severity of decision making implications, the available resources, and so forth. From a biodiversity perspective two scale aspects are important:

- The *extent* of the study, in terms of size of the area and duration of time under consideration. Physical, biological, or social processes work on different scales in time and space. The extent of the study is not necessary limited by the geographical limits or by the time horizon of the policy or plan under assessment. It is important to know the relevant process to be studied and define the extent of the study accordingly.
- The *level of detail*, in ecology often referred to as grain size, of the study. An important determinant of the required level of detail is the level of decision making. Looking at the idealised tiered structure of SEA, in general it can be stated that a high level of decision making, such as policy decisions, usually requires low level of detail. Descending from policy to programmes and plans the required level of detail increases while in some cases (but definitely not always) the extent of the study area is reduced. The availability of information and financial resources, and the priorities expressed by stakeholders during the scoping process will further define the level of detail at which the study needs to be carried out.

Biodiversity has fine grain and large extent. In studying biodiversity a fine grain has to be sacrificed for a large extent, or reciprocally, a requirement for fine-grain information often limits the extent of the study. Some practical examples in Box 7.11 show how the dilemma of large extent and fine grain of biodiversity can be addressed in different situations. They show that the biodiversity aspects composition, structure, and key process provide a good means to focus the assessment.

Box 7.11. Examples of focussed biodiversity studies

Limited extent with high level of detail, and focus on a key aspect of biodiversity (species composition) to reduce information requirements. The dominant biophysical change caused by forestry activities, that is, selective removal of valuable tree species, primarily affects species composition. SEAs for district forestry plans in Nepal consequently concentrated on the effects of forestry on forest composition and looked at species level information only. The extent of the study was limited, so species level information could be obtained (Uprety, 2004).

Very large extent and low level of detail, with focus on key processes as determinants of impacts. An SEA for a 600 km road in Bolivia concentrated on main ecosystems and hydrological processes (apart from social aspects not elaborated here). Road construction potentially affects the hydrology of the area. Because the road crosses wetlands of international importance, this key wetland process was the focus of study. The extent of the study area was of such magnitude that further detailed biodiversity analysis was not feasible (Consorcio Prime Engenharia, 2004).

Medium extent and sufficiently reduced level of detail by focussing on ecosystem structure: An SEA for the siting of a nuclear power plant in India focussed on the connectivity of tiger habitats. The highly endangered and strictly protected tiger triggered the study, but the study focussed on ecosystem structure, thus avoiding unnecessary detailed surveys (Rajvanshi and Matur, 2004a).

Large extent, with high level of detail, but strongly focussed on one key process and the use of indicator species: An SEA for a National Drinking Water Policy in the Netherlands concentrated on the main biophysical effects of water extraction (hydrological change). The extent of the study was large (the entire nation); defining a limited number of vegetation indicators for impact determination provided the required level of detail for policy decisions. The availability of detailed vegetation inventories facilitated the use of computer technology to highlight areas that are sensitive to hydrological changes (van Schooten, 2004a).

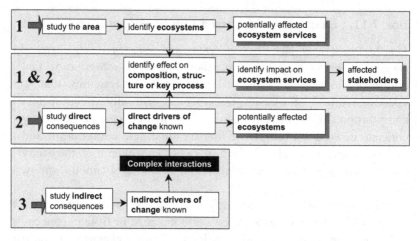

Figure 7.2 Summary overview of analytical steps to define impacts on biodiversity starting with one or a combination of biodiversity triggers.

A general complaint in SEA is the lack of clearly defined activities and the lack of information on biodiversity. In our view lack of clarity and data is a standard condition to virtually all environmental assessments. Although we acknowledge the difficulty presented by lack of information and clarity, we are convinced that relevant information with respect to impacts on biodiversity can be obtained in such situations. The biodiversity triggers provide the most concrete starting point when defining the kind of information of relevance to biodiversity that can be obtained during the SEA study. They also provide insight in the type of information that can NOT be obtained.

Conclusions

Figure 7.2 provides a summary overview of the way in which potential impacts on biodiversity of a policy, plan, or programme can be identified. It starts with the identification of potential biodiversity triggers in the policy, plan, or programme to be analysed, including: (i) a geographically defined area with valued ecosystem services; (ii) activities affecting direct drivers of change; (iii) activities affecting indirect drivers of change; or a combination of (i) and (ii) where activities with known drivers of change influence a known area with valued ecosystem services. If one of these triggers is present in the policy, plan, or programme, the flow chart shows the type of information that can and should be obtained in the SEA process. The link between indirect and direct drivers of change is characterised by complex interactions, many of which are presently subject to research efforts worldwide.

The generic and conceptual nature of this chapter implies that further elaboration of its practical application is needed to reflect the ecological, socioeconomic, cultural, and institutional conditions for which any SEA system is designed. The elaboration can be done on a country or sectoral level and is a task to be taken up by government agencies and the SEA practitioners. From a scientific point of view the research needs encountered during the Millennium Ecosystem Assessment have been summarised by Carpenter *et al.*, (2006). Given the prominent role of ecosystem services in the approach taken in our book, their observations are of great relevance to the world of SEA:

Basic theory. From the evidence we provided in this chapter it is obvious that with the combined knowledge of experts and stakeholders many biodiversity issues in SEA can be addressed satisfactorily. Nevertheless, 'we still lack a robust theoretical basis for linking ecological diversity to ecosystem dynamics and, in turn, to ecosystem services underlying human well-being. We lack the ability to predict thresholds for catastrophic changes, whether or not a change may be reversible, and how individuals and societies will respond'.

Local to global scales. 'Most ecosystem services are delivered at the local scale, but their supply is influenced by regional or global-scale processes. Additionally, there is a mismatch between the scales at which natural and human systems organize.' The conceptual framework we introduced in Chapter 2 clearly addresses this issue. The CBD ecosystem approach is an attempt to circumvent this problem. Nevertheless, it remains difficult to implement such framework in day-to-day reality of SEA.

Monitoring and indicators. 'Trends in ecosystem services are often most effectively communicated through indicators that simplify and synthesize the underlying complexity. There is no consensus on a manageably small set of indicators that can be consistently applied and serves the needs of decision-makers and researchers.' 'Attributes used for monitoring social and economic variables, such as gross domestic product or population, have an established role in decision-making, but their spatial resolution is coarse. Biophysical observations typically have great spatial detail, but little political traction. Integrating both types of data into policy discussions is a key challenge'.

Policy assessment. 'We need to understand how the effects of response strategies vary among ecological and social contexts. We don't know what conditions must be met or how to tailor planning and decision-making to local circumstances. Understanding of the costs and benefits of alternative management approaches for the entire range of ecosystem services is essential. The few examples that assess the bundle of ecosystem services provided by a region show that a single-service analysis misses key trade-offs'.

Linking social to ecosystem change. 'Most research related to ecosystem services focuses on direct drivers, such as land use change or invasive species. Yet, effective management requires more attention to indirect drivers such as demographic, economic, socio-political, and cultural factors. In some cases, indirect drivers may provide better leverage points for policy than the direct drivers'.

Economic instruments and valuation: 'Valuation translates ecosystem services into terms that decision-makers and the general public can readily understand. At present, most ecosystem services are not marketed. The resulting lack of information about prices that reflect social value is an impediment to design and implementation of economic policy instruments. The gap is particularly acute for 'regulating services', such as disease and flood regulation and climate control, which are rarely priced, yet have strong effects'.

In this chapter we have provided clues how to identify potential impacts on biodiversity through the identification of biophysical effects with know impacts on biodiversity composition, structure, or key processes. Knowledge on the area where these impacts occur further provides information on affected ecosystem services, which allows for identification of stakeholders and a complete representation of biodiversity in the SEA process. A question of great concern to all those involved in environmental assessment is when NOT to study certain issues further. Environmental assessment can only be effective if it focusses on real issues of societal concern; it should not end up in endless data gathering exercises with little added value to decision making. Each human activity leads to biophysical effects (by our very existence we continuously change our environment), but not all biophysical effects lead to relevant impacts on biodiversity.

A recent questionnaire among 44 impact assessment professionals from 14 Southern African countries revealed that 'in the context of biodiversity considerations, impact assessments seldom answer strategic questions, seldom deliver relevant information and provide little inspiration to decision makers' (Brownlie *et al.*, 2006). The authors make the important observation that a focus on ecosystem services provides a means to address biodiversity in SEA from a 'biodiversity FOR development' point of view, in stead of the polarising message of 'biodiversity OR development' (Brownlie *et al.*, 2006). Especially in the developing world the linking of biodiversity to people and poverty is a prerequisite.

Part III
Emerging issues

8 · Reconciling conservation and development: the role of biodiversity offsets

Asha Rajvanshi and Vinod B. Mathur

Background

Many of the efforts of the global conservation community are directed to achieve two seemingly incompatible goals – biodiversity conservation and economic development. The synchronization of efforts for achieving these conflicting goals is often hampered by multiplicity of factors that threaten biodiversity. Those threats to biodiversity that are driven by an increasing array of homogenizing forces include at least three factors. First, there is the spread of introduced species (Mack *et al.*, 2000). Second, there are increasing demands on biodiversity resources due to dominance of humans as principle components of natural ecosystems (Putz, 1998; Sanderson *et al.*, 2002) in most biodiversity-rich countries. Third, there is the rising impact of competing land uses in wilderness areas for meeting contrasting objectives of satisfying the socioeconomic needs of the human population and conserving the fast declining diversity and decreasing sizes of populations of species. The increasing trends of habitat fragmentation, modification, and loss of forests, wetlands, coral reefs, and other ecosystems resulting from unabated and accelerating transformation of the earth for urban development, perhaps remain the single most pervasive threat to biodiversity resources (Sala *et al.*, 2000; McNeely, 2006).

Biodiversity conservation has a number of distinguishing features which differentiate it from more conventional resource management issues and which must be taken into account in implementing development projects. Some of the biodiversity losses may be irreversible, and once lost, a species is gone forever. Also, many species, especially invertebrates, microbes and viruses, have yet to be discovered. Therefore much of the biodiversity loss that is presently occurring is in the form of loss of species we have yet to discover (Young *et al.*, 1996). Stopping developments cannot always be an answer. Moving beyond the environment versus development debates and making a fresh resolve to mainstream biodiversity in development may perhaps lead to more positive efforts towards a sustainable future. Finding innovative ways to link biodiversity conservation with development becomes a unique challenge and urgency for conservation organisations and businesses as well as voluntary bodies, governments, and civil societies.

Biodiversity offset: Concept and definition

Biodiversity-inclusive impact assessment has steadily emerged as a useful planning tool to manage development and to restoring biodiversity as a means to address the impacts of our expanding footprint. The mitigation step in EIA frameworks provides options for preventing and minimizing the impacts of development projects on biodiversity by utilizing an array of strategies, policy instruments, economic incentives, and market solutions for compensating the residual impacts. The concept of a 'biodiversity offset' as a compensation measure is relatively new. Comprehensive and universally acceptable definitions for 'biodiversity offsets' that would be useful for encouraging new policy directions for conserving biodiversity are therefore currently lacking. The review of the terminology use in different parts of the globe provides several working definitions that help in understanding of the concept, design, and conservation benefits of offsets (See Box 8.1).

Box 8.1. Defining 'biodiversity offsets'

Biodiversity offsets are:
- Environmentally beneficial activities undertaken to counterbalance an adverse environmental impact to achieve 'no net environmental loss' or a 'net environmental benefit' (Western Australia EPA, 2004).
- One or more appropriate actions that are put in place to counterbalance (offset) the impacts of development on biodiversity (NSW, 2002).
- Conservation actions intended to compensate for the residual, unavoidable harm to biodiversity caused by development projects to ensure 'no net loss' of biodiversity (Ten Kate et al., 2004).
- A form of mitigation used to address net biodiversity loss after all other mitigation measures have been taken (EBI, 2003a).

Policy context

Interest and practical experience of using biodiversity offsets in making development and conservation mutually supportive is gradually building up in response to various policy directives across different countries and economic sectors (Johnston and Madison, 1997; NRC, 2001; Wilkinson and Kennedy, 2002). The concept is also increasingly appearing on the international agenda as a potential mechanism for securing conservation outcome in the face of growing development pressures. One of the most far-reaching provisions of the Convention on Biological Diversity (CBD, 1992) requires the Conference of Parties (CoP) to integrate the conservation and sustainable use of biodiversity into relevant sectoral or cross-sectoral plans, programmes, and policies. The relevance of offsets as conservation actions at regional and national level is being increasingly realised 'to help reduce the current rate of biodiversity loss at the global level by 2010 with an objective to alleviate poverty and benefit all life on earth' as proposed by the CBD Conference of Parties (CBD, 2002). The principles of in situ and ex situ conservation advocated in Articles 8 and 9 of CBD and in Article 14 of the CBD that promotes biodiversity-inclusive impact assessment together provide the impetus for encouraging onsite and offsite activities as 'green development mechanisms'.

Targeting for 'no net loss' or 'net biodiversity gain' as opposed to mitigation of negative impacts on biodiversity is being strongly recommended by voluntary organisations and professional bodies (IAIA, 2005; IEEM, 2006; Slootweg et al., 2006) as best practice principles and approaches. Such best practice benchmarks guiding the integration of biodiversity in impact assessment of developments in different economic sectors has created tremendous scope and opportunities for designing biodiversity offsets to ensure that development makes a positive contribution to local and regional biodiversity resources. The findings of the recent Millennium Ecosystem Assessment (2005) are most significant in determining drivers of global biodiversity losses and in providing useful guidance on promoting market-oriented approaches (use of incentives, easements, and tradable development permit programmes) to integrate protection of biodiversity and ecosystem services in the development planning (see Millennium Ecosystem Assessment 2005a). This guidance has been adequately captured in many of the biodiversity offset approaches, including conservation banking, development of tradable rights and biodiversity credits, direct payments for access to habitats and use of biodiversity resources and ecological services, and creation of trust funds and monetary bonds for financing mitigation of impacts. These enabling mechanisms of mainstreaming biodiversity conservation in business plans are aimed at creating mutually beneficial opportunities for both business and biodiversity. Business groups are beginning to take a lead to demonstrate that responsible biodiversity stewardship is a fundamental business issue for managing risks, capitalizing on opportunities, and improving the corporate performance in environmentally and socially responsible manners (Anonymous, 1997, 2002b). Oil and gas industries are already beginning to explore the use of offsets (EBI, 2003a; IPIECA, 2003; IPIECA and OGP, 2005).

A number of donor organisations, governments, and intergovernmental institutions are studying offsets to develop enabling policies and setting standards for promoting offsets as a biodiversity management and conservation tool to benefit different sections of society. For governments, offsets offer opportunities to drive the implementation of national policies to be able to achieve national conservation goals and fulfil the obligations and commitments under Millennium Development Goals and CBD and to allow for better balancing of costs and benefits of conservation and economic development. For developers, offsets offer means to secure new market opportunities and provide cutting edge advantages of improving performance and 'reputational benefits' by caring for nature. For both custodians of natural resources and the conservation community at large, offsets are mechanisms to mainstream biodiversity in development plans. The offsets offer opportunities to enter into negotiations on the conservation of high priority biodiversity values and habitats instead of highly compromised sites to achieve better conservation outcomes from development. For resource economists, offsets present a new approach to the financing of conservation and achieving greater economic value for biodiversity. For society, offsets ensure the perpetuity of the benefits from healthy and productive ecosystems for sustenance, livelihood security, and well-being.

Policy makers wishing to promote biodiversity offsets draw on a range of underlying benefits to make enabling policy interventions, thus building the case for the promotion of biodiversity offsets. In Uganda, the policy framework for biodiversity offsets requires the employment of offsite measures when the impacts cannot be mitigated by onsite remediation. Offsets are already accepted as an additional environmental management tool to contribute to achievement of sustainable development in the Western Cape Province of South Africa (Department of Environmental

Affairs and Development Planning, 2006), Sweden, and the United Kingdom (Petterson, 2004). The World Bank has a long history of financing regional/multinational projects and national investments. The Bank's Safeguard policy, that is, the Operational Policy on Natural Habitats, is being updated (World Bank, 2001) to incorporate performance standards for conservation of biological diversity and sustainable natural resource management. It states that 'the client will not significantly convert or degrade natural habitats unless there are no technically and economically feasible alternatives, the overall benefits of the project substantially outweigh the social and environmental costs; and any conversion is appropriately mitigated' (www.ifc.org/ifcextpolicyreview.nsf/content/home). The Ecological Compensation Programme in the Netherlands incorporates the condition that developers who damage habitats are required to offset this damage through protection of three times the original area in the same zone (OECD, 1996; Landell-Mills and Porras, 2002). In many cases, Government policies requiring or supporting offsets lead to the establishment of national systems to define and administer biodiversity offsets and develop institutional frameworks for the consideration of offset proposals and their implementation. The amended EU Habitats Directive (European Commission, 2000) that requires provisions for recreation or replacement of a habitat to mitigate adverse effects on the integrity of the 'Natura 2000' sites is an example of government-led support to implement offset schemes.

Legal and regulatory provisions

Biodiversity offsets are already a part of the legal framework in many countries (Table 8.1). Many of these regulations, requiring the implementation of offsets along with protection of species or ecosystems, have motivated the development of offset schemes in these countries and have created a still greater incentive for offsetting activities.

The experience from the United States, which has one of the longest standing biodiversity offset schemes, is the Wetland Banking System that was developed in response to provisions under its Clean Water Act. The offset activity has been most robust in United States, and its implementation has increased markedly in recent years as a result of two decades of experience. As a result, the wetland offsets in the United States accounted for about 6,000 hectares per year in the early 1990s, further increasing to an average of over 16,000 hectares per year since 1995 (Fox and Nino-Murcia, 2005).

Table 8.1 *Regulatory provisions for the adoption of biodiversity offsets.*

Legislation	Provisions
United States	
Endangered Species Act (1973) U.S.C.1531 et. Seq.	Mandates preservation of critical habitat on private land and the implementation of a species recovery plan (Ten Kate *et al.*, 2004). Provides for an 'incidental take' of enlisted species if a landowner provides a long term commitment to conserve species through the development of a Habitat Conservation Plan (HCP) that ensures recovery of species survival through the hectare for hectare compensation (McKenny, 2005).
Clean Water Act (Section 4B)(amended Federal Water Pollution Control Act of 1972) and US Army Corps of Engineers Regulations (33CFR320.4(r))	Encourages the Wetland Banking for ensuring no net loss of wetlands by mandating creation or restoration of wetlands of comparable value through mechanisms of purchasing 'credits' in an alternative specified area determined by the Army (McKenny, 2005).
Canada	
Fisheries Act under R.S.1985, c, F-14 Policy for Management of Fish Habitat (1986)	Encourages compensation by increasing the productive capacity of existing habitat for a different stock or species on or offsite to ensure 'no net loss' of fisheries habitat in Canada (www.dfo-mpo.gc.ca/canwaters-eauxcan/infocentre/legislation-lois/policies/fhm-polocy/index_e.asp).
Habitat Conservation and Protection Guidelines (1998)	Encourages developers to relocate or redesign their activities, mitigate where relocation and redesign is not possible and compensate for unmitigated damage.
European Union	
UK Electricity Act 1989 Schedule 9	Stipulates that electricity suppliers and generators must mitigate any adverse effect and produce a statement confirming actions taken to ensure mitigation (Ten Kate, 2003).
Habitat Directive and implementing regulations (Council Directives 92/43/EEC and 79/409/EEC.)	Provides a step-by-step scrutiny of development plans affecting the European designated sites to secure compensatory measures to replace the affected habitat (European Commission, 2000). Directives for protection of sites under Natura 2000 network.

Table 8.1 (*cont.*)

Legislation	Provisions
European Union Environmental Liability Directive (2004):	The EU directive makes specific reference to biodiversity and operates on the 'polluter pays principle' requiring companies to undertake compensation for environmental damage.
Switzerland Federal Law for Protection of Nature and Landscape (1983)	Mandates reconstitution and replacement of protected biotopes where impacts are unavoidable (www.admin. ch/ch/f/rs/451/a18html).
Brazil Protected Areas Law (#9985) Decree 4340	Regulation requires rural property to maintain a forest reserve of at least 20 percent (Government of Brazil, 1965, 2000).
Forestry Code (#4771) Provisional Measures 2166/67	Where a development has a significant environmental impact, it must compensate for this by supporting a unit within a National System of Conservation Units (SNUC). The sum paid depends on the degree of environmental impact of the project but must be at least 0.5 percent of the total investment costs and in the rainforest areas may be above 6 percent. The law requires that the landowners must maintain a fixed minimum percentage of natural vegetative cover on their property. The requirement can be satisfied through the use of offsite conservation offsets (McKenney, 2005).
Australia (states and territories) Environment Protection and Biodiversity Conservation Act (1999)	Establishes that Commonwealth approval may be required for native vegetation clearing that is likely to have significant impacts on those aspects of the environment that are of national significance (Commonwealth of Australia, 1999).
New South Wales Environmental Planning and Assessment Act (1979)	Grants power of negotiating biodiversity offset programs for licensed premises as part of their development consent (McKenney, 2005; NSW, 2006b).

Table 8.1 (cont.)

Legislation	Provisions
Environmental Protection Act (1986) 511(2)(b)	Regulatory provisions for abating, mitigating, or offsetting the loss of the cleared vegetation and allowing for making monetary contributions to a fund maintained for the purpose of establishing or maintaining vegetation.
The Sydney Water Catchment Management Act (1998)	Development must demonstrate a 'neutral or beneficial effect' on the water quality.
Native Vegetation Act (2003) No. 103	The legislation seeks to prevent broad scale clearing and encourages the revegetation or rehabilitation of land. Supports biodiversity certification process and provides regulatory procedures and tools for developing biobanking schemes to promote biodiversity offsets.
Victoria	
Threatened Species Legislation Amendment Act (2004).	Establishes legal and administrative structure to enable and promote the conservation of Victoria's flora and fauna and to provide a choice of procedures which can be used for the conservation.
Flora and Fauna Guarantee Act 1988 No. 47/1988 version No. 30.	

The recently amended Environmental Protection Act in Australia, relating to the clearing of native vegetation, makes specific reference to environmental offsets. A range of state and territory level schemes has been developed in Western Australia, Victoria, and New South Wales (NSW). These link environmental offsets into the state's sustainable development strategies and encourage the use of biodiversity offsets as part of a company's development requirements. The NSW Government's 'green offsets' scheme is one of the new economic tools to address the cumulative environmental impacts of development (NSW, 2002) under the Threatened Species Legislation Amendment Act (2004). The successful implementation of this scheme subsequently led to the introduction of other market-based approaches such as Biodiversity Certification and Biodiversity Banking Schemes (NSW, 2006b) for conservation of threatened species on private lands. These new schemes would provide landholders with financial incentives to 'maintain or improve' the biodiversity values in the area and offset any losses that may occur as a result of the development process.

Voluntary approaches

Apart from policy interventions and regulations requiring impacts on biodiversity and ecosystems to be offset, voluntary approaches promote offsets as a business case beyond legislative compliance. Literature on voluntary schemes designed for forest services, water services, and carbon sequestration (Landell-Mills and Porras 2002; Ten Kate and Laird, 2002) have significantly encouraged the use of biodiversity offsets. Many professional networks have already taken a lead in institutionalizing the practices of developing and designing offsets (Box 8.2) as these have considerable virtues of low administrative costs, high community acceptability, and minimal equity implications. Companies such as Newmont, Rio Tinto, and Shell are also pioneering innovative approaches to integrate biodiversity offsets into their business plans (Bertand, 2002; EBI, 2003d; Fish *et al.*, 2004).

Box 8.2. Voluntary initiatives for biodiversity offsets

- The **Business and Biodiversity Offset Program** (BBOP) (www.forest-trends.org/biodiversityoffsetprogram) Forest Trends and Conservation International recently established the Business and Biodiversity Offset Program. This BBOP Learning Network is a partnership of 50 institutions representing companies, scientists, NGOs, government agencies, and research institutes to pilot projects which compensate for the residual, unavoidable harm to biodiversity caused by major development projects. The objectives of BBOP are to (i) demonstrate conservation and livelihood outcomes in a portfolio of biodiversity offset pilot projects; (ii) develop, test, and disseminate best practices on biodiversity offsets; and (iii) influence policy and corporate developments on biodiversity offsets so they meet conservation and business objectives.
- **Biodiversity Neutral Initiative** (www.biodiversityneutral.org). Biodiversity Neutral Initiative (BNI) is a nonprofit organisation that researches and promotes best practices for corporate biodiversity management. The organisation's long-term goal is to develop guidelines for measuring, communicating, and offsetting impacts

on biodiversity through compensatory conservation projects so that the leading companies can become biodiversity 'neutral'.

- **Energy and Biodiversity Initiative** (www.theebi.org). The EBI is a partnership between four energy companies (BP, Chevron, Shell, and Statoil) and five international conservation organisations (Conservation International, Fauna and Flora International, IUCN, Smithsonian Institution, and The Nature Conservancy) to produce valuable products to ensure better integration of biodiversity considerations into oil and gas operations. EBI is developing practical guidelines, tools, and models to improve industry's environmental performance, reduce impacts of development projects on biodiversity, and maximize opportunities for conservation. A report of EBI (2003a) focuses on biodiversity offsets.
- **International Council on Mining and Metals** (www.icmm. com). ICMM, formed in October 2001, represents the leading international mining and metals companies. ICMM members' mission is to offer strategic industry leadership towards achieving continuous sustainable developments in the mining, minerals, and metals industry. ICMM's environmental stewardship work programmes are aimed at promoting science-based regulations and material-choice decisions that encourage market access and the safe production, use, reuse, and recycling of metals. One of its key work programmes is the development of biodiversity offsets and their field testing for positive contributions to biodiversity conservation and the environmental performance of the industry.

Objectives of implementing biodiversity offsets

The concept of biodiversity offsets is premised on a 'no harm' principle. Primarily, offsets aim to ensure that all residual impacts on biodiversity which are considered significant but not severe enough to hold back the proposed development, are counterbalanced by gainful compensatory conservation measures.

Accordingly, the objectives for implementing offsets can vary from 'no net loss' to 'net gain' in biodiversity values (Ten Kate et al., 2004; Western Australia EPA, 2004). The objective of 'no net loss' of values of ecological functions of ecosystem is inherent in the Memorandum of Agreement that is laid down between the US Army Corps of Engineers and the

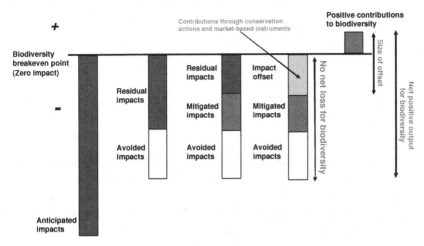

Figure 8.1 Role of offsets in influencing the project's net impact at the site level (Modified from ICMM, 2005b and Ekstrom, 2005).

Environmental Protection Agency for Wetland Mitigation Banking (US Army Corps of Engineers Regulations (33CFR320.4(r)). This objective was later rewritten to advocate the policy of 'net gain' in wetland areas in the United States (Coyne, 2004). The 'better' and 'more' conservation goals proposed by Goodland (2003) suggest that the offsets should be extensive in area, yield greater conservation value in perpetuity (i.e. be less disturbed, more varied in biodiversity, and more efficient in productive functions for greater benefits of environmental services) and should be better protected. In this context, offsets may be seen as a means to achieve a 'net positive' contribution to biodiversity conservation. These measures go beyond just achieving the thresholds of 'minimum requirement', 'no net loss' or 'no net gains' (EBI, 2003b) and can be in any of the forms leading to generating benefits greater than the project's net impact at the site level (Figure 8.1).

Biodiversity offsets virtually represent the last line of defence for the natural ecosystems. They serve as a final option to achieve the 'biodiversity breakeven point', that is, the point where no net loss of biodiversity is ensured through comprehensive biodiversity management responses to developmental impacts. The challenge to recognise a point where and when the breakeven point would occur is conditioned by the location, time and scale of offset, and the type of offset activity. The following are the most recognizable forms of offsets:

- *Onsite offset*: where a developer secures and improves biodiversity values within the same development zone.
- *Offsite offset*: where the developer secures and improves biodiversity values in another piece of land, for example, creation of an alternate habitat for endangered species.
- *Offsite offset through a third party*: when a developer purchases biodiversity credits from a third party to provide an offset ex ante or at the time of the development and subsequently maintains the offset on its behalf.

The activities intended to help counterbalance the environmental impacts with the aim of achieving no environmental difference are considered as **direct offsets** (e.g. restoration of biodiversity corridors, rehabilitation, re-establishment, and sequestration). The **secondary offsets** represent the selected complementary activities which along with the direct offset meets the offset principles of 'no harm'. These include protection mechanisms (e.g. fencing and buffering); management initiatives (e.g. monitoring, education and research); removal of threats (e.g. eradication of exotic species); activities having a proven environmental benefit (e.g. watershed management); or contributions to an approved 'bank', credit trading scheme or trust fund.

Relevance of biodiversity offset as a mainstreaming instrument

The process of integrating biodiversity in EIA offers adequate ground to stimulate biodiversity offsets as mainstreaming instrument. The inclusion of offsets in the generic EIA process represents a possible scenario for visualisation of conservation actions to be explored and recommended in the mitigation step of any impact assessment exercise. Mitigation in an environmental context refers to a sequence of options designed to help manage adverse environmental impacts. These may include:

- *avoidance* (avoiding the adverse environmental impacts all together);
- *minimization* (limiting the degree or magnitude of the adverse impacts);
- *rectification* (repairing, rehabilitating or restoring the impacted site as soon as possible);
- *reduction* (gradually eliminating the adverse impacts over time by better maintenance of operations during the life of the project); and
- *compensation* (undertaking such activities that counterbalance/offset an adverse, residual environmental impact).

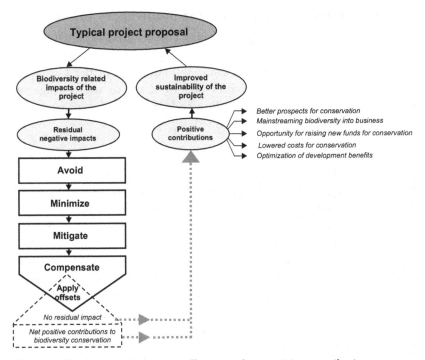

Figure 8.2 The role of biodiversity offsets in making positive contributions to biodiversity through sustainable development.

Although conservation of existing biodiversity values should be upheld as a priority above the use of biodiversity offsets, the onsite adverse biological impacts must follow the ground rule for mitigation sequence, that is, avoid, minimise, rectify, reduce, compensate, or offset, in that order. The offsite offsets should then be used to address any significant residual environmental impacts subsequently (Figure 8.2).

No single approach or instrument can serve to be a magical wand that would be sufficiently powerful to solve all threats to biodiversity in any country. As a general rule, it is better to explore a mix of instruments to meet the complex objective of offsetting the impacts on biodiversity of development projects. The choices can be made from a range of conservation-oriented activities that can qualify as compensation actions or from the various market-based instruments that can represent satisfactory compensations in economic terms for biodiversity losses.

Conservation-oriented actions for biodiversity offsets

Prescriptions and guidance for developing conservation-oriented biodiversity offsets is fast accumulating from the experience of countries such as Australia, Brazil, the United States, and countries in the European Union. While appropriate activities vary from site to site, a range of different terrestrial and aquatic, offsite and onsite ecosystem management interventions can be distinguished.

Establishing habitat networks

As the size and spatial relationships between habitat patches greatly influences biodiversity, this approach of establishing habitat networks essentially involves identifying and securing the management of lands that offer greater utility as biological corridors between Protected Areas (PAs). The fragmentation of habitats by urbanization and linear development projects (e.g. network of canals, roads, railway lines, or power lines) has highlighted the need to restore habitat networks to provide better functional attributes and maximise opportunities for biodiversity. Ecological engineering solutions such as road passages for animals and restoration of biological corridors certainly represent an important strategy to offset impacts of species isolation, mortality, and habitat fragmentation by existing roads (McKinney and Murphy, 1996; Cuperus *et al.*, 1999; Van Bohemen, 2004). New guidance is now also becoming available on how to master plan green infrastructure for a sustainable community. The initiatives of encouraging 'biodiversity by design' into new urban developments to offset climate change effects already offer a strong business case in many countries. (TCPA, 2004)

Upgrading protection in nondesignated areas

This approach follows the principle of conservation science that advocates strengthening of conservation efforts by placing land into protected areas and nature reserves in order to reduce its vulnerability to threats. Some examples of offsets resulting in enhancement of conservation of protected ecosystems and species are presented in Box 8.3.

Demarking sites of conservation importance to ensure protection

Acquisition of land for the creation of alternate habitat, or the fencing of areas of conservation importance, such as sites of reintroduction or

Box 8.3. Examples of offsets involving conservation enhancement activities

- The **Chad–Cameroon Petroleum Development and Pipeline Project** involved construction of a 1,070 km pipeline to transport crude oil from three fields in southwestern Chad to a floating facility 11 km off of the Cameroon coast. Cameroon has some of the most biologically diverse and important forests in Africa. The project threatened valuable ecosystems, particularly in Cameroon's coastal rainforest where the corridor of the pipe cuts straight across these sensitive ecosystems. The World Bank Group applied its safeguard policies to the project and related infrastructure, and worked with the sponsors to ensure that whereever possible, the pipeline avoided areas of high biodiversity. Other conservation efforts included the setting up of two new large national parks in Cameroon to offset a small but unavoidable loss of forest. The parks, which will help protect biodiversity, are being independently managed. (www.ifc.org/ifcext/africa.nsf/AttachmentsByTitle/ChadCam ProjectOverview/$FILE/ChadCamProjectOverview.pdf).
- The construction of **Indira Sagar Hydropower Project** in India presented an opportunity to create three Protected Areas in forested habitats including the riparian stretches outside the zone of submergence. These new Protected Areas are being set up to offset the loss of habitats of several ungulate species (four-horned antelope, spotted deer, sambar), endangered carnivores (tiger and leopard), the endangered smooth-coated Indian otter, and several bird species. As part of the mitigation proposal (WII, 1994) the costs of the preparatory feasibility studies, the land transfer and the subsequent management of the protected areas are to be borne by project authorities. The anticipated offset benefits are: (i) securing the contiguity of the habitat; (ii) the enhancement of status and conservation value of habitats outside the project area, currently having low anthropogenic pressures; and (iii) the opportunities of conducting studies on newly created islands within the future reservoir area.
- As part of the **Pacific Highway upgrade program**, the Roads and Traffic Authority (RTA) of New South Wales proposed to construct a 9.8 km section of dual carriageway around the town of Karuah. Although the preferred route for the bypass was selected

to avoid environmental impacts, the environmental assessment identified that the removal of 47 hectares of vegetation, including 16 hectares from the Karuah Nature Reserve, would have major ecological impacts. The RTA acknowledged that it would not be possible to avoid all the impacts on habitat or threatened species and a compensatory habitat package would have to be developed. The National Parks and Wildlife Service (NSW, 2006a) now the Department of Environment and Conservation (DEC) sought an offset that would deliver an outcome of overall ecological gain rather than applying specific habitat ratios. An 89-hectare block of privately owned land was identified near the proposed road alignment. It contained similar vegetation and many threatened species affected by the road upgrade. The DEC agreed to incorporate the land into the adjacent Karuah Nature Reserve. The RTA acquired the land to transfer it to DEC for which the approval of the Parliament was granted. Legislative provisions for land transfer require the new nature reserve to be of equivalent or better value than the habitat being lost from the reserve. The RTA also agreed to contribute $15,000 towards initial management costs, such as weed control and active rehabilitation. The offset benefit in this case was the added area that created a larger contiguous block of habitat that had significant benefits for biodiversity.

feeding and breeding grounds of endangered species, can sometimes help to compensate impacts of a development activity. Such activities as part of a mining operation or the construction of a dam and associated human displacement are examples of secondary offset activities. Secondary offsets can only be accepted after attempts have been made to address the impacts of the project through primary offset activities.

Removal of threats

This involves initiatives to remove potential and existing threats to biodiversity at the direct offset site, thereby preventing it from being potentially damaged in the future. Examples might include eradication of feral animals or exotic flora, removing pollutants, preventing livestock entry, and controlling the spread of diseases in areas maintained as natural ecosystems.

Building partnerships to enhance biodiversity conservation in habitats on private land

Many conservation organisations have shown strong interest in pursuing partnerships with the private sector to capture opportunities for the conservation of high-priority areas. The Land for Wildlife Programme is run by the Department of Conservation and Natural Resources and the Bird Observers Club of Australia with the objectives to foster attitudinal change of private landholders and to encourage them to conserve flora and fauna habitat. As a result, currently more than 40,000 hectares of land is being voluntarily managed as wildlife habitat without offering any direct financial assistance (Victorian Environment Protection Authority, 1994).

In a hypothetical case from 'Arboria', Kumari and King (1997) calculated the incremental cost for logging companies to extend the existing system of commercial inventory (which constitutes the baseline for planning logging operations) to include a comprehensive inventory of the remaining biodiversity. This was intended to assist the Forest Department in profiling the biodiversity of one of the mega diversity 'hot-spot' countries and to provide the basis for designing medium- and long-term conservation measures for natural biodiversity.

Address underlying causes of biodiversity loss at offset sites: biodiversity-related sustenance needs of local communities

This approach is based on the assumption that if the stakes of the local communities are nurtured and secured, they may relate and respond more readily and ably to 'on-the-ground' biodiversity issues, such as wildlife conservation, than to other less tangible environmental impacts, such as carbon dioxide emissions and ozone depletion. For example, in Papua New Guinea, the Government has established Wildlife Management Areas, where local communities comanage resources. The management committees, consisting of representatives from local communities, can apply measures, such as asking royalties on the taking of game and fish by outsiders, putting restrictions on some hunting or fishing techniques, prohibiting the collection of crocodile eggs, or imposing restrictions on logging (Eaton, 1985).

Market-based approaches for biodiversity offsets

The premise for promoting market-based approaches to mainstream biodiversity in mitigation planning is that ecological values on the site of a proposed development can be translated in economic values. Market-based approaches are receiving increased attention as they possibly offer more effective alternatives and complementing options to regulatory approaches (Young *et al.*, 1996; Tietenburg and Johnstone, 2004). The primary motivation for incentive-based approaches is that if environmentally appropriate behaviour can be made more rewarding to land managers, the best business choice will correspond to the best social and environmental choices. Economic instruments have enormous potential as a source of funding for ecosystem restoration and for conservation of biodiversity. These instruments can guide the choice between alternatives that influence quantities of resources used. We will explore some of the options for offsetting impacts on biodiversity using economic instruments below.

Taxes, fees, and charges

Direct charges and taxes are the most common instruments used to align private and social incentives, promote environmentally sound behaviour, and raise funds to protect biodiversity or reduce likely negative impacts of development proposals. Collectively these approaches imply that, wherever possible, the costs of providing access to biodiversity should be recovered from the direct beneficiaries of biodiversity conservation. Taxes can follow the 'polluter pays principle' by charging those who cause environmental damage. Taxes are levied on the extraction of resources, such as sand and gravel (e.g. in Eastern Europe, taxes are applied to extracting minerals from river beds), on the discharge of effluents into water sources, on logging, or on visitor use (e.g. a fee is charged to tourist visiting the Greek Island of Zakynthos in order to reduce the pressures on the sea turtle *Caretta caretta* (Bräuer, 2006)). Charges can also include entrance fees for protected areas, payments for water services, and schemes to internalise the costs of pesticide or fertiliser use.

Subsidies, grants, and funds

Direct payments through subsidies, support schemes, grants, and funds are market-based instruments which help to establish a direct link between economic incentives and conservation actions. Funds can be used to target the protection of particular species or the conservation of valued ecosystems in protected or other wilderness areas (Box 8.4).

Box 8.4. Conservation funds to offset impacts on biodiversity

- The Monarch butterfly conservation fund was established by WWF in Mexico to conserve the habitat of the Monarch butterfly. This involved paying communities within the biosphere reserve to conserve forest by forgoing logging permits in conservation actions.
- A revolving fund for biodiversity is administered by The Victorian Trust for Nature in Australia. It is used to buy up unsustainably managed land from landowners. The land is put under a covenant covered by Australian law, specifying activities that may or may not be carried out on the site. It is then sold on to landowners who are bound by the covenant (Missrie and Nelson, 2005).
- Kutai National Park (Bontang, East Kalimantan, Indonesia) is one of ten protected areas to be formally approved as a National Park. Covering about 200,000 hectares, it has immense international significance for the conservation of lowland tropical rainforest and contains a high diversity of wildlife including all five primate species endemic to Borneo. At the same time, Kutai National Park faces many pressures from at least eight large industries, including oil, gas, and coal mining, fertilizer production, and logging operations within or in close proximity of the Park. Since 1996, these industries have contributed funding to support the conservation activities in the PA. An association formed especially for this purpose, the 'Friends of Mitra Kutai', has channelled contributions for Park management and community development activities. Cash contributions go directly to the Park's budget, while noncash contributions include fire-fighting equipment, trucks, personnel, fuel and food, as well as tree nurseries established by the industry owner. Between 1996 and 2000, participating companies invested more than $300,000 in Kutai National Park. Corporate donations have been used on training and to map the Park, to develop ecotourism, to prevent forest fires, and to support community development activities in four villages located within the PA (Suratri, 2000).

Property rights

Property rights mechanisms seek to compensate for, or reverse market failures through mechanisms which make resource use opportunities consistent with social values. As with other market-based instruments, the aim is to alter private costs and benefits so that unaccounted social costs (and benefits) can be 'internalised' to ensure the desired environmental improvement. Most property rights begin by defining what may be done and then restricting what may be done by using covenants and conditions to specify what may not be done. A conservation covenant, for example, might prohibit clearing. The experience of a number of countries suggests that covenants and easements offer considerable scope for the establishment of buffer zones and wildlife corridors, and for protected area management (Gunningham and Young, 2002). Property rights–based approaches include the establishing of clear ownership rights, conservation easements, and communal property rights.

Payments for environmental services

In this model, 'consumers' of environmental services are taxed or are made to contribute voluntarily to the generation of funds to help maintain those services that are threatened. One of the best-known developing country examples is the Environmental Services Payment Program in Costa Rica, which pays landowners in key watershed areas to maintain forest in degraded areas for their four basic services: (i) watershed protection, (ii) mitigation of greenhouse gas emissions, (iii) biodiversity protection, and (iv) ensuring natural scenic beauty. Part of the funding comes from developers. For example, in La Esperanza Hydropower Project a contract between the hydropower project and the Monteverde Conservation League (MCL), owning most of the 3,000 hectares watershed, required payment of $10 per hectare for watershed services as a means of assisting the MCL to protect the forest effectively (Rojas and Aylward, 2001).

Bond and funds for mitigating anticipated impacts

Bonds and deposits are product surcharges which shift the responsibility for the mitigation of project impacts on to individual producers and consumers. They are applied to natural resource-based industries, such as forestry, mining, fisheries, and other extractive activities. They are also highly relevant to tourism, and urban, industrial and residential developments that may harm biodiversity. By charging in advance for any anticipated impact on biodiversity, restoration and assurance bonds provide conditional security of funds to meet the costs of anticipated damage. At the same time, such bonds and funds also ensure that producers or consumers cover the cost themselves. This approach thus presents an incentive to avoid impacts on biodiversity.

Trading rights

Tradable permits and biodiversity credits are trading instruments to counterbalance the harm to endangered species and habitats of high conservation significance. This approach involves the creation of ecologically comparable area(s) that are managed for biodiversity and the purchase of species-specific credits from what has become known as 'conservation banks'. The wetland and conservation banking schemes, developed in the context of the U.S. Endangered Species Act and the Clean Water Act are perhaps the best known examples of trading instruments. The wetland mitigation banks and the conservation banks are essentially private (usually for profit) entities that protect specific species with a view to selling species' mitigation credits to future needy developers. The potential benefits of such conservation banks is that they allow advance mitigation at a single large site for multiple future projects that would otherwise be mitigated at several smaller sites. The state of California pioneered the approach in 1990 as a creative way of financing the conservation of gnatcatcher habitat. Since then, private companies have been setting up wetland banks to create wetlands to serve as 'wetland credits' to be sold out to developers (Shabman, 2004). Estimates indicate that these trading schemes have created 72,000 ha of wetland and endangered species habitat in more than 250 approved habitat 'credits' in more than 45 states in the United States (Wilkinson and Kennedy 2002; Fox and Nino-Murcia, 2005).

Acquisition of land with high conservation values through open land markets

As the environmental benefits of acquiring a 'critical asset' for conservation may greatly outweigh the overall environmental loss, acquisition of land for conservation can be considered as a direct offset. Examples

---

Box 8.5.   Land acquisition to improve conservation opportunities for developers

• The Roads and Traffic Authority (RTA) purchased a property with a known population of Striped Legless Lizards. The protection of this population to compensate for unavoidable impacts of a highway upgrade helped to maintain the overall viability of the species (NSW, 2006).

• An offset scheme was developed for a bentonite mine in Western Australia which was proposed to occupy about 10 hectares in Watheroo National Park, 200 km north of Perth. In addition to rehabilitation of the mine site after extracting the resource, the company agreed to provide an offset for the residual impacts on a Eucalypt. *Eucalyptus rhodantha* (*Rose mallee*) occur only as a few remnant populations in Midwestern farmland with fewer than 500 plants recorded. The company purchased 50 hectares of remnant native vegetation that contained the largest stands of the tree and donated it to the State Government. The area was then managed as a conservation reserve, maintaining a nationally significant population of rose mallee. The offset proposed for the project was of significant value to biodiversity conservation as it provided more than equivalent values to the estate than those that were lost through mining (CBD, 2005).

---

in Box 8.5 illustrate that conservation through a combination of land acquisition, protection, and ongoing management can become viable offset packages.

Payments for conservation and management of biodiversity

Biodiversity presents great opportunities for developing links to achieve best biodiversity performance between a business and its stakeholders. In situations where a company's business operations lead to unavoidable negative impacts on biodiversity, key elements of a corporate action plan could be the incorporation of restoration and compensation actions to reduce species and habitat loss through responsible investments in conservation actions. Examples of business initiatives to offset impacts are presented in Box 8.6.

Box 8.6. Examples of business cases for conservation and management of biodiversity

• At its Kennecott Utah mine site, Rio Tinto faced a permit require-ment for an additional storage area of tailings produced by copper ore milling. A wetland area was chosen that had to be offset. Rio Tinto purchased 1,000 hectares of wetland to replace the lost wetland, in a location fairly remote from the mine. This wetland provided significant shorebird habitat. Subsequent surveys indicated a many fold increase in use of the new wetland area by birds, justify-ing its expansion by addition of another 450 hectare. (*Source*: www. biodiversityeconomics.org/offsets).

• The red cockaded woodpecker is listed under the US Endangered Species Act (ESA), which mandates the development and imple-mentation of a 'species recovery plan' including obligatory protec-tion of critical habitat on private land. Conservation efforts in forests owned by the International Paper Company (IPC) led to excess breeding pairs of woodpecker. This became an offset opportunity for IPC. Each excess pair became a 'bank' used by the company to offset impacts on endangered species elsewhere and also a valuable commodity that could be sold out to other land owners to offset their requirements (Heal, 2000).

Supporting biodiversity conserving business

This option can promote opportunities of cost sharing and promot-ing business shares in biofriendly products. Examples of partnerships between the conservation and business communities and of cost sharing include many situations where private entities have voluntarily assumed certain management responsibilities or have funded conservation activi-ties that could be used to promote sustainable production and consump-tion practices. Financial assistance includes targeted grants to promote sustainable livelihoods and conservation, bounties, or other cash rewards, conservation leasing, and soft credits and loans designed to encourage conservation activities.

Creating markets for biodiversity conservation

A number of specific markets have already been developed around biodiversity-related activities. Examples include: organic agriculture, sustainable forestry, nontimber forest products, genetic resources, and ecotourism. Two highly successful examples where the instruments themselves created the market are: (i) the trading in access to fishing rights and (ii) transferable development rights to land. The emergence of private parks in many regions of the world also demonstrates that there is the scope to capture public values in private markets. The development of new markets enhances the capacity of interested parties to delineate attributes of biological resources and may also trigger the creation of new products, services, and corresponding markets.

Carbon sequestration offsets encourage landowners to conserve natural vegetation and to reforest land by providing a market that allows them to be compensated for their costs and forgone profits. Ecolabelling schemes and environmental certifications, mostly voluntary and often created by private agents, have gained significant market importance, particularly in the area of natural resource extraction and management. These labelling schemes seek to increase incentives for environmentally sound production by enabling consumers to differentiate between production techniques, product quality, or producing organisations. A somewhat similar approach involves the creation of offset certificates. These certificates are granted to business groups and individuals who are able to demonstrate that by purchase of offsets they have been able to balance activities that create greenhouse gases.

Framework and ground rules for implementing offsets

Offsets are a relatively new conservation concept. There is no universal 'how to use' tool kit to plan, design, measure, and monitor the offsets. This creates an urgency to develop a generic framework for the improvement of the existing practice of using offsets for conservation. The advantage of a uniform framework lies in the ease of minimum requirements which can be set pursuant to the needs of different organisations that have to contemplate offsets. A framework presented in Figure 8.3 represents a generic approach for planning biodiversity offsets. This incorporates good practice guidance understood from extensive review of literature on offsets (Anonymous, 2002b; Ten Kate et al., 2004; McKenny, 2005; BSR, 2006; WWF, 2006) and the lessons drawn from the application of offsets in countries, such as Australia, Brazil, Canada, the UK, and the United States.

There is universal agreement that even although biodiversity offsets have been successful in many cases, they are far from being a universal

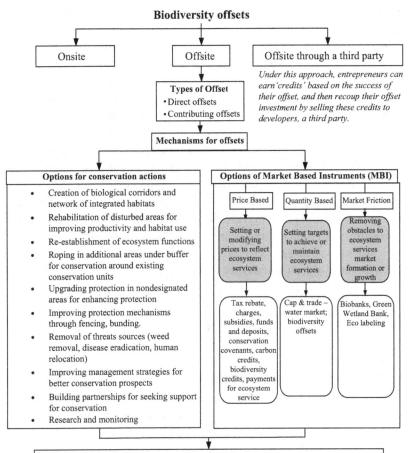

The following content is part of Figure 8.3:

Onsite

Offsite

Offsite through a third party

Under this approach, entrepreneurs can earn 'credits' based on the success of their offset, and then recoup their offset investment by selling these credits to developers, a third party.

Types of Offset
- Direct offsets
- Contributing offsets

Mechanisms for offsets

Options for conservation actions

- Creation of biological corridors and network of integrated habitats
- Rehabilitation of disturbed areas for improving productivity and habitat use
- Re-establishment of ecosystem functions
- Roping in additional areas under buffer for conservation around existing conservation units
- Upgrading protection in nondesignated areas for enhancing protection
- Improving protection mechanisms through fencing, bunding.
- Removal of threats sources (weed removal, disease eradication, human relocation)
- Improving management strategies for better conservation prospects
- Building partnerships for seeking support for conservation
- Research and monitoring

Options of Market Based Instruments (MBI)

Price Based	Quantity Based	Market Friction
Setting or modifying prices to reflect ecosystem services	Setting targets to achieve or maintain ecosystem services	Removing obstacles to ecosystem services market formation or growth
Tax rebate, charges, subsidies, funds and deposits, conservation covenants, carbon credits, biodiversity credits, payments for ecosystem service	Cap & trade – water market; biodiversity offsets	Biobanks, Green Wetland Bank, Eco labeling

Generic guidance for offset application

- Positive offset ratios should be used where there is a reasonable risk that the offset will not fully succeed over the long term. That is, the size of the offset to impact ratio should be larger than 1:1
- The risk that the offset action will fail to deliver the expected credits should also be considered as part of the mitigation plan.
- Mitigation replacement ratios should adjust to reflect a "temporal premium" for full benefits.
- Offsets 'for un-tradable' values at stake must not be an option and in areas of high biodiversity such as protected areas, offsets should be allowed only in exceptional circumstances.
- Offsets must recognize the dynamism of ecosystems, their complexity and uniqueness to sites.
- Project-by-project development of offsets in the same locality should be discouraged as far as possible as these limit opportunities for larger gains *(e.g. improved viability of populations, larger buffer zone to guard against potential disturbances from adjacent land uses)*.
- The biodiversity outcome is more important than ensuring equivalence of the actual ecosystem

Figure 8.3 Generic framework for the application of biodiversity offsets.

antidote. The outcome of offsets is likely to vary significantly depending on many different factors. These may include regulatory controls in place, status of other conservation measures already being implemented in natural areas, magnitude and significance of anticipated threats from upcoming development proposals, and the economic flexibility in the implementation of offsets. The practical experience of developing, implementing and evaluating the success and drawbacks of pilot offset projects (Pagiola *et al.*, 2002; Jenkins *et al.*, 2004; Ten Kate *et al.*, 2004; ICMM, 2003 and 2005a; NSW, 2006a) have brought planners, business groups, policy makers, governments, and the conservation community to a point of building a general consensus on guiding principles and ground rules for a globally acceptable framework for offset schemes. The key principles recommended by IUCN, BBOP, EBI and ICMM are based on practical experience (NSW, 2002; Ten Kate *et al.*, 2004; EPA, 2006). They provide a strong basis to the framing of the following ground rules, which are now almost universally accepted:

- The application of offsets should be necessitated in the context of only those developments that are legally appropriate and federally authorized, and where the developer has first used best practices to avoid and minimize harm to biodiversity.
- Offsets are no substitute for 'no go' areas.
- Offsets are not a project negotiation tool.
- Offsets must not reward ongoing poor environmental performance.
- Offsets should follow the principle of 'like for like or better' and therefore must result in a net conservation benefit.
- Offsets should follow the mitigation hierarchy.
- An environmental offset package should address both direct offsets and contributing offsets.
- Biodiversity offset should represent a conservation benefit that would not be possible without the investment companies' contributions and must overcome the impacts of a temporal gap between project impacts and offset benefits.
- Offsets must have local context and must be sensitive to indigenous people's rights.
- Offsets should be convincing and the impacts should be quantifiable.

Challenges in implementing biodiversity offsets

Despite the growing interest in offsets and their increasing application in diverse situations and locations (Anonymous, 2002b; IUCN and ICMM

2005a; World Bank, 2003; Energy and Biodiversity Initiative, 2004a; Ten Kate *et al.*, 2004), biodiversity offsets raise many scientific, social, political, legal, and economic questions, to which there are no easy answers. Regardless of the approach, the effectiveness of biodiversity offsets as a compensation tool is constrained by a number of methodological challenges and practical difficulties linked to their design and implementation. Some of these are discussed here.

Defining the nature of offsets

Although it is critically important to establish what would make an ideal offset package, in reality it is often difficult to make firm decisions on alternatives of in- kind and out-of-kind offsets. The argument for in-kind compensation favours that the best means of ensuring full and equivalent replacement of losses is to compensate with the same type of habitat, functions, and values. The 'out-of-kind' offsets allow stakeholders to experiment with different forms of compensation that may also result in new opportunities for conservation. Making choices about activities that offer exclusive benefits as direct or secondary offsets or a combination of these often pose a dilemma for planners and conservation community.

Identifying conservation strategies that make an offset

Biodiversity offset as a mitigation option, is seemingly a controversial approach and raises complex questions about the value, role, and replaceability of biodiversity. Most professionals working in areas of mainstreaming biodiversity in impact assessment feel that natural ecosystems are too complex for an assessment of the full range of impacts on their function, role, and richness that need to be compensated. Others argue that offsets can never adequately compensate for the immediate loss of unique habitats and species. The ability to choose from a portfolio of offsets that includes conservation-oriented actions and incentive-driven tools is often constrained by the lack of clarity about the conservation values of the site of proposed development and the national conservation priorities that should guide levels of compensation. Most countries have established their conservation priorities in National Biodiversity Strategies and Action Plans, but these have not been fully mainstreamed in environmental decision making policies and processes, leaving the evaluation processes open ended and dependent on judgment of EIA professionals.

Limitations of economic instruments in offsetting impacts on biodiversity

Lack of experience with economic instruments has also led to uncertainties about whether pure economic and development benefits can represent satisfactory compensation for biodiversity losses. This inexperience is compounded by the fact that biodiversity, with its various meanings for different people, has a strong public goods character that inhibits the development of markets for its products and services. Most biodiversity values are implicit rather than explicit and thus are often not captured in economic evaluation that form the basis of incentive-based offsets (OECD, 2003).

Defining the appropriateness of scales and location

Varied complexities of ecological processes and systems preclude simple broad-brush solutions and 'one size fits all' approaches to designing offsets. To achieve equivalent gains to losses caused by a particular project, the planning of offsets need to consider the best distance to the size of the impact offset ratio. Yet, determining the ratio is never easy as measuring the impacts of land use change on biodiversity is the most daunting task. The problem becomes further compounded when the planning of offsets is based on snapshot observations through rapid EIA studies. If such one-time assessments fail to capture the 'bigger picture' the offset ratio is not likely to yield 'no net loss' benefit.

The geographic flexibility offered by offsets enable conservation efforts to be focused on areas where the long-term conservation benefits are more likely. The identification of substitute land at a distance from the project site offers the greatest dilemma of ensuring how best the new site replaces or complements the intrinsic values of nature, functionality, conservation significance, endemism, and integrity of lost or degraded habitats. The issue of scale is similarly significant for offsets. Ten Kate *et al.*, (2004) observed that there are cases where a restored mine site, for instance, appears as a small oasis in highly degraded surrounding area. On the other hand the conservation goals in some sites are achievable by increase in mere numbers of breeding pairs of endangered birds species (e.g. the red cockaded woodpecker listed under the US Endangered Species Act) or improving the regenerative capacity of forest stands to protect individual plant species (e.g. *Rose mallee*, a species of Eucalyptus that is nationally important for New South Wales).

Timing of the offset

In dynamic ecosystems cause and effect relationships are not always easy to understand and establish. This leads to uncertainties in establishing the appropriate timing of applying offsets (before, after or along with the project implementation cycle) and the duration (extending to the term of the project or perpetuity) of offset activities. There are still no clear answers to when an offset should be applied in different situations. Some argue that damage to biodiversity should be allowed only after the offsetting activities are operational and have proven their effectiveness. This would ensure that there is no net loss at any period of time and the risk of the project failing to deliver the desired conservation outcomes is reduced. Others have argued that having the offsetting project operational before the damage is inflicted is essential for the viability of the model. This view takes into account the implications of aggravated damage due to time lags when the full effects of decisions on environment and biodiversity are not known. However, some strongly believe that offsets should be applied in perpetuity (Ten Kate et al., 2004).

The problem of defining currency to assess gains from offsets

Determining the unit of conservation – the 'currency' is almost a universal challenge. Without a common currency, it is difficult to ensure that a like is exchanged for like (or an acceptable equivalent) to assess the potential costs and benefits of using offsets. As no two areas are ecologically identical, the fundamental issue in designing offset is, 'whether we can confidently trade x for y'. According to Salzman and Ruhl (2002), unless the currency captures what we care about, we can end up trading the wrong things. This points out that 'trading' of project impacts for offset benefits' begs the questions of what the relevant values are, how we measure them and how we reflect them in a conveniently traded currency. A more commonly used currency is a measure of hectare for hectare. The quality measure is combined with a measure of area to create a measure for the offsets called 'Habitat hectares' (habitat score × area). The number of Habitat hectares needed for a given offset become dependent on the conservation significance of the area to be affected. Habitat Evaluation Procedures (HEP), a modelling system developed by US Fish and Wildlife Service that results in Habitat Units (HU) relate closely to the hectare for hectare approach as the currency for project/mitigation exchange. This approach has been extensively used in United States and also adapted for applications in many other countries. The constraints in extended use of this approach are that Habitat Suitability Index models of different species which actually form the basis of habitat modelling may not be available at the time of planning offsets. Developing HSI models also requires extensive research efforts.

Furthermore, if an offset is conducted in a similar ecosystem to that affected by development, the development may affect relatively 'mature' habitat, while offsets may involve rehabilitating or restoring habitat on comparatively degraded lands. In such scenarios, more biodiversity will be lost per hectare in the site developed than conserved in the offset site. This is likely to undermine the potential benefit of 'no net loss' and may necessitate that the offset area is larger or ecologically 'richer'. These obvious inconsistencies in the scale and currency are the inherent problems that make designing offsets a real challenge.

Barriers for implementing offsets

Some biodiversity experts, conservation groups, local communities, and other key stakeholders are of the view that biodiversity offsets will not genuinely result in 'no net loss' and are therefore generally suspicious about the intent and commitments made by business groups and government for offsetting. This invariably breeds fear among developers who start refraining from revealing the status of endangered species and their habitat and the actual benefits of environmental services provided by the biodiversity in the proposed development site or its surroundings. The planners fear that if they reveal the true conservation value of the development site, they would be subjected to more regulations and scrutiny by the public, decision makers, NGOs, and the media. For example, Shell Oil had invested in a carbon sequestration project in Indonesia but was more worried about being held accountable for the fate of the Orangutan population that lived there (Ten Kate et al., 2004). Many mining companies in India follow a similar kind of practice of excluding 'risk prone species' from being listed in the EIA reports of mining projects that are to be located in biodiversity rich areas. The obvious intent is to avoid rigorous scrutiny and huge conditional investments in conservation planning (based on the author's personal experience of appraisal of projects on behalf of the Ministry of Environment and Forests, Government of India).

Some argue that environmental services represent an attempt to translate every living thing into an 'own-able' commodity (BSR, 2005). Some environmental services that may not be very obviously recognisable (e.g. nitrogen fixation by bacteria) do not get accounted in economic valuation. As a result, the trading instruments used for offsetting biodiversity losses can often lead to misinterpretations, increased scrutiny and distrust of public who feel that the real benefits will always be short of promised benefits.

Designing or implementing biodiversity offsets can also raise unforeseen legal liabilities and consequently, make developers incur additional investments that may be several times more in proportion of the size of the offset. This becomes a major dampener for business groups in designing offsets that can offer effective mitigation.

Offsets pose the greatest risk of becoming the 'gate passes' for encouraging the approval of development projects that should not take place (e.g. destruction of unique habitats in protected areas or irreversible loss of ecosystem processes). Decision makers often fear that promises of compensating biodiversity losses by 'bigger and richer' offsets may become a 'permit' for trash EIA reports and may even become precedence for overselling of untested offsets.

Conclusion

Although the aforementioned barriers do pose genuine constraints in popularising the concept and practice of offsets, there can be no denial that the objective of offset is ideologically sound. There is a clear need to overcome the various challenges and barriers to achieve better levels of success with offsets for more and better biodiversity conservation outcomes. A greater effort on the part of developers, policy makers, and the conservation community is needed to strike synergies in the development of innovative approaches.

Learning from the growing volume of literature should be utilised to encourage practical experience through pilot projects. These will enable different players to demonstrate their results and to bring real experience to serve as an impetus to the discussion on the role of offsets. One of the recent and most active initiatives in this direction is the Business and Biodiversity Offsets Program (BBOP). BBOP is a partnership between companies, governments and conservation experts to explore biodiversity offsets. Through several pilot projects, BBOP partners are testing and disseminating best practice on biodiversity offsets through their website (www.forest-trends.org/biodiversityoffsetprogram/).

Building such partnerships will help business groups achieve landscape level conservation outcomes by pooling resources for the design of 'bigger and richer' offsets which will have greater benefits for the restoration of fragmented and isolated habitats.

For environmental professionals, EIA practitioners and business groups, the design canvas will always become larger as the retooling of offsets for newer models will have to remain an ever going effort.

9 · Valuation of ecosystem services: lessons from influential cases

Pieter van Beukering and Roel Slootweg

Introduction

The concept of ecosystem services has received significant attention since the appearance of the Millennium Ecosystem Assessment (2003). Ecosystem services are the benefits people obtain from ecosystems (Box 9.1). A growing body of knowledge is developing on ecosystem services. Knowledge institutes around the world have worked with the concept of ecosystem services for years already. Environmental economics have produced an impressive collection of valuation studies (more than 3,000 have been reported by Environmental Valuation Reference Inventory (EVRI)),[1] applying valuation techniques with ever increasing sophistication and reliability. Gradually the approach is being applied in practice, to support decision making and to guide development into a more sustainable direction. Yet, cases where economic valuation of ecosystem services has actually contributed to or exerted influence on strategic decision making on real-life policies, programmes, or plans remain scarce (Van Beukering et al., 2008). As Ehrlig already stated,

a general problem is the failure of ecological economists adequately to communicate their results and concerns to the general public and to decision makers. In view of the demonstrable failure of traditional economics to focus its attention on what will be the central issues of the twenty-first century, it is clear that ecological economics is in a position to become the central subdiscipline of economics. (Ehrlig, 2008)

So far, the Strategic Environmental Assessment (SEA) community has used the opportunities provided by ecosystem services as a means to translate the environment into societal benefits and link these to stakeholders even less. Even although Van Beukering et al. (2008) seriously looked for good SEA case material, only few SEA cases were available with a clear recognition of ecosystem services. In other words, it was

---

[1] Environmental Valuation Reference Inventory (EVRI): www.evri.ca/.

Box 9.1. Ecosystem services

The Millenium Ecosystem Assessment (MA) has subdivided ecosystem services into four categories: (i) *provisioning*, such as the production of food and water; (ii) *regulating*, such as the control of climate and disease; (iii) *supporting*, such as nutrient cycles and crop pollination; and (iv) *cultural*, such as spiritual and recreational benefits. Although not described as such by the MA, other categories have been recognised in scientific literature such as 'carrying' services (providing a substrate or backdrop for human activities) or 'preserving' services, which includes guarding against uncertainty through the maintenance of diversity (see Chapter 2 for more information).

Source: Millennium Ecosystem Assessment (2003).

extremely difficult to find good practical evidence that application of the ecosystem services concepts 'works' in the context of SEA. Yet, from personal experience in a number of cases we know it does work well in SEA. Therefore, 20 influential cases have been documented where the recognition, quantification, and valuation of ecosystem services have significantly contributed to strategic decision making. In all cases, the use of the ecosystem services concept supported decision making by providing better information on the consequences of new policies or planned developments. Ten cases have been elaborated in detail and are reproduced in the Annex of this book; ten additional cases providing further supporting evidence are reproduced in boxes in the Annex. In several cases SEA or a process similar to SEA was followed. Yet, in all cases, valuation of ecosystem services, in one form or another, resulted in major policy changes or decision making on strategic plans or investment programmes.

The aim of this exercise is to contribute to closing the gaps between the three main communities involved in the use of biodiversity in environmental assessment: (i) the ecologists and environmental economists predominantly based within knowledge institutes; (ii) the strategic environmental assessment community, consisting of competent authorities, consultants, and environmental agencies; and (iii) the decision makers at all levels of government. This chapter does not attempt to provide an exhaustive overview of all available approaches to (economic) valuation of ecosystem services, nor does it provide a scientific discussion on the pros and cons of various valuation techniques. This knowledge is readily accessible in many good publications, the most important of these is briefly summarised and cited in this chapter. What this chapter does is provide a number of cases where the valuation of ecosystem services has had a marked influence on concrete decision making; in other words, where the worlds of ecology, economy, environmental assessment, and decision making have actually met and produced tangible results. In a sense this is only a first attempt to collect and analyse such cases; given the limited available time and the limited availability of good cases we can only touch upon a number of striking issues. Because of the practical relevance of our initial observations we would like to invite the scientific community to search for more cases and apply in-depth analytical tools to learn much more from practice. This will reveal relevant research questions to direct future research efforts.

This chapter is structured somewhat differently than the other chapters. The main chapter text is relatively short and contains the results of the analysis of ten influential example cases. These cases are briefly introduced in the next section and make reference to the full case descriptions that are in the Annex of this book. The description of cases in the Annex constitutes the bulk of text of this chapter. It is, however, not necessary to read the cases to be able to understand the messages of the main text. As we have mentioned, the Annex further provides ten boxes with additional cases. We decided to present all this case material because of the great demand for practical evidence. Moreover, the cases provide hands-on experience of many of the issues introduced in this book.

This chapter provides a minimum background on valuation of ecosystem services, including a short 'how-to' description providing minimal requirements for the implementation of a valuation study. Based on the analysis of the case studies we have deliberately expanded the term 'valuation' to 'noneconomic quantification and societal valuation of ecosystem services'. As we will show, simple quantification or noneconomic valuation of ecosystem services can provide relevant information for decision making. The main messages obtained from the cases are presented in this chapter.

This chapter is a reworked version of two technical reports commissioned by the Netherlands Commission for Environmental Assessment, that is, Slootweg and Van Beukering (2008) and Van Beukering et al. (2008).

Influential cases

Because the potential of using ecosystem services as a means to translate the environment into societal benefits is not yet recognised fully by the

SEA community, there is a need for convincing evidence that such an approach is the right way to go. In our search for influential examples of this approach, we started with the creation of a long list of potentially relevant cases, all recognising ecosystem services and all having resulted in concrete decision making at the strategic level (above the project level). From this long list, ten cases were selected for further detailed analysis. This selection aimed at an even distribution over geographical regions and among different sectors, with a preference for cases from nonindustrialised countries. Because the most relevant material comes from industrialised countries, these are still overrepresented. It is also evident that cases linked to water or 'wet' environments are very dominant in the list of cases. Apparently, the multifunctional character of water triggers the need for an ecosystem services assessment. And, of course, the community of wetland experts has long promoted the multifunctional character of wetlands – for two decades the Ramsar Wetlands Convention has promoted the notion of *wise use* of wetlands, even before *sustainable use* became a commonly used term (see, for example, David, 1993).

Below, summaries of the main cases provide a minimum of background information to be able to position the studies; Table 9.1 provides an overview of these cases, with page references. Table 9.2 provides an overview of these supporting cases with page references.

Summary 1. *West Delta Water Conservation and Irrigation Rehabilitation Project (WDWCIRP) (Egypt, 2006).*

Ecoservices	Related to ground and surface water in desert area, Nile Delta, and coastal zone.
Valuation	Financial gains and losses linked to agricultural water supply quantified; other services quantified in terms of numbers of jobs or people affected.
Assessment	Voluntary SEA during planning phase of a public-private investment programme.
Decision	Magnitude, technical design, and conditions for resulting projects influenced.
Scale	West Delta region: investment initially planned for approximately 100,000 hectares.
Planning level	Private–public investment programme.
Sector	Water resources management and irrigation.

In the desert area west of the Nile Delta, groundwater-based, export-oriented agriculture has developed, with an annual turnover of about

Table 9.1 *Case studies summarised below, elaborately explained in the Annex.*

No.	Page	Study	Ecosystem	Country	Type	Valuation
1	334	Water Conservation and Irrigation Rehabilitation	Reclaimed desert and river delta	Egypt	Voluntary SEA	Financial gains and losses linked to agricultural water supply quantified; other services quantified in terms of labour or people affected.
2	343	Wetland Restoration Strategy	Wetlands	Aral Sea	SEA-like	Participatory MCA with semiquantified services for 6 alternatives. Full CBA of resulting pilot project.
3	352	Strategic Catchment Assessment	Watersheds	South Africa	Part of SEA process	Annual value of key ecosystem services quantified at the level of the municipality
4	358	Making Space for Water in Wareham	Coastal wetlands	United Kingdom	Experimental SEA	Quantification of impact on services, followed by valuation. Two sets of results: absolute values, and relative differences between baseline and alternatives and sensitivity analysis.
5	361	Climate policies and the *Stern Review*	Global	Global	Inform policy making	Comparison of the social costs of carbon emissions to society as a whole for a business as usual and a stabilisation scenario. Excess of benefits over costs, in net present value terms, would be US$2.5 trillion if strong mitigation policies were implemented this year.

(cont.)

Table 9.1 (*cont.*)

No.	Page	Study	Ecosystem	Country	Type	Valuation
6	368	Natural gas extraction in the Wadden Sea	Wetlands	Netherlands	Inform EIA and SEA process	Various cost–benefit analysis techniques, partly using contingent valuation.
7	374	Management of marine parks	Coral reefs	Dutch Antilles	Sustainable financing	Contingent valuation to determine the willingness to pay among reef users
8	380	Watershed rehabilitation and services provision	Forest	Costa Rica	Payments for Env. Services	Basic economic valuation techniques, such as the replacement cost method
9	387	Water scarcity and transfer	Rivers	Spain	Advocacy	Various valuation techniques combined in an extended cost–benefit analysis. Proposal compared to alternative desalinisation and water saving scenario.
10	387	Exxon Valdez oil spill in Alaska	Coastal resources	United States	Damage assessment	Travel cost methods, hedonic pricing, contingent valuation methods

Table 9.2 *Supporting cases summarised in boxes in the Annex.*

No.	Page	Study	Ecosystem	Country	Policy context	Valuation
1	342	Impact of dams on wetlands and livelihoods	Wetlands	Mali	Investment decision	Various techniques to determine external costs downstream, including production function approach and contingent valuation
2	342	Livelihood and conservation of Korup National Park	Tropical forest	Cameroon	Nature conservation	Potential loss in income of 187 villages when access to a park is closed
3	350	Large scale wetland restoration	Wetlands	Everglades	Nature conservation	Multiattribute choice model to evaluate preferences for restoration of the Greater Everglades ecosystem
4	393	Management of Durban's open spaces	Open spaces	South Africa	Environmental planning	Annual value of key ecosystem services quantified at the level of the municipality
5	364	Cost of policy inaction for biodiversity	Biodiversity	Global	Awareness raising	Economic impact of worldwide biodiversity loss resulting from policy inaction; approach similar to the *Stern Review*

(*cont.*)

Table 9.2 (*cont.*)

No.	Page	Study	Ecosystem	Country	Policy context	Valuation
6	367	Carbon offset investments in Iwokrama National Park	Tropical forest	Guyana	Investment decision	Total economic value of the entire forest, leading to a payment for ecosystem services scheme
7	371	Mangrove rehabilitation	Mangroves	Philippines	Nature conservation	Comparison of total value of conserved mangroves and alternative uses, taking in to account equity and sustainability objectives
8	379	Voluntary user fee system for divers	Coral reefs	Hawaii	Sustainable financing	Contingent valuation method (CVM)
9	381	Watershed rehabilitation for drinking water	Rural areas	New York	Payments for Environmental Services	Various valuation techniques
10	394	Penalty system for coral reef injury	Coral reefs	Florida/Hawaii	Damage assessment	Contingent valuation in the context of avoided losses of coral reefs through improved conservation

€500 million (US$750 million). However, the rate of groundwater exploitation by far exceeds the rate of renewal. Groundwater is rapidly depleting and turning saline. To reverse this situation the Government of Egypt has proposed a plan to annually pump 1.6 billion cubic metres of fresh Nile water from the Rosetta Nile branch into an area of about 45,000 hectares.

The use of SEA at the earliest possible stage of the planning process has guaranteed that environmental and social issues beyond the boundaries of the project area were incorporated in the design process. Valuation of ecosystem services focussed on the services linked to water resources under influence of the major driver of change, that is, transfer of water from the Nile to the desert area. Simple quantification techniques provided strong arguments for decision makers at the Ministry of Water Resources and Irrigation and the World Bank to significantly reduce the scale of the initial phase.

The diversion of water from relatively poor smallholder farmers in the Nile Delta to large investors in the desert west of the delta poses unacceptable equity problems. It was decided to follow a phased implementation of the plan, providing time for the National Water Resources Management Plan to be implemented, including its water savings programme. Short-term measures can produce necessary water savings to allow for the first, relatively small pilot phase of the WDWCIRP plan. Further water saving measures will provide room for further expansion.

Summary 2. *Aral Sea Wetland Restoration Project (Uzbekistan, 1996).*

Ecoservices	Restoration of wetland services for local livelihoods and health.
Valuation	Participatory MCA of strategy based on semiquantified ecosystem services for six alternatives. Full CBA of pilot project based on provisioning services.
Assessment	SEA integrated in a water resources management strategy development process.
Decision	Resulted in decision making by regional government and donor. One component successfully implemented.
Scale	Regional: Amu Darya Delta – approximately 12,000 km².
Planning level	Both plan (strategy) and project (pilot project).
Sector	Water resources and wetland management.

Intensification and expansion of irrigation activities in Central Asia led to the shrinking of the Aral Sea and the degradation of the Amu Darya Delta south of the sea. Loss of biodiversity, loss of vegetation and fisheries, the occurrence of salt and dust-laden winds, and salination of groundwater led to deteriorating living conditions. About 10 percent of the original wetlands remained in the delta, largely maintained by a mix of incidental floodwaters and saline drainage water flowing into constructed water reservoirs.

The Interstate Committee on the Aral Sea in consultation with the World Bank requested the development of a coherent strategy for the restoration of the Amu Darya Delta, broadly accepted by local stakeholders and government authorities and an investment programme of priority pilot projects. One pilot project, the restoration of the Sudoche wetlands, was designed in detail, and which, as of 2008, has been successfully implemented.

Valuation of ecosystem services was used in an SEA-type approach, as a means to structure the decision making process on a future development strategy of the delta. Valuation was instrumental in changing the course of development from technocratic and unsustainable interventions, towards the restoration of natural processes, which are much better capable of creating added value to inhabitants under the dynamic conditions of a water-stressed delta. The process created a strong coalition of local stakeholders and authorities, resulting in necessary pressure to convince national government and the donor community to invest in a pilot project.

Summary 3. *Strategic Catchment Planning at uMhlathuze municipality (South Africa, 2006).*

Ecoservices	Ecosystem services of subcatchments in hilly region under urbanisation pressure.
Valuation	Annual value of key ecosystem services quantified at the level of the municipality.
Assessment	Integrated Development Planning (legal requirement) must 'contain a strategic assessment of the environmental impact of the spatial development framework'.
Decision	Strategic Catchment Assessments were undertaken by the uMhlathuze Municipality to avoid conflict and time delays arising during EIAs.
Scale	Municipality.
Planning level	Plan.
Sector	Spatial planning.

Biodiversity issues in the South African City of uMhlathuze have led to various conflict situations. The classic 'development' versus 'conservation' situation exists, with the local municipality mostly in favour of development as a result of the poor socioeconomic climate. The area has, however, been identified as a biodiversity hotspot, and in order to alleviate the conflict and time delays that arise during Environmental Impact Assessments, the uMhlathuze Municipality opted to undertake a Strategic Catchment Assessment.

Instead of identifying and declaring conservation-worthy areas as 'no-go', the study stresses the ecosystem services that the environment provides free of charge to this Municipality. Nutrient cycling and waste management, water supply, water regulation, and flood and drought management are some of the most highly valued services. Wetlands have a particularly high value, relating to the high costs of trying to replace a vital but finite resource. The value of environmental services provided by all catchments was estimated at R1.7 billion (nearly US$200 million) per annum.

Politicians, known to be 'biodiversity averse', reacted positively once they realised that ecosystem services have an economic value. The Municipality embarked upon a negotiating process to identify (i) sensitive ecosystems that should be conserved, (ii) linkages between ecosystems, and (iii) areas that could be developed without impacting on the area's ability to provide environmental services. More importantly, (iv) it would identify the management actions that need to be implemented in the area in order to ensure not only the survival for key biodiversity assets, but also the sustainable use of biodiversity resources to benefit all residents of uMhlathuze.

Summary 4. *Wareham Managed Realignment case study (UK, 2007).*

Ecoservices	Estuarine tidal area: flood defence measures prevent flood damage or loss of land, and also create new habitats with multiple services.
Valuation	Quantification of services, followed by valuation: absolute value and relative differences between baseline and alternatives and sensitivity analysis.
Assessment	Experimental, government initiated study to enhance initial policy appraisal.
Decision	Need to decide on the cause of action in flood defences. Policy appraisal asked for changes in the flood risk-management regime (in progress).
Scale	Regional.
Planning level	Policy.
Sector	Flood defence.

This case study describes an analysis of the way ecosystem values are monetised, absolutely and relatively, in the Wareham Flood and Coastal Erosion Risk Management study. Economic values are applied to ecosystem service changes under different scenarios. The results (aimed to be practical guidance on how to conduct valuation of ecosystem services) will be used as input to a handbook on Economic Valuation of Environmental Effects in flood and coastal erosion risk management.

The main conclusion is that economic valuation of ecosystem services, even when a policy framework for incorporation of ecosystem services in a cost–benefit analysis is present, in daily practice still is difficult. Many uncertainties exist concerning scientific data, human economic behaviour, values, and methodological issues rising when transferring data from existing knowledge.

The case shows that even in situations with great potential for valuation of ecosystem services (a cost–benefit analysis is required for all coastal defence projects), practical implementation is difficult. However, the case also shows that valuation contributes to identification of a most favourable option and to reject other options.

Summary 5. *Climate policies and the Stern Review (UK, 2007).*

Ecoservices	Climate regulation and impact of global warming on all ecosystem services.
Valuation	Cost of climate change to society as a whole. Excess of benefits over costs, in net present value terms, would be US$2.5 trillion if strong mitigation policies were implemented this year.
Assessment	UK government initiative (the Chancellor of the Exchequer) to solve the UK's divide on the position regarding the Kyoto Protocol and the Intergovernmental Panel on Climate Change.
Decision	The UK Climate Change Bill introduced in Parliament; contains legally binding target for a significant reduction on UK carbon dioxide emission. Large impact beyond UK.
Scale	Global.
Planning level	National climate policy, but study led to many new initiatives around the globe.
Sector	Energy generation based on fossil fuels.

Changes in the global climate lead to fundamental changes throughout the world's ecosystems and therefore also affect the economic sectors that depend on these ecosystems. The *Stern Review* is one of the best-known assessments to estimate the economic impact of climate change. The 700-page report was prepared by a team of economists at

HM Treasury upon a request from the Chancellor of the Exchequer (the present PM Gordon Brown) to (i) address the lack of political consensus on climate change in the UK, (ii) to fill the gap in knowledge on the economics of climate change, and (iii) to resolve UK's divide on the position regarding the Kyoto Protocol and the Intergovernmental Panel on Climate Change (IPCC).

The main message of the *Stern Review* is that what we do now can have only a limited effect on the climate over the next 40 or 50 years, but what we do in the next 10–20 years can have a profound effect on the climate in the second half of this century. In other words: the benefits of strong, early action considerably outweigh the costs. Each tonne of carbon dioxide emitted causes damages worth at least US$85. At the same time, emissions can be cut at a cost of less than US$25 a tonne. Shifting the world onto a low-carbon path could eventually benefit the economy by US$2.5 trillion a year. Stern characterises climate change as 'the greatest and widest-ranging market failure ever seen'. The *Stern Review* has been heavily criticised by some economists but is supported by many others. The low discount rate, causing future economic losses to way heavily in net present values terms, was one of the main points of criticism.

The *Stern Review* attracted more attention than any other economic valuation study in history. Influential people from all over the world were inspired by the *Stern Review* to stress the urgency of immediate action. The most significant impact of the *Stern Review* was seen in the policy arena. A number of governments responded by announcing expansion of their climate policies. In the UK, the Climate Change Bill was introduced in Parliament in 2007. It will shortly go to the House of Commons. The Bill contains provisions that will set a legally binding target for reducing UK carbon dioxide emissions by at least 26 percent by 2020 and at least 60 percent by 2050, compared to 1990 levels.

Summary 6. *Extraction of natural gas from the Wadden Sea (the Netherlands, 2006).*

Ecoservices	Risks for biodiversity, fishery, recreation versus revenues from natural gas.
Valuation	Various CBAs, also using contingent valuation techniques.
Assessment	CBAs, EIA for gas exploitation and SEA for planning decision.
Decision	Gas can be extracted under strong precautionary conditions.
Scale	National.
Planning level	Mega project, within boundaries of planning process (key spatial planning decision).
Sector	Energy.

The Dutch Wadden Sea is a shallow, semienclosed tidal flat, part of the largest tidal wetland area in Europe and bordering the North Sea. An estimated 200 billion cubic metres of gas are located below the Wadden Sea. The Wadden Sea is a wetland of international importance under the Ramsar wetland convention, part of European Nature 2000 network, and a National Park.

Opponents to the exploitation of gas argued that the proponent in its EIA did not take into consideration the effects on ecosystem services, such as water regulating, drinking water supply, tourism, fisheries, and so forth. They pointed out that the economic value of these services had been underestimated in previous studies. Therefore, they conducted an economic valuation study of the Wadden Sea, including a cost–benefit analysis (CBA) of gas exploitation. Estimations of damage to ecosystem services, in case serious effects would occur as a result of gas exploitation, were estimated at €1.1 billion.

In December 1999, the government eventually decided, based on the precautionary principle, not to give permission for gas exploitation. However, research and discussion on the effects of gas exploitation on soil subsidence continued. In 2003, the government appointed an advisory committee. The committee concluded that there are no ecological reasons to prohibit gas exploitation. Due to natural dynamics and the supply of sand and mud from the North Sea, the effect of the main driver of change, that is, soil subsidence resulting from gas exploitation, will be balanced by increased sedimentation and soil accretion. The committee therefore recommended that gas exploitation from the Wadden Sea could take place under strict conditions. Gas has been extracted since February 2007.

Summary 7. *Self-financing of marine protected areas in the Netherlands Antilles (2005).*

Ecoservices	Supporting and cultural services of coral reefs.
Valuation	Significant willingness to pay among users of reefs for better conservation of marine areas.
Assessment	Economic valuation study played crucial role in policy design decision making.
Decisions	Establishment of self-funded management system for marine parks.
Scale	All Netherlands Antilles islands.
Planning level	Policy.
Sector	Tourism/nature conservation.

Bonaire and its marine park are representative of the issues facing many marine protected areas in the Caribbean. The case explicitly combines analysis of ecological and economic factors. Bonaire's coral reefs, humid elfin forests, and semidesert scrublands represent an irreplaceable tourism resource – the most important source of income of the Caribbean island. Good management requires funds, but funding has in the past been plagued by instability and deficits. Economic valuation studies helped to establish an effective and sustainable revenue generation system. Bonaire's marine park is now among the best-managed park in the region.

A contingent valuation survey was conducted to establish willingness to pay user fees for the marine park resulting in an average value for willingness to pay (WTP) of US$27.40. This exceeded the relatively modest US$10 fee instituted in 1992. The difference between what people would be willing to pay for an ecosystem service and what they actually paid amounted to US$325,000 annually.

With the introduction of new legislation all the users of the Bonaire National Marine Park, not solely the divers, pay a user's fee. The most significant changes include admission fees to the marine park also admit entrance to land-based Washington/Slagbaai National Park. Price tags for divers changed to US$25 for a year pass or US$10 for a day pass. Swimmers, board sailors, and all other users of the park are required to pay US$10 for a year pass. Recently, it was decided that tag receipts go directly to the park management organisation and are used entirely for the management of Bonaire's National Parks.

Summary 8. *Payments for Environmental Services in Costa Rica (1997).*

Ecoservices	Forests guaranteeing stable water supply (provisioning service).
Valuation	Basic economic valuation techniques, such as replacement cost method.
Assessment	Valuation studies showed economic feasibility of a Payments for Ecosystem Services (PES) scheme through a change in tax policy.
Decisions	Costa Rica pioneered the development of PES as formal government policy.
Scale	National.
Planning level	Tax policy.
Sector	Forestry sector.

In the last two decades, Costa Rica transformed from one of the most rapidly deforesting countries in the world to one of the foremost pioneers in reforestation, forest management, and forest protection. One of the driving forces was the Payments for Environmental Services (PES) programme, initiated in 1997, becoming the first country-wide PES programmes in the world, and the first to adopt the terminology of environmental services and PES. Since its inception, it has become a point of reference for environmental authorities and practitioners around the world, as well as becoming one of the pillars of Costa Rica's image as a 'green' country that is a model for sustainable development.

The programme was fostered by the 1996 changes in the Forest Law that created the legal framework to pay landowners for the provision of four types of ecosystem services: (i) carbon sequestration, (ii) watershed protection, (iii) scenic beauty, and (iv) maintenance of biodiversity. Later public water supply was added to these. The primary funding source for the original PES programme was a 15-percent consumer tax on fossil fuels. Later, 3.5 percent of the tax revenue was directly assigned to the PES programme. As of 2003, such tax revenues provided an average of US$6.4 million per year to the PES programme.

In several studies the value of Costa Rican forests has been calculated. These studies showed that in the most pessimistic distribution of benefits (from the Costa Rican perspective) 66 percent of the environmental services are enjoyed by the global community (US$137 million) and only 34 percent by Costa Rica (US$71 million). Conclusion: the value of environmental services is high, the global community receives the major benefits of these services, and owners of the resources that provide these services are not compensated for their full value.

Summary 9. *National Hydrological Plan/Ebro water transfer works (Spain, 2006).*

Ecoservices	Wetland biodiversity, fisheries, aquaculture, groundwater supply in Ebro Delta.
Valuation	Various valuation techniques in an extended cost benefit analysis, comparing the proposed plan with an alternative, more sustainable scenario.
Assessment	Independent valuation study, responding to serious societal concerns.
Decisions	Financing by EU rejected; after elections alternative plan launched.
Scale	Water transfer between river basins (national).
Planning level	Mega infrastructure plan.
Sector	Water/agriculture.

The Spanish National Hydrological Plan (SNHP) was passed into law in July 2001. The chief objective of this €4.2 billion (US$6.3 billion) plan was the transfer of water from the Ebro Basin to four other river basins in the east of Spain. These water transfers would lead to serious impacts on the Ebro River. Ecosystem services in the Ebro Delta produce an annual turnover of €120 million (US$180 million) from fisheries, aquaculture, agriculture, and tourism. A part of the Ebro Delta is an important wetland designated as Natura 2000 and Ramsar site. The Plan merely stated that the transfer would not have any impacts on the economic activities of the donor basin, nor would it have any negative consequences on population distribution in the regions within the donor basins.

The Plan claimed to comply with the requirements of the European Water Framework Directive. However, extensive analyses indicated that on economic and environmental terms the Plan was not compatible. Aragón and Cataluña, two regions in the Ebro basin, strongly opposed the Plan. In terms of sustainability, numerous analyses indicated that the environmental and the economic principles were mostly ignored. The Plan was also questioned because of its lack of assessment of social issues. The University of Zaragoza showed the real costs of the SNHP were highly underestimated, in fact the SNHP made a negative contribution to economy of €3.5 billion (US$5.3 billion).

The lack of proper estimates of the real costs and benefits associated with affected ecosystem services strongly influenced decision making with regard to the plan. Critics agreed that additional studies were needed for a proper economic evaluation of the impacts of the water transfer. Before the European Commission could take a (probably negative) final decision on its support, Spain's newly elected socialist government cancelled the SNHP and launched a new water policy, strongly recognising the economic value of ecosystem services of rivers and wetlands.

Summary 10. *Compensation payments after Exxon Valdez oil spill (Alaska, USA, 1991).*

Ecoservices	Services supporting marine and coastal biodiversity, tourism, and fisheries.
Valuation	Travel cost methods, hedonic pricing, contingent valuation methods.
Assessment	The use of survey research (e.g. CVM) became a well accepted appraisal method as a result of the complex valuation problems associated with contamination.
Decision	Court awarded US$287 million actual damages and US$5 billion
(1991)	punitive damages.
Scale	Considered one of the most devastating environmental disasters ever at sea.
Planning level	State and national regulations.
Sector	Nature conservation, tourism, fisheries.

On 24 March 1989, the oil tanker Exxon Valdez ran aground near the coast of Alaska. Approximately 38,800 tonnes of oil were spilled on 9,000 miles of shoreline. It is one of the most-studied environmental tragedies in history and can be considered extremely influential in changing policies. The accident also led to the ultimate recognition of the validity of economic valuation studies in environmental damage assessments.

Immediately after the oil spill the U.S. government and Alaskan government began a series of studies – the Natural Resource Damage Assessment – to determine the effects of the oil spill on the environment. The studies were designed to support: (i) the development of restoration plans to promote the long-term recovery of natural resources and (ii) the determination of damages to be claimed for the loss of services of the natural resources.

Ultimately, five ecosystem services were valued in economic terms: (i) replacement costs of birds and mammals, (ii) losses in recreational fishing, (iii) sport fishing losses, (iv) tourism industry, and (v) contingent valuation of lost passive-use values. The contingent valuation measured the loss of option values, existence values, and other nonuse values. Respondents were then asked their willingness to pay for a realistic programme that would prevent with certainty the damage caused by a new oil spill. The median household willingness to pay for the spill prevention plan was found to be US$31. Multiplying this number by an adjusted number of U.S. households resulted in a damage estimate of US$2.8 billion dollars.

On October 8 1991, Exxon agreed to pay the United States and the State of Alaska US$900 million over ten years to restore the damaged resources and the reduced or lost services (human uses) they provide. Exxon was fined US$150 million, the largest fine ever imposed for an environmental crime. The court forgave US$125 million of that fine in recognition of Exxon's cooperation in cleaning up the spill and paying certain private claims.

Valuation of ecosystem services

A major worry among planners and decision makers is the time and costs involved in environmental assessment; similarly so for valuation studies. Full-fledged valuation studies are thought to be time consuming, as large amounts of data need to be collected. The practice of EIA and SEA has shown that environmental assessment can be done at any required level of detail, varying from a 'back-of-an-envelope' assessment to a comprehensive *Stern Review*–like evaluation. Moreover, approaches have been developed to be able to support decision making, even in cases where data are scarce or incomplete. More strongly stated, environmental assessment, by definition, has to deal with incomplete information that must be collected in a limited amount of time, within the limits of a budget more or less defined by the magnitude of the project under study.

The analysis of cases in this study has produced results similar to experience from the field of environmental assessment. Valuation studies can be done in great detail and at great length and costs (such as the Exxon Valdez case and the *Stern Review*), but they can also be applied in a very rapid and cost effective manner (most of the other cases). Full information and knowledge is not always needed to be able to provide relevant information for decision making. When comparing alternatives it usually is sufficient to know relative values: what alternative performs better in comparison (qualitative)?; does an alternative perform much better, or only a little better (semiquantitative)? Absolute values are not always needed.

By and large, there are four reasons to value ecosystem services (Van Beukering *et al.*, 2007):

(1) Advocacy: economic valuation is often used to advocate the economic importance of the ecosystem services, with the ultimate purpose of encouraging sustainable development. For example, by demonstrating that the economic values of threatened ecosystem services have previously been underestimated, it can be argued that the ecosystem should receive more attention in public policy.
(2) Decision making: valuation can assist the government to allocate scarce resources to achieve economic, environmental, and social goals. Decision makers constantly operate under short time frames, their windows of opportunity are limited by the election cycle, and they often have to take decisions without full information. Economic valuation studies are critical to assist decision makers make fair and transparent decisions.
(3) Damage and risk assessment: valuation is increasingly used as a means of assessing damage inflicted on an ecosystem, and the risk thereof. Damage assessment has been used in many cases to asses the compensation owed after oil spills by large ships and after accidents in mining companies that lead to tailings dam leakages or other toxic waste spills. Risk assessment is increasingly used to determine the value of risks related to a recognised threat such as climate change.

(4) Sustainable financing: valuation of ecosystem services can be used to set taxes or charges for the use of those goods and services. Setting taxes or charges have a double role in terms of environmental management. They help to control the extent to which environmental resources are exploited (i.e. the more a resource costs the less it is used) and simultaneously generate revenue that can be used to pay for management, protection and restoration of the ecosystem. Valuation results can be used to set taxes or charges at the most desirable level.

The presented case studies fall within one or several of the categories above. They show a wide variety of forms in which ecosystem services can be recognised, quantified, and valued and represent most of the commonly applied valuation techniques. We have created a rather straightforward classification of ways in which ecosystem services are represented or valued in the cases, ranging from simple recognition to full fledged economic valuation.

Identification and recognition

The simplest way of paying attention to ecosystem services is the qualitative listing of services in studies to support decision making. It raises awareness on issues that may not have been thought of before. Most studies paying attention to ecosystem services start with a listing of services. More often than not the actual quantification and valuation of services is done only for the easiest and/or the most important services. Others simply remain listed.

How: Identification of ecosystem services involves experts with knowledge of the area, whose preliminary identification of potential ecosystem services is checked with local stakeholders or representative bodies for these stakeholders.

Who: Most important is to have people with the right 'mind set' to recognise ecosystem services. More often than not, sector-oriented experts tend to overlook the effects their plans may have on ecosystem services linked to other sectors. A mix of natural resources management experts and ecologists with good local knowledge works well.

Data needs: maps indicating main ecosystems and types of land use; overview of main economic activities in the area; population data; field reconnaissance.

Time required: for the actual study only several days. The decision to actually spend attention to ecosystem services may take longer as competent authorities or proponents need to be convinced of its usefulness (see Ebro case).

Quantification of ecosystem services

Ecosystem services can be quantified in units of measurement directly linked to the service. Units of measurement have a very broad range. Some examples: quantity of renewable water supply for an aquifer, annual sustainably harvestable fish or timber or fruits in certain area, amount of agricultural produce per hectare, amount of carbon stored per hectare of forest, number of species occurring in certain area, and so forth.

Based on the ecosystem services identified in Step 1, a selection of the relevant ecosystem services to be quantified can be made. Selection highly depends on the purpose of the study and can be part of a scoping process, where also the required level of detail can be defined. An impact-oriented assessment will focus on the main drivers of change resulting from an activity and highlight potentially affected ecosystem services (see the Wadden Sea and Egypt cases). A spatial planning–oriented type of assessment may try to identify ecosystem services with opportunities for development or relevant services with major constraints (see the South Africa and Aral Sea cases). Management planning focusses on the purpose of management (see the Costa Rica – forest management for water supply and Antilles – coral reef management for tourism cases).

How: quantify an ecosystem service in units of measurement relevant to the service. Some examples: the amount of sustainably harvestable fish from a water body; the number of scuba divers a coral reef can handle without unacceptable damage; the amount of renewable water to be extracted from an aquifer; the percentage of the world population of a threatened bird species making use of a wetland area; amount of agricultural produce per hectare; amount of carbon stored per hectare of forest, and so forth.

Who: full quantification may involve experts supported by computer models (hydraulic, population, harvest, preferences). Proxies can be obtained from national or regional statistics, local stakeholders, narrative information, and data from similar services elsewhere.

Data needs: national or regional statistics often provide good information; remote sensing information may provide relevant information on surface areas and productivity. Research institutes may provide access to computerised models. A reality check with people on the ground is always recommended.

Time required: from a week to several months, depending on the level of detail required, number and complexity of the services to be assessed, the surface area, availability and reliability of statistical data, and presence of local (scientific) information. See Boxes 9.2 and 9.3 for two practical examples of SEAs in which ecosystem services were quantified and

Box 9.2. Practical aspects of Egypt SEA

Duration: three months.
Time expenditure: three expatriate and two local consultants for one month each and farm surveys by local agricultural extension workers. *Cost of SEA study*: approximately US$80,000, on a total estimated plan budget of around US$100 million.

As a result of good coordination the study was fully integrated in the planning process which did not experience any delays. Data were obtained from project planning documents, government statistics, farm surveys, two existing computational groundwater models and surface water models and a number of additional field visits and on-farm interviews for verification. Two stakeholder workshops provided relevant scoping information and discussion on the outcome of the study. The level of detail and reliability of information was sufficient to guide the planning process. Where links between hydrological changes and impacts were very difficult to quantify in economic terms, the impact description was limited to the identification of numbers of affected people.

The subsequent detailed technical design was subject to a full fledged ESIA, which could at a later stage zoom in on a limited number of issues to provide more detailed information.

Box 9.3. Practical aspects of integrated SEA in the Aral Sea case

Duration: strategy development, including all preparatory studies, participatory process, and environmental assessment – 12 months.
Time expenditure: 1 permanent expatriate project leader; 3 permanent local experts; 6 expatriate experts – 2 visits of 1 month each; 12 hired local scientists, 3 months each.
Total costs: US$1 million (impossible to separate the SEA components). Investment cost for the proposed programme of projects was US$20 million. The Sudoche pilot project was implemented at an approximate cost of US$4 million.

Ecosystem services were quantified in semiquantified terms; some were valued in societal terms. Level of detail was sufficient for MCA exercise. Discussing values expressed in their own terms, and more importantly, recognising stakeholders for each ecosystem service did not distract the discussion to aggregated figures on money.

In a later stage, when concrete investment projects were proposed, cost–benefit analysis was the proper tool to provide sufficient and convincing arguments that the investments are justified.

valued. Study duration, number of people involved, and available budget greatly varied, because the Egypt example represents only a small part of the planning process, while the Aral case was a complete strategy development process, with SEA integrated into the process.

Societal valuation

Society places a value on ecosystem services. The quantities in which ecosystem services are expressed can be translated into values for society. This does not necessarily mean values have to be directly expressed in monetary terms. Values can also be expressed in social or ecological terms, represented in the conceptual framework of Chapter 4. The Ramsar Wetland Convention provides a similar approach: three main types of values are defined which together determine the Total Value (or importance) of wetlands. These are (i) ecological, (ii) sociocultural, and (iii) economic values. Each type of value has its own set of criteria and value units (de Groot et al., 2006).

Examples of social values are the number of households depending on a service, the number of jobs related to a service, and the number of people protected against forces of nature. Ecological values can relate to the number of threatened (red-listed) species in an area; the importance of an area as living repository of wild ancestors of agricultural crops; or the contribution certain area makes to the maintenance of other areas (e.g. marine fish reproducing in coastal wetlands; the importance of wetlands as stopover locations for migratory birds). Some values may be difficult to quantify in their own terms; examples are the religious value or the historical value of certain ecosystem features. Contingent valuation may in such cases provide estimates of economic value (see the next section).

How: quantify the societal value of an ecosystem service in units of measurement relevant to the value. Examples of social values are the number of households depending on a service, the number of jobs related to a service, and the number of people protected against forces of nature. Ecological values can relate to the number of threatened (red-listed)

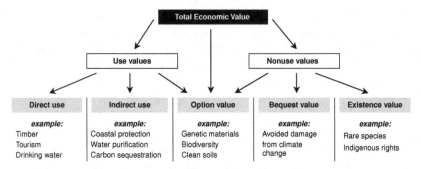

Figure 9.1 Composition of the Total Economic Value (TEV) of ecosystems into use and nonuse values (source: Beukering et al., 2007).

species in an area, or the number of wild ancestors of agricultural crops for which an area serve as living repository, or the contribution certain area makes to the maintenance of other areas (e.g. marine fish reproduce in coastal wetlands).

Who: for full quantification, labour-intensive questionnaires may be needed. Sampling with good statistical analysis provides a means to reduce workload.

Data needs: proxies can be obtained from national or regional statistics on population size, economic activities, agricultural outputs, fisheries and forestry productivity, and so forth.

Time required: from a week to several months, depending on the level of detail required, number and complexity of the services to be assessed, the surface area, availability and reliability of statistical data, and presence of local (scientific) information.

Economic valuation

Different economic valuation methods exist to value the range of benefits provided by different ecosystem services (see Figure 9.1). The selection of which method to use depends on a number of aspects. For instance, when planning a valuation study, it is necessary to balance the benefits of using the best scientific and analytic techniques with the financial, data, time, and skills limitations to be faced. Realise that no single method is necessarily the best; for each application it is necessary to consider which method(s) is the most appropriate. Sometimes a number of different methods is to be used in conjunction in order to estimate the value of different services from a single ecosystem.

Table 9.3 *Methods for estimating values (based on Mitchell and Carson, 1989: 75).*

	Observed behaviour	Hypothetical
Direct	Undistorted market prices Simulated markets	Contingent valuation Choice experiments
Indirect	Hedonic pricing Travel cost method Production function approach Avoided damage approach	Contingent ranking

Advancements in environmental economics have provided tools to monetise the values of ecosystem services, even in absence of a functional market for services. Table 9.3 shows a classification of valuation methods (Mitchell and Carson, 1989). It distinguishes two dimensions. For the first dimension, the columns make a distinction between values derived by observing people operating in the real world where the consequences of their choices are felt by these same people and preferences revealed by raising hypothetical questions. The second dimension is whether monetary values are estimated directly or whether these values are inferred through indirect valuation techniques.

The valuation techniques in three out of the four quadrants in Table 9.3 are known to be the most commonly applied techniques. (The quadrant containing 'contingent ranking' is not included.) These categories are also labelled as:

(a) *Market-based valuation*: goods traded in an open market have a price, which serves as the basis for valuation. Similarly, the effect of services can be priced using market prices. For example, coastal mangroves or dunes protect the inland and thus avoid damage to infrastructure and economy. A valuation technique that commonly applies market values is the 'net factor income approach'.

(b) *Revealed preference methods*: people's behaviour can reveal the value attached to a service. For example, waterfront houses in the Nether-lands are significantly more expensive than comparable houses elsewhere; or people spend money to travel to certain places that have something special to offer, such as national parks. Examples of commonly used revealed preference techniques are the 'hedonic pricing', 'avoided damage approach', and 'travel cost method'.

(c) *Stated preference methods*: value nonmarket resources, such as environmental preservation or the impact of contamination. Although these resources do give people utility, certain aspects of them do not have a market price as they are not directly sold. For example, people receive benefit from a beautiful view of a mountain. Contingent valuation and choice modelling are techniques used to measure these aspects.

A special case of valuation is the value transfer. Values obtained from studies in comparable areas and/or comparable situations can be transferred to another situation. Although value transfer avoids time-consuming data collection efforts, the accuracy of the estimates is generally limited. Valuation transfer is typically applied to determine the value of particular ecosystems (e.g. wetlands, coral reefs), as well as the economic importance of specific ecosystem services (e.g. provision of drinking water, flood protection).

How: in the context of ecosystem services, it is crucial to start identifying the providers and the beneficiaries of the relevant ecosystem services. Next, valuation techniques need to be selected. This choice is context specific and dependent on a number of factors, including whether or not the environmental service is traded directly or indirectly in a market, the stakeholders that hold values for the service, the available budget for conducting a valuation study, and the availability of existing information on the value of similar resources.

Who: it is advisable to have at least one environmental economist in the team who is properly trained to conduct economic valuation studies. The actual implementation of surveys and interviews can be conducted by noneconomists as well. However, for the design and analysis of the data, thorough economic knowledge is essential.

Data needs: in economic valuation, there are broadly three main types of data that will be used: (i) market prices that can be found from private sector sources, government statistics or international organisations; (ii) local social, environmental, and economic information that can be found through local surveys or government statistics where they exist; and (iii) preference data that are generated by asking people through questionnaire surveys. The categories are described in detail in Van Beukering *et al.* (2007).

Time required: depending on the comprehensiveness of the study, a valuation exercise may vary from a few months to two years or more. Obviously, the data availability present at the start of the study is a major factor in this regard. An illustration of the time and budget needed for economic valuation is provided in Box 9.4.

Box 9.4. Examples of planning and budget for valuation studies

To provide a sense of how long studies can take (from the shortest to the longest) some of the time taken to complete a variety of studies and the resources used to complete them are shown below (from Van Beukering *et al.*, 2007: 113)

Examples of case studies conducted for Hawaii and the Philippines

	Case study 1	Case study 2
Type of valuation exercise	WTP for conservation among 750 visitors	Total Economic Value (TEV) study on mangrove rehabilitation
Location of valuation exercise	Hawaii	Philippines
Type of activities	Survey at dive shops and on tour boats	Surveys, country statistics, scientific literature
Number of people involved	One economist, four interviewers, one data-enterer	Three economists, one social scientists, one biologist, four interviewers
Total human resources used	80 man days	300 man days
Total cost	Total US$30,000[a]	Total US$100,000[b]
Time taken	4 months	16 months

[a] Questionnaire US$5,000, Interviewers US$8,000, Data-entry and cleaning US$1,000, Analysis US$7,000, Report writing US$4,000, Travel costs US$5,000.

[b] Questionnaires US$7,500, Interviewers US$20,000, Data-entry and cleaning US$21,500, biodiversity assessment US$10,000, Data purchase US$2,000, Analysis US$20,000, Report writing US$15,000, Travel costs US$15,000, Policy brief US$5,000.

Economic valuation techniques

Market prices

The most commonly used method for valuing goods and services is to look at the market price of each. In a competitive market without market failures prices reflect their true marginal value (i.e. the value of a small change in the provision of that good or service). *Market prices* are therefore useful for valuing ecosystem services that are directly traded in markets, for example, products such as timber, fuel wood, fish, and other foods.

The major advantage of this technique is that it is relatively easy to apply, as it makes use of generally available information on prices and only requires simple modelling and few assumptions. A major disadvantage is that many environmental goods and services are not traded directly in well-functioning markets and so readily observable prices for them are not available. If markets for environmental goods and services do exist but are highly distorted, the available price information will not reflect true social and economic values and cannot be used. The main sources of market distortion are taxes and subsidies, noncompetitive markets, imperfect information, and government-controlled prices (Krugman and Wells, 2006). The market price method is therefore straightforward and inexpensive to apply and is particularly relevant for valuing ecosystem services when market prices are available in 'non-distorted markets'.

Net factor income

The *net factor income method* estimates the value of ecosystem services as an input in the production of a marketed good. It estimates the value of an ecosystem input as the total surplus between revenues and the cost of other inputs in production. For example, the value of a coral reef in supporting reef-based dive recreation should be calculated as the revenue received from selling diving trips to the reef, minus the labour, equipment and other costs of providing the service. The net-factor income method is likely to be useful for valuing many recreational ecosystem services such as the support of tourism. It is a simple method to apply and uses generally available data.

Replacement cost

The *replacement cost method* estimates the value of ecosystem services as the cost of replacing them with alternative man-made goods and services (Freeman, 2003). Basically, it is assumed that the amount of money society spends to replace an environmental asset is roughly equivalent to the lost benefits that asset provides to society. For example, the value of a wetland that acts as a natural reservoir can be estimated as the cost of constructing and operating an artificial reservoir of a similar capacity.

The replacement cost method is particularly useful for valuing ecosystem services that have direct man-made or artificial equivalents, such as water storage or wastewater processing. The method is also relatively simple and inexpensive to apply. It does not require the use of detailed surveys or complex analysis. However, the replacement cost method does not produce a strictly correct measure of economic value. After all the measure is not based on people's preferences for the goods and services being valued, but on the assumption that if people pay to replace a lost ecosystem service, then that service must be worth at least the cost of replacement. This method is therefore most appropriately applied in cases where replacement expenditures have been, or will be, made. A key weakness of this technique is that it is often difficult to find exact replacements for ecosystem services that provide an equivalent level of benefits; moreover, ecosystems often provide multiple services simultaneously. The replacement cost method is a useful tool for valuing ecosystem services, such as water storage and purification and coastal protection in a straightforward way.

Damage cost avoided

Ecosystems frequently provide protection for other economically valuable assets. The *damage cost avoided method* uses either the value of property and assets protected, or the cost of actions taken to avoid damages, as a measure of the benefits provided by an ecosystem. For example, if a mangrove forest provides protection to coastal areas from storm damage, the value of the coastal protection function of the mangrove forest may be estimated as the damages avoided.

The damage cost avoided method is particularly useful for valuing ecosystems that provide some form of natural protection. A potential weakness of the method is that in most cases estimates of damages avoided remain hypothetical. They are based on predicting what might occur under a situation where ecosystem services decline or are lost. Even when valuation is based on real data from situations where such events and damages have occurred, it is often difficult to relate these damages to changes in ecosystem status. The damage cost avoided method provides an effective approach to estimate the value of protection services by ecosystems.

Production function

The *production function method* estimates the value of a nonmarketed ecosystem service by assessing its contribution as an input into the production process of a commercially marketed good. A production function estimates the functional relationship between inputs and outputs in production. For example, the production of fruits and nuts from a forest may be described as a function of hours spent harvesting (labour) and the area and quality of the forest. A change in the availability of an ecosystem service may result in both a change in total output and a change in the use of other inputs. For example, a reduction in the area of forest may result in either a reduction in the harvest of fruit or/and an increase in the number of hours spent harvesting a given quantity. One study found that forest-based pollinators increased coffee yields by 20 percent within 1 km of forest (Ricketts *et al.* 2004). In theory, the production function method is well-suited to value ecosystem services, because it is based on the notion that ecosystem services and economic benefits are strongly linked. However, in practice the production function valuation method is technically difficult to apply and has substantial data requirements.

Hedonic pricing

The basic premise of the *hedonic pricing method* is that the price of a good is related to its characteristics, including its environmental characteristics. The hedonic pricing method should therefore be used to estimate economic values of ecosystem services that directly affect the price of marketed goods. The hedonic pricing method is often used to value environmental amenities that affect the price of residential properties. For example, a house that is close to an aesthetically pleasing natural area is often worth more than a similar house that is further away. Such differences in house characteristics and prices may be used to identify the value of natural amenities by employing statistical methods. The first economist to attempt to estimate potential links between real estate values and environmental quality was Ridker (1967) finding evidence that air pollution negatively affects property values. One year later, Strotz (1968) also found evidence that (environmental) land improvements can benefit property values.

Hedonic property value studies use statistical regression methods and data from real estate markets to examine the increments in property values associated with different attributes. Structural attributes (e.g. number of bedrooms and age of house), neighbourhood attributes (e.g. population demographics, crime, and school quality), and environmental attributes (e.g. air quality and proximity to hazardous waste sites) may influence property values. When assessing an environmental improvement, it is essential to separate the effect of the relevant environmental attribute on the price of a housing unit from the effects of other attributes. Similar to the production function approach, the hedonic pricing method may be difficult to apply for valuing ecosystem services in poorly documented areas due to its high complexity of analysis and large data requirements.

Travel cost method

The *travel cost method* is used to estimate the value of ecosystems or sites that are used for recreation. The premise behind this method is that the travel expenses that people incur to visit a site represent the 'price' of access to the site. Travel expenses include the actual travel costs (e.g. price of using public transport, petrol and maintenance for travel by private car, aeroplane ticket etc.), time costs, and admittance fees. For example, for a forest that is used for recreation, information on the number of people that visit the site and the time and cost they spend travelling to reach it can be used to estimate the economic value of the recreational service that is provided. Freeman (2003) provides an overview of studies that measured the value of site quality through recreational activities, such as angling, beach visits, swimming, and skiing.

The travel cost method is also dependent on a relatively large data set. Data are usually collected through visitor interviews and questionnaires, which require sampling to cover different seasons or times of the year, and to ensure that various types of visitors from different locations are represented. Complex statistical analysis and modelling are required in order to construct information on visitor demand. As a result, travel cost surveys are typically expensive and time consuming to carry out. An additional source of complication is that several factors make it difficult to isolate the value of a particular ecosystem in relation to travel costs, and these must be taken into account in order to avoid overestimating ecosystem values. Visitors typically have several motives or destinations on a single trip, some of which are unrelated to the ecosystem being studied. The travel cost method is particularly useful for valuing recreational ecosystem services that are visited by tourists (e.g. coral reefs, national parks).

Contingent valuation

The contingent valuation method is a stated preference method and involves directly asking people, in a survey, how much they would be willing to pay for specific environmental services (Mitchell and

Carson, 1989). The contingent valuation method can be used to estimate economic values for all types of ecosystem services. The term 'contingent' denotes that valuation is based on a specific hypothetical scenario and description of the ecosystem service. For example, in the case that a wetland provides a habitat for a popular species of animal, respondents to a survey might be asked to state how much additional tax they are willing to pay to preserve the wetland in order to avoid a decline in the population of that species. The first practical application of contingent valuation was done by Davis in 1963 to estimate the value that hunters and tourists placed on a particular wilderness area (Davis 1963).

The idea is that a hypothetical, yet realistic, market for buying or selling the use and/or preservation of a good or service can be described in detail to an individual, who then participates in the hypothetical market by responding to a series of questions. These questions relate to a proposed change in the quality or provision of the good or service. The responses to these questions are then analysed to estimate the average value the respondents associate with the proposed change. This value can subsequently be aggregated over the affected population to derive a measure of total benefit (or cost).

An advantage of the contingent valuation method is that it can be applied to estimate values for all types of environmental goods and services, including nonuse values, and also changes in ecosystem services that have not yet occurred. A disadvantage of this method is that responses to willingness to pay questions are hypothetical and may not reflect true behaviour. Hypothetical scenarios described in contingent valuation questionnaires might be misunderstood or found to be unconvincing to respondents, leading to biased responses (Hanemann, 1994). Another disadvantage of the contingent valuation method is that it requires complex data collection and sophisticated statistical analysis and modelling. The large-scale surveys that are necessary for contingent valuation can also be expensive to conduct.

Choice modelling

Like contingent valuation, *choice modelling* is also a stated preference method and can be used to estimate economic values for virtually any ecosystem good or service. Choice modelling is generally regarded as one of the most suitable method for estimating consumers' willingness to pay for quality improvements. The Nobel Prize for economics in 2000 was awarded to a principal exponent of the Choice Modelling Theory – Daniel McFadden.

Choice modelling is a hypothetical method – it asks people to make choices based on a hypothetical scenario. Choice modelling is based around the idea that any good can be described in terms of its attributes or characteristics. Values are inferred from the hypothetical choices or tradeoffs that people make between different combinations of attributes. Choice modelling is different from contingent valuation in that it asks respondents to select between a set of alternatives, rather than asking directly for values.

The choice modelling valuation method addresses a number of the difficulties associated with traditional valuation methods. For example, rather than simply asking respondents how much they are willing to pay for a single improvement in an ecosystem service, a choice model forces respondents to repeatedly choose between complex, multiattribute profiles which describe various changes in ecosystem services at a given cost (e.g. a change in tax paid).

Because it focusses on tradeoffs among alternatives with different characteristics, choice modelling is especially suited to policy decisions where a set of possible actions might result in different impacts on ecosystem services. For example, a restored wetland will improve the quality of several services, such as floodwater storage, drinking water supply, on-site recreation, and maintenance of biodiversity. In addition, while choice modelling can be used to estimate dollar values, the results may also be used simply to rank options, without focussing on dollar values. A further advantage of the choice model approach is that research is not limited by preexisting market conditions, because the levels used in a choice experiment can be set to any reasonable range of values. As such, the choice modelling is useful to use as a policy tool for exploring proposed or hypothetical futures (e.g. for the impact of climate change on ecosystem services). Finally, choice experiments allow individuals to evaluate nonmarket ecosystem services described in an intuitive and meaningful way, without being asked to complete the potentially objectionable task of directly assigning dollar figures to important values such as culture.

Choice modelling is therefore a useful tool to value ecosystem services given its flexibility for valuing different environmental goods and services in different contexts (see Table 9.4). However, this method involves complex data analysis and relatively expensive data collection. This method is therefore only applicable when the necessary expertise and budget are available.

Table 9.4 *Ecosystem services and commonly applied valuation methods (source: Van Beukering et al., 2007).*

Ecosystem service	Valuation method
Food, timber, fuel wood	Market prices
Water filtration	Replacement cost, net factor income, production function
Water storage	Replacement cost, net factor income, production function
River flow control	Replacement cost, damage cost avoided, production function, net factor income
Coastal protection	Replacement cost, damage cost avoided, production function, net factor income
Support to fisheries	Net factor income, production function
Recreation site	Market prices, contingent valuation, travel cost, hedonic pricing, choice modelling
Visual aesthetics	Contingent valuation, hedonic pricing, choice modelling
Nature conservation	Contingent valuation, choice modelling
Nonuse/existence values	Contingent valuation, choice modelling

Main messages from case studies

The case studies presented in this report provide a rich source of information. We try to highlight the messages from these cases by providing the main message and illustrate the message with prominent examples. Other cases may also provide the same lessons, but for reasons of presentation we have chosen to link the messages to fewer cases where the issue is most prominent.

Recognising ecosystem services: a first step towards more transparent and engaged decision making

It is generally accepted that quality of SEA and transparency of decision making is greatly enhanced if stakeholders are at least informed about, or preferably invited into a planning process. The recognition of ecosystem services facilitates the identification of relevant stakeholders – the word service by definition links an ecosystem (the supply side) to stakeholders representing the demand side. In the Aral Sea Wetland Restoration Project an inventory of wetland related ecosystem services pointed towards the economic and social interests of these services and the associated groups in society. By inviting these stakeholders into the process of defining alternative restoration strategies it was possible to make an estimate of the former level of service delivery, its presently degraded state, and the desired future level of ecosystem service delivery. The assessment also revealed the geographical distribution of the ecosystem services. Similarly, in the West Delta Water Conservation and Irrigation Rehabilitation Project in Egypt, the identification of ecosystem services linked to surface water from the Nile River and to groundwater from the underlying aquifers facilitated the identification of relevant stakeholders to be invited into the SEA process.

When it is obvious that a plan leads to significant impacts on ecosystem services, ignoring such impacts may lead to opposition and ultimately the cancellation of the plan. Not studying (the impacts on) ecosystem services and their respective ecological, social, and economic importance thus can have serious repercussions. The case on planned water transfer from the Ebro River in Spain provides a clear example. The proposed water transfer would seriously affect water flow into the Ebro Delta. The delta combines multiple ecosystem services, such as maintaining internationally important biological diversity, and providing suitable conditions for rice cultivation, aquaculture, and fisheries. The protected status and the economic importance of the delta have been highlighted by independent studies. Ignoring these tangible ecosystem services and their beneficiaries by the authorities has contributed greatly to the failure of the water transfer plan to get approval.

Economic valuation increases the transparency of complex systems; the *Stern Review* provides one of the most convincing cases in this respect, addressing an issue with global consequences over a very long period of time. By explicitly highlighting the crucial uncertainties of certain economic activities, environmental conditionality for continuation of projects can be defined in the approval procedure. Economic valuation does not intend to prevent actual implementation of projects with impacts on ecosystem services, but it may affect the design of the intervention such that costs and benefits are traded off in a rational manner.

Methodological complexities do not necessarily hinder influential decision making

Due to the complex links between ecosystems and society, economic valuation of ecosystem services is often faced with methodological difficulties. The Wareham study from the UK was specifically designed to make an inventory of such difficulties in a real-life case, a regional flood control plan. The conclusion of this study was that reliable monetary values of ecosystem services are difficult to establish when depending on metadata or data transfer from other areas. Local data collection is needed, but is laborious. Nevertheless, the same study concluded that for comparison of alternatives, absolute valuation figures are not necessarily needed; a relative value measure provides enough information for decision making.

In spite of methodological difficulties, economic valuation of ecosystem services provides acceptable clues for legal procedures and fines. The Exxon Valdez oil spill is probably the most widely publicised case. Exxon was fined with the largest fine ever imposed for an environmental crime. Valuation studies covered various types of ecosystem services, most of these based on market prices. A significant part of the losses, however, was based on contingent valuation of lost passive use values linked to maintenance of biodiversity. The case shows that this technique based on stated preference of respondents is a legally accepted technique. The Exxon Valdez case set an example for liability claims for damage inflicted upon biodiversity. Some other examples are provided in the Annex where fines are based on contingent valuation, relating to damage inflicted upon coral reefs.

Of course, in cases where uncertainty about the (impact on the) value of ecosystem services is significant and the service itself is considered of great societal importance, the precautionary principle should be applied. The SEA for gas exploitation under the Dutch Wadden Sea is a classical case. The Wadden Sea provides multiple ecosystem services of economic importance (fisheries, tourism), and is an internationally important biodiversity conservation area. The main driver of change was soil subsidence by gas exploitation. There was uncertainty about the rate of sediment accretion, which would counteract the subsidence. The combination of important ecosystem values and uncertainty led to significant further research on this theme before a decision could be reached. Gas exploitation now is subjected to strict monitoring and can be forced to stop if impacts are larger than expected.

The *Stern Review* also urges the world to take a precautionary approach, but in a very particular manner. Instead of doing more research before taking action, Stern advises us to take action in response to potential climate change as soon as possible, and not wait for further evidence of climate change to emerge. In spite of the methodological complexities of calculating economic consequences of potential climate change, the *Stern Review* presents a convincing case that action now will prevent considerably larger future costs. Acting now is the best precautionary measure.

Apart from the need to do additional research as a result of a precautionary approach, there may also be methodological reasons to do so. Sensitivity analysis is an important tool to avoid the risk on major errors and to focus efforts for further research on most relevant issues. The Wareham case highlighted the need for sensitivity analysis to identify those factors where small changes in values have great influence on the outcome.

Insight in the distribution of ecosystem service benefits, highlight, poverty, and equity issues

In early planning stages, recognition of ecosystem services and identification of stakeholders can provide important clues on winners and losers of certain changes, and thus provides better understanding in poverty and equity issues. In the Egypt case, the diversion of Nile River water is proposed to enhance agricultural output of a desert area where large investors have created an economy with annual value of €500 million, producing agricultural outputs for the European market. If unmitigated, the withdrawal of water would go at the cost of ecosystem services in the downstream Nile Delta where poor smallholder farmers and fishermen would suffer from deteriorating water quality and supply. Even although the investments would make economic sense, the social consequences were considered unacceptable. The SEA study thus recommended adjusting the timing of the water diversion plan to the implementation of the national water resources management plan, in order to avoid the equity problems.

Another lesson from the Egypt case is that benefits and costs associated to ecosystem services can occur in geographically completely separate areas and affect different stakeholders, belonging to different divisions of society. In the Egypt case the 'winners' were large investors practising high-tech agriculture in the West Delta, while the potential 'losers' were relatively poor inhabitants of the Nile Delta living hundreds of kilometres away from the plan area. A similar spatial distribution effect was observed in the economic valuation study in Mali where the hydrodams transferred welfare from the poor downstream communities to the wealthier urban population in the capital.

A manner to overcome distributional effects as described above is provided by payments for ecosystem services (PES). Costa Rica provides an example where the existing inequity in distribution of costs and benefits between providers of an ecosystem service and the ones benefiting is solved by a legally embedded PES scheme. PES facilitates market processes between individual landowners, urban water consumers, and the world carbon market. For the protection of water resources the upstream landowners receive a payment if they leave their forest untouched, while the downstream urban inhabitants benefit from a secured source of drinking water. Similarly, the benefits of carbon sequestration accrue to the global community, while the opportunity cost of not converting a forest lies with local landowners.

SEA and planning processes are enhanced by the identification and quantification of ecosystem services

The Aral case represents a strategy development process for a large region, where reliable quantitative data were scarce. After the collapse of the Soviet Union, research and data collection efforts in Uzbekistan came to a standstill. Yet, this did not hinder the effective comparison of alternative restoration strategies for the Amu Darya Delta, based on ecosystem services assessment. The participatory multicriteria analysis involving both local scientists and stakeholders was a guarantee that all relevant local knowledge was represented in the process. Linking ecosystem services to stakeholders provided a good approach to involve relevant actors. By using the MCA tool it was possible to compare the performance of ecosystem services under different alternatives in semiquantified manner. 'Currencies' to compare values for different alternatives ranged from simple five-point scales (much more, more, neutral, less, much less) to actual quantification of societal values (such as income, number of jobs, number of inhabitants receiving good drinking water). At a higher strategic level this provided enough information for effective decision making. The Wareham case where different coastal flood management options were compared in terms of their impacts on ecosystem services came to a similar conclusion – relative differences in values provided a good basis for comparison. Full quantification and monetisation is not needed in early planning stages or at higher strategic levels.

In South Africa a spatial planning approach based on a SEA-like strategic catchment assessment provided a way out in a situation where biodiversity issues repeatedly caused discussion and delays in decision making at EIA/project level. Identification and valuation of ecosystem services and identification of stakeholders put biodiversity in the perspective of social and economic development needs of the municipality. Some services were under critical pressure and in need of conservation, not only because of biodiversity per se but also because of essential services for human well being. Other services are performing well and may provide a development potential when underexploited. Such a constraints and opportunities approach resulted in an open and better platform for discussion.

SEA provides a platform to put valuation results in a societal context

A general observation on the available literature on ecosystem services valuation is the lack of knowledge on the actual effects of the studies in planning and decision making processes. Moreover, there is a general feeling that the great potential of such studies to have an impact is not used to the full benefit. This is to a large extent caused by the divide between the worlds of environmental economy and environmental assessment. Economists often are not aware of the SEA instrument and the opportunities provided by this instrument to embed their methods and knowledge in a planning context and decision making process.

The case studies in this document provide evidence that economic valuation tools can easily be integrated in the SEA process, providing information much wanted by decision makers. Of course, the cases also show that SEA is not necessarily needed to make effective use of valuation tools for decision making. In cases where money was the key issue, economic valuation, of course, was the most preferred tool available. Examples are the penalties in the Exxon Valdez case, the compensation payments in the Costa Rica PES case, and the management fees in the Antilles case. In other cases, the use of valuation tools was not the obvious choice but played an important role in final decision making. In the South Africa case, valuation provided the necessary vocabulary to convince decision makers; in the Wadden Sea case, it contributed to the recognised need for a precautionary approach and a strict environmental management plan. In both cases, SEA or SEA-like processes supported decision making, and provided the platform to merge the valuation results with the decision making process.

Decision making supported by relevant information

The authors of the South African case clearly state that monetisation of ecosystem services has put biodiversity considerations on the decision makers' agenda. Instead of identifying and declaring conservation-worthy areas a 'no-go', the study stresses the ecosystem services that the environment provides free of charge to the Municipality. The use of ecosystems services and focus on the value of these services for society was of key importance to convince local councils that biodiversity conservation makes economic sense. Politicians reacted negatively to the term 'biodiversity' but more positively once they realised that environmental services have an economic value.

Presentation of results is an important aspect of environmental assessment. All too often assessment reports are voluminous and filled with jargon, rendering these reports inaccessible for decision makers and the public at large. Some lessons can be drawn from the case studies. In the Aral case the construction of an 'ecosystem services – values' table provides a good visualisation of the variety of services and their stakeholders. It served as a good communication tool. For the strategic catchment assessment in uMhlathuze Municipality, a status quo report on the condition of ecosystem services was presented in four poster-like pages for each catchment. This communication-oriented output was ideal to rapidly inform planners and decision makers. The thought behind this was that 'planners are in the best position to influence sustainable development, so they should also be educated'.

Similarly, the *Stern Review* case teaches us that the one who conveys the message also makes a difference in the impact of the study. This case shows that the most far-reaching policy changes for improving the functioning of ecosystem services can be achieved by making the Treasury the champion of the economic valuation study. They have both the authority and the means to follow up on the recommendations. In general, the case teaches us that boundary conditions such as timing, communication and ownership can be more important in terms of generating societal impact than the quality of the study only. The *Stern Review* was published shortly after the world famous *Inconvenient Truth* by former U.S. Vice-President Al Gore. The documentary paved the way for the more complex message of the economics of climate change.

Valuing ecosystem services directly facilitates sustainability

The Exxon Valdez case has confronted oil companies with severe financial consequences of oil spills. Undoubtedly, this has contributed to the ever-increasing safety norms for oil transport, thus reducing such mishaps in future. On the other hand it provides a mechanism for the financing the clean-up operations of environmental damage for which a party can be held accountable. In a strange manner this generates financial 'sustainability' of clean-up operations; of course, an environmental disaster can never be considered environmentally sustainable.

The introduction of a payment for ecosystem services scheme (PES) in Costa Rica has played a major role in changing Costa Rican destructive and rapid deforestation into forest restoration efforts and more sustainable management, with tangible and convincing results.

Similarly, contingent valuation of coral reefs has effectively been applied in the Netherlands Antilles case where it has lead to the implementation of measures guaranteeing better management of national parks and financial sustainability of the management operations. In other cases, valuation of ecosystem services has resulted in more sustainability-oriented decision making (i.e. South Africa, Aral, Egypt, Wadden Sea), although it cannot be judged how decision would otherwise have been taken.

The Ebro case shows the power of valuation tools in the hand of opponents of an obviously unsustainable project. Although environmental assessment never has the intention to hinder or to stop development, in this case the use of independent assessment and simultaneous pressure on the main funding agency has avoided great damage. In the end it has resulted in a much better plan, although a change of government was needed for this major step.

In summary, the cases provide evidence that the recognition and valuation of ecosystem services within the context of well-informed strategic decision making, facilitates a better representation of the three pillars of sustainability:

(1) Financial sustainability of environmental and resource management;
(2) Social sustainability by facilitating participation of stakeholders and by highlighting and addressing equity issues; and
(3) Environmental sustainability by providing better insight in the long- and short-term trade offs of investment decisions.

Epilogue – Topics in need of further elaboration

Roel Slootweg, Asha Rajvanshi, Vinod B. Mathur, and Arend Kolhoff

A number of topics have not been addressed in a separate chapter in this book. If we had covered everything then the book was in danger of becoming an encyclopaedia. The aim of this epilogue is to raise some of these additional topics in the anticipation and hope that others will address them – each deserves a book in its own right!

Climate change

Despite the rapidly expanding state of knowledge about climate change impacts, it is strange not to address this issue in a separate chapter of this book. Many environmental professionals view climate change as a symptom of our unsustainable use of the biophysical environment. A clear distinction is maintained between efforts to address the symptoms of climate change from those dealing with the causes. These distinctions are underpinned by the United Nations Framework Convention on Climate Change (UNFCCC) which identifies two responses to climate change: (i) mitigation of climate change by reducing greenhouse-gas emissions and enhancing sinks and (ii) adaptation to the impacts of climate change. In response to the issues pertaining to climate change, most research on adaptation and mitigation so far has been disparate, involving largely different communities of scholars who take different approaches to analyse the two responses. This dichotomy of objectives poses a significant challenge for the impact assessment community which is trying to see a potential for creating synergies between adaptation and mitigation. Impact assessment professionals have a responsibility to link climate, development, and environmental policies by, for example, linking energy efficiency (related to mitigation) to sustainable communities or poverty reduction (related to adaptation). Yet, in most impact assessments climate change is an externality to be taken into account when describing expected future developments in the description of autonomous developments. Due to their limited scale of influence, most projects or plans under study cannot influence or mitigate climate change itself and can only take climate change scenarios into account when these are expected to have influence on the outcome of the project or plan (adaptation). Mitigation of climate change is the ultimate policy challenge as it can only be addressed globally. For example, agreements on a global scale lead to national standards on emissions reduction. These agreements set the boundary conditions for planned activities and can be taken into account in environmental assessment, in the same way as noise or water quality standards have to be taken into account. The impacts of climate change are only recently becoming visible. How these processes of change have to be weighed against changes induced by other human activities, and how to deal with potential cumulative or synergistic effects are questions we are only starting to recognise. We feel there is a need to use environmental assessment to improve resilience of ecosystems through the adoption of biodiversity-based adaptation and mitigation strategies. We also suggest that impact assessments can be thought of as "trading tool" through which actors, such as developers, economists, EA experts, or local communities, can negotiate data needs and research priorities, participation, and methodological issues. This theme, by any means, is too big for this book, even though Chapter 3 does pay attention to the linkages between climate change and ecosystem services. Additionally, we have included the highly publicised Stern report on the economic consequences of climate change as an example case in the Annex.

Stakeholder participation: the ecosystem approach in practice

Stakeholder participation appears prominently in most chapters of this book. Yet, we feel that it merits more attention, especially in relation to the implementation of the ecosystem approach. Chapter 2 stated that 'discussing biodiversity is very much about discussing people's behaviour and interests'. The main threats to global biodiversity are associated with human activities. It is argued that the value of biodiversity is best guaranteed among people by means of a participatory discussion of environmental goals. The ecosystem approach of the CBD explicitly states that humans, with their cultural diversity, are an integral component of many ecosystems. The ecosystem approach is based on the application of appropriate scientific methodologies but also states that ecosystem management is a social process. The need for stakeholder participation in decision making appears prominently throughout the approach. The most prominent principles highlighting the role of stakeholder participation are cited below (see Box 2.4, Chapter 2, for a complete overview).

Principle 1 states that the objectives of management of land, water, and living resources are a matter of societal choice. There are many interested communities, which must be involved through the development of efficient and effective structures and processes for decision making and management. The ecosystem approach should, according to principle 10, seek the appropriate balance between, and integration of, conservation and use of biological diversity. Biological resources provide goods and services on which humanity ultimately depends. There has been a tendency in the past to manage components of biological diversity either as protected or nonprotected. There is a need for a shift to more flexible situations, where conservation and use are seen in context and the full range of measures is applied in a continuum from strictly protected to human-made ecosystems. In doing so, the ecosystem approach should consider all forms of relevant information, including scientific and indigenous and local knowledge, innovations, and practices (principle 11). Information from all sources is critical to arriving at effective ecosystem management strategies. Sharing of information with all stakeholders is equally important. Therefore, principle 12 logically argues that the ecosystem approach should involve all relevant sectors of society and scientific disciplines. The integrated management of land, water, and living resources requires increased communication and cooperation, (i) between sectors, (ii) at various levels of Government (national, provincial, local), and (iii) among Governments, civil society, and private sector stakeholders.

The public participation discussion in environmental assessment has evolved along with the emergence of more strategic level assessment. The most simple level of participation in EIA allows the general audience to react to the Terms of Reference of an EIA and to provide comments on the Environmental Impact Statement resulting from an EIA. Public review in EIA is considered too little and too late. SEA has emerged to provide an earlier possibility to influence the planning process. Increasingly, a more active role of both proponent (finding and inviting stakeholders) and stakeholders is promoted to be able to come to sustainable plans and projects. Yet, at strategic level public participation is more complex to organise. Plans are not concrete yet, locations may still be unknown, and government may not be willing to publicise it plans in early stages. Yet, in a recent book, Dietz and Stern (2008) conclude that, when done correctly, public participation improves the quality of federal agencies' decisions about the environment. Well-managed public involvement also increases the legitimacy of decisions in the eyes of those affected by them, which makes it more likely that the decisions will be implemented effectively.

With respect to biodiversity the distributional or equity effects of ecosystem services come into scope: who is most vulnerable, who is most resilient, and how do we tackle this aspect in planning and environmental assessment? Biodiversity is providing a resource base for the livelihoods of many. These stakeholders can easily be identified, although their participation in a planning process may not be as easy (literacy, language, education, access to information, etc.). Biodiversity is also considered to be a life insurance to life itself; in other words, it has to be maintained for future generations. How can the interests of future generations be represented in the assessment process? Issues of scale present further complexities in the public participation discussion. Stakeholders in the climate change debate are global – how can these stakeholders be represented in a process? Similarly, other global commons such as the oceans do not have easily identified representatives. Here the role of global conventions comes into the picture.

Capacity development and institutional issues

Environmental assessment can only be effectively used when the instrument is used in a setting with certain characteristics. For example, the ecosystem approach requires stakeholder participation in decision making and management. Environmental assessment has always strongly emphasised the role of stakeholders and consequently provides an important vehicle for the implementation of the ecosystem approach. But, true stakeholder participation requires a society with certain institutions which guarantees that the views of stakeholders are not disregarded. In general terms, environmental assessment is often considered to require a minimal level of democracy and transparency to be effective.

From a biodiversity perspective this is extremely relevant as we position biodiversity in terms of ecosystem services, which translates biodiversity into stakeholder values. If these values cannot be taken into account as a result of the lack of necessary institutions to govern the environmental assessment process, the basis for good assessment is jeopardised. In other words, biodiversity is an issue for which certain governance mechanisms have to be in place in order to be effectively maintained for human well-being, now and in future.

Issues that need to be addressed include access to information, equity in expressing interests, respecting human rights, and transparency. New developments need attention; for example, how can the rapid development of new communication tools (e.g. the Internet) contribute to the process.

Biodiversity and the law

Increasingly impact assessment is becoming a legal battle. Legal protection guarantees minimum protection but may lead to a free-for-all in the remaining area. Nonprotected areas or species may go unnoticed in environmental assessment. Environmental assessment has always been positioned as a tool to provide insight in all relevant consequences of human activities; 'relevant' is interpreted as relevant to decision making. However, 'relevant' is often interpreted as 'regulated by law'. It can be argued that with a strong focus on legally required issues, many other relevant aspects (of biodiversity) may be lost and out of sight and may even lead to a loss of independent thinking. A possible way to better harmonise the need for legal 'clarity' with the diverse character of biodiversity is the recognition of ecosystem services. This may provide another way to put unprotected biodiversity in the picture. In short, we would highly welcome good contributions on this aspect. An issue of interest may be the implementation experience of some European directives, such as the EU habitat and birds directives and the water framework directive. While the habitat and birds directives are designed to provide minimal protection to threatened habitats without jeopardising economic development, they are considered by many countries as too prescriptive. The water framework directive better recognised that different countries do things in different ways to achieve the same outcomes. The implementation of these (framework) directives provides a wealth of experiences when different countries have different interpretations and implementation mechanisms. Evaluation studies within the broader context of the objectives and approaches from the CBD could provide relevant practical clues.

Sustainability assessment, integrated assessment, and more

In this book we have addressed environmental assessment from the perspective of real-world decision making. For this, the instruments of environmental impact assessment (EIA) and strategic environmental assessment (SEA) have been developed, and in most countries regulated by law. In many countries these are the only regulations providing some minimal form of transparency and participation in governmental decision making. In practice EIA as well as SEA are legalised battle grounds in which stakeholders fight out their interests in a formalised 'arena'. This explains the preoccupation of the impact assessment community with procedural aspects. In this book we have tried to provide some contents to these largely procedural instruments. However, there is a world beyond EIA or SEA, where many different tools and approaches are being developed around the key words 'sustainability' and 'integrated'. Sustainability is based on three pillars, a social, an ecological and an economic pillar. In order to address these in a balanced manner, the worlds of sociology, ecology, and economics have to be integrated into one coherent instrument. Sustainability assessment is often described as a process by which the implications of an initiative on sustainability are evaluated, where the initiative can be a proposed or existing policy, plan, programme, project, piece of legislation, or a current practice or activity. This generic definition covers a broad range of different processes, many of which have been described in the literature as 'sustainability assessment' (Pope *et al.*, 2004). Integrated assessment (including social cost–benefit analysis) has largely similar characteristics, but has been developed by economists. These largely scientific endeavours have produced relevant new insights and tools. Yet, the methods are 'content driven' and lack a procedural framework comparable to EIA or SEA. Consequently, there is neither a standard approach (yet), nor a standard terminology – many different approaches with different applications coexist. In recent years the world of sustainability or integrated assessment, dominated by scientists, and impact assessment (EIA(SEA), dominated by practitioners, are nearing each other. There are obvious lessons to be learned for both sides. The scientists can enrich and improve the present practice of impact assessment with better approaches to the integration of different scientific disciplines, while the impact assessment practitioners can indicate what works in practice and what does not work. In the end, both scientists and practitioners all want to provide relevant and timely information to decision makers, leading to decisions that contribute to a more 'sustainable' world. In this book the focus is on maintaining biodiversity for human well-being, now and in future.

Annex: valuation of ecosystem services: influential cases

Pieter van Beukering, Roel Slootweg, and Desirée Immerzeel

Introduction

For reasons of consistency the ten main cases (see Table A.1) have been written with a (more or less) fixed format. As much as possible the following six items have been addressed in the analysis of cases:

(1) *Introduction to the case*: description of the issue, social and environmental setting, sector, and location.

(2) *Context of the case study*: where and how was the valuation study used in the planning process? (or what was the policy context of the study?) Where in the process did the study fit?

(3) *Assessment context*: was the study carried out as (part of) a formal SEA (or EIA) procedure?

(4) *Ecosystem services*: the type of ecosystem services, the way in which ecosystem services were included in the assessment, the type of valuation applied, and the role of stakeholders in the process.

(5) *Decision making*: in what way did valuation of ecosystem services influence decision making? What constraints where encountered in using ecosystem services to inform decision making?

(6) *SEA boundary conditions*: relation between study effort and magnitude of the decisions involved; source of data; the level of detail required at which level of planning; timing of the assessment in the process.

Ten further cases will appear as Cases A.1–A.10 throughout this chapter. These cases have not been analysed in detail but merely provide further illustration of the messages derived from the main cases. These additional cases are listed in Table A.2 with a reference to the cases that they support.

West Delta Water Conservation and Irrigation Rehabilitation Project, Egypt

Main messages

• In early planning stages, recognition of ecosystem services and identification of stakeholders can provide important clues to poverty and equity issues.

Table A.1 *Case studies elaborately explained in separate chapters.*

#	Study	Ecosystem	Country	Type of study
1	Water Conservation and Irrigation Rehabilitation	Reclaimed desert and river delta	Egypt	Voluntary SEA
2	Wetland Restoration Strategy	Wetland	Aral Sea	SEA-like
3	Strategic Catchment Assessment	Watersheds	South Africa	Part of SEA process
4	Making Space for Water in Wareham	Coastal wetlands	United Kingdom	Experimental SEA
5	Climate policies and the *Stern Review*	Global	Global	Inform policy making
6	Natural gas extraction in the Wadden Sea	Tidal wetlands	Netherlands	Inform EIA and SEA process
7	Management of marine parks	Coral reefs	Dutch Antilles	Sustainable financing
8	Watershed rehabilitation and services provision	Forest	Costa Rica	Payments for Ecosystem Services
9	Water scarcity and transfer	Rivers	Spain	Advocacy
10	Exxon Valdez oil spill in Alaska	Coastal resources	United States	Damage assessment

- Benefits and costs associated with ecosystem services can occur in geographically completely separate areas and affect different stakeholders belonging to different divisions of society.

Introduction to the case

Since the 1980s Egypt continues to expand its groundwater-based agriculture on the desert plains west of the Nile Delta, an area with the confusing name 'West Delta'. A highly productive and economically important, export-oriented agriculture has developed, based on modern irrigation technology and advanced agricultural practises. However, the rate of groundwater exploitation by far exceeds the rate of renewal and thus is not sustainable. Groundwater is rapidly depleting and in some places already turning to saltwater. In order to reverse the deteriorating situation, to save the economic potential (about US$500 million annually) and the many jobs in 'on' farm and 'off' farm activities, the Government of Egypt has proposed the West Delta Water Conservation and Irrigation Rehabilitation Project (WDWCIRP) to supply Nile water to the area. The Government of Egypt is preparing a public–private partnership project to pump fresh Nile water from the Rosetta Nile branch

Table A.2 *Case studies briefly addressed in boxes.*

#	Study	Ecosystem	Country	Type of study	Link
1	Impact of dams on wetlands and livelihoods	Wetlands	Mali	Investment decision	Case A.1. in Case 1
2	Livelihood and conservation of Korup National Park	Tropical forest	Cameroon	Nature conservation	Case A.2. in Case 1
3	Large scale wetland restoration	Wetlands	Everglades	Nature conservation	Case A.3. in Case 3
4	Management of Durban's open spaces	Open spaces	South Africa	Environmental planning	Case A.4. in Case 3
5	Cost of policy inaction for biodiversity	Biodiversity	Global	Awareness raising	Case A.5. in Case 5
6	Carbon offset investments in Iwokrama National Park	Tropical forest	Guyana	Investment decision	Case A.6. in Case 5
7	Mangrove rehabilitation	Mangroves	Philippines	Nature restoration	Case A.7. in Case 6
8	Voluntary user fee system for divers	Coral reefs	Hawaii	Sustainable financing	Case A.8. in Case 8
9	Watershed rehabilitation for drinking water	Rural areas	New York	Payments for ecosystem services	Case A.9. in Case 8
10	Penalty system for coral reef injury	Coral reefs	Florida/Hawaii	Damage assessment	Case A.10. in Case 10

into the project area and distribute it over 40,000 hectares of farmland in the West Delta area on a full cost recovery basis.

Context of the case study: the planning process

Egypt's National Water Resources Management Plan (NWRP, 2000) is based on a strictly defined amount of available water, agreed upon among the Nile Basin states. Within this limitation, the NWRP describes measures to save water in the existing water resources management system to facilitate expansion of irrigation works in desert areas. Water-saving measures include the ongoing urbanisation on farmland (thus saving on irrigation water), waste water treatment, shifts in the cropping pattern (restrictions on rice and bananas), and irrigation improvement projects. At the time of the study a timetable for the implementation of these water-saving measures still needed to be devised.

The West Delta is one of the identified areas for further land reclamation for irrigation development. The first planning step was a study on a 'Conceptual Framework and Transaction Model for a Public-Private Partnership in Irrigation in the West Delta'. This study provided a conceptual design based on public–private water management partnership and the willingness of the beneficiaries to connect on a full cost recovery basis. The study was conducted under the condition that 1.6 billion cubic metres (BCM) of Nile water would be available which amounts to about 16 percent of the total flow in the Rosetta branch.

Assessment context

The creation of a public–private partnership is a relatively new procedure. How to deal with environmental and social impact assessment in such circumstances had yet not been clearly defined. Slootweg *et al.* 2007 provides a simplified overview of the steps in the planning process and the points where impact assessment played and still has to play its role. The WDWCIRP started with a preliminary technical design providing a general framework on how to address the predictable future problems of groundwater availability. This preliminary study provided the basis for a Drainframe assessment, that is, an SEA-like assessment of the provisional plan, following an approach developed by the World Bank Agriculture and Rural Development Division (Abdel-Dayem *et al.*, 2004).

Drainframe is a water resources planning approach that ensures the integration in the planning process of the multiple services provided by natural resources, taking into account the interests of stakeholders. It has the characteristics of an integrated SEA. Integrated in the sense that economic, social, and environmental aspects are taken into account; strategic in the sense that it offers options for decision making in early stages of planning (Slootweg *et al.*, 2007). At the time of the study no decisions had been taken yet on the exact location and size of the project intervention area, irrigation technology, or institutional arrangements. The Drainframe assessment is subject of this case description.

The preliminary study was based on a number of stakeholder workshops and interviews throughout the project area. The focus was on identifying the needs and aspirations of farmers in the West Delta area. The Drainframe SEA study has extended participation to other stakeholders who, based on the assessment of affected ecosystem services linked to surface and groundwater, could be identified as potentially affected by the project. The outreach study made all farmers in the area aware of the process (instead of sampled groups) to guarantee broad knowledge of and contributions to the planning process. Drainframe and outreach studies have determined the scope of a further detailed technical study. This technical study, however, is NOT a final design, but it provides the boundary conditions for the bidding process.

The environmental and social impact assessment (ESIA), required in the project preparation cycle of the World Bank, moved from the broad overview provided by

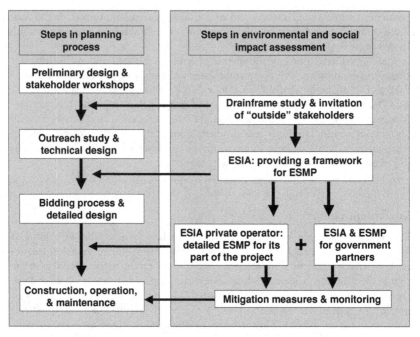

Figure A.1 Project planning process including environmental and social assessment (Source: Slootweg *et al.*, 2007).

the Drainframe study, to more detailed project-level impact assessment. However, because the final design of the project is not available yet and many issues remain unresolved, the impact assessment in many instances could not go further than defining the various tasks and responsibilities of both the private and public partners in the remaining phases; the ESIA provided a framework for further work. The final design will be made by the private service provider who wins the bid for the PPP project. The process, at the time that this chapter was written, did not yet reach this phase.

Ecosystem services and valuation

A first round of qualitative analysis resulted in an overview of affected ecosystem services through the identification of main drivers of change:

- *Withdrawal of water from the Rosetta Nile branch* – The reduction of water availability downstream of the intake by 16 percent affects water supply to tens of thousands of smallholder farms (many of these being among the poorest of Egyptian society), public water supply in Beheira Governarate and the city of Alexandria, the ecological status of coastal lagoons (one Ramsar site), and their fisheries productivity.
- *Surface water supply to West Delta* – Transferring this water to the project area can lead to reduced exploitation of groundwater in the project area, but it can also lead

Table A.3 *Economic performance of three alternatives (not taking into account offsite and indirect impacts).*

	A	B	C
Benefit–cost ratio in 2017	0.99	1.07	1.05
Income per feddan in 2017 (LE/feddan)	−70	521	398
NPV ($r = 10\%$) million LE	975	1013	588

to intensified agricultural exploitation by jointly using imported surface water and local groundwater. This results in complex groundwater level fluctuations in the aquifer underlying the entire West Delta.
• A permanent supply of water will induce increased social and economic development in the West Delta Region. This aspect is left untouched in this case study description.

Stakeholders in ecosystem services were invited, in a workshop, to make an assessment of the relative importance of the affected ecosystem services. This resulted in the identification of main issues. A second round of analysis included a comparison of alternative project concepts based on quantified impacts. Relationships between interventions, the changes that were expected, their effects on ecosystem services, and the impact on societal values of these services were first described. These relationships, in most cases, were modelled in simple mathematical equations. The team also took advantage of two existing computational models for (i) simulation of water availability and yield relations in the Nile Delta and for (ii) simulation of groundwater behaviour in the West Delta region. The results were presented and discussed in a second workshop with about 60 stakeholders from both the private sector and the government.

The Drainframe study considered three alternative strategies for water supply to the project area, with various subalternatives. *Strategy A* represents the case of doing nothing – pumping of groundwater continues at unsustainable levels. *Strategy B* uses surface water for irrigation in conjunction with groundwater use for peak demands; the water conveyance infrastructure is modest and the surface area potentially cultivated is the largest. *Strategy C* considers no groundwater use at all; the capacity of the conveyance system has to be significantly larger to meet peak water requirements.

The evaluation exercise concentrated on the following main impacts expected to result from the considered alternatives in the study area:

• *Net economic benefits of an average farm in the project area (quantitative)*: As shown in Table A.3, conjunctive use of surface and groundwater gives highest net present value. At a discount rate of 10 percent the continuation of the present practise of groundwater pumping remains more beneficial than investing in a system that fully depends on Nile water (Strategy C). This is explained by the fact that rapid deterioration of groundwater only takes place after 2013, while investment costs for the Strategy C alternative are huge.

Table A.4 *Performance of strategies on job creation.*

Number of jobs	A	B	C
Seasonal jobs	−29,809	273,255	210,070
Permanent jobs	0	54,607	41,897
Total jobs	−29,809	327,862	251,966

Table A.5 *Hypothetical annual production loss in the Nile Delta as a result of 1.6 BCM annual water withdrawal (in millions of US$).*

	Impacts distributed over entire Nile Delta	Impacts on Rosetta branch only
Cropping pattern unchanged	263.731	132.678
Cropping pattern adapted	77.353	78.563

- *Numbers of permanent and seasonal jobs in the project area (quantitative)*: As shown in Table A.4, the different strategies generate different levels of employment, with Strategy B being the most labour-intensive approach.
- *Impact on the production in the downstream area of the Nile Delta (quantitative)*: As shown in Table A.5, the potential losses in production value in the Nile Delta were calculated for two situations: the water is only taken from the Rosetta branch, or the water is taken from the entire delta. Two scenarios where evaluated: (i) the loss of net return of agriculture when the cropping pattern remains unchanged and (ii) the loss when farmers cope as well as possible with increasing water scarcity. The alternatives, Strategies B and C would both take the same amount of water.
- *Fishery benefits*: Fisheries productivity of the coastal lakes amounts to 152,295 tonnes annually, caught by 18,000 boats, providing employment to 48,000 persons. No attempt was made to calculate production losses.
- *Impact on drinking water availability (qualitative)*: Beheira governorate depends on groundwater for 60 percent of its population (2.4 million people). The remaining 40 percent of the population (i.e. 1.6 million people) depends on surface water. Mahmoudia canal, taking water from Rosetta branch, is the only source of public water supply to Alexandria, serving between six million inhabitants in the winter and eight million in the summer. Any reduction in water supply will have severe consequences, because the water supply is already under stress.

Decision making

The use of strategic environmental assessment at the earliest possible stage of the planning process guarantees that environmental and social issues beyond the boundaries of the project area are incorporated in the design process. Valuation of ecosystem services focussed on the services linked to water resources in the area under the influence of the major driver of change, that is, the transfer of water from the Nile Delta to the West Delta desert area.

Very simple quantification techniques, in terms of net present value and benefit/cost ratio of investments at farm level, job creation, numbers of people negatively affected, and overall production losses in the Nile Delta, provided strong arguments for decision makers at the Ministry of Water Resources and Irrigation and the World Bank to significantly reduce the scale of the initial pilot project. The diversion of water from relatively poor smallholder farmers in de Nile Delta to large investors in de West Delta poses equity problems unacceptable to stakeholders as well as government decision makers.

All experts and stakeholders agreed that water withdrawal from the Rosetta branch should be fully compensated by measures to save water in the entire irrigation system. Water quality in Rosetta branch is already below the needs of the command area, water quality in the coastal lakes is similarly under serious stress, agriculture in the Nile Delta would face serious losses under reduced water availability, and public water supply to Alexandria is of such overwhelming importance that any reduction in water supply to Rosetta branch must be avoided.

However, the National Water Resources Plan does not give a timetable of water-saving measures and therefore does not provide any clues to the timing of water savings. At present an implementation programme for water-saving measures is developed. It was considered important to harmonise implementation of the West Delta project with the necessary measures to save the required amount of water.

It is decided that the WDWCIRP project will have a phased approach, providing room to implement the water-saving programme. Short-term measures can produce necessary first savings to allow for the first, relatively small pilot phase of the WDWCIRP project. Further water-saving measures will provide room for further expansion of the WDWCIRP project.

SEA boundary conditions

The Drainframe assessment has been carried out over a period of three months. Time expenditure included hiring of three expatriate and two local consultants for one month each. Furthermore, farm surveys were carried out by local agricultural extension workers. The study was carried out in close collaboration with the persons responsible for project planning, at the Ministry and the World Bank. The cost of the study was approximately US$80,000, on a total estimated project budget of around US$100 million. The study was well-coordinated, was fully integrated in the planning process, and did not experience any delays.

Data were obtained from project planning documents, government statistics, farm surveys, two computational groundwater and surface water models, with a number of additional field visits and on-farm interviews for verification. Two stakeholder workshops provided relevant scoping information and discussion on the outcome of the study. The level of detail and reliability of information was sufficient to guide the planning process. Where links between hydrological changes and impacts were very difficult to quantify in economic terms, the impact description was limited to the identification of numbers of affected people. The subsequent detailed technical design was subject to a full-fledged ESIA, which could zoom in on a limited number of issues to provide more detailed information.

Case A.1. More sustainable management of the Niger River in Mali

One million people in the Inner Niger Delta make a living from arable farming, fisheries and livestock. Upstream dams (one built for electricity generation and one for irrigation) affect this downstream multifunctional use of water. Additionally, the Inner Niger Delta, which is one of the largest Ramsar sites in the world, is a hotspot of biodiversity and accommodates two of the largest known breeding colonies of large wading birds and staging water birds, residents and migrants from all over Europe and western Asia. The hydrological and related ecological conditions in the Inner Delta largely determine the health of the ecology as well as the economy.

The major aim of a three-year study was to develop a decision-support system for river management in the Upper Niger, in which ecological and socioeconomic impacts and benefits of dams and irrigation systems are analysed in relation to different water management scenarios. The study involves various components: hydrology, arable farming, livestock, fisheries, ecology, and socioeconomics (Zwarts *et al.* 2005).

An economic analysis has been conducted to determine the role of dams in the economy of the Inner Niger Delta and the Upper Niger region. By innovatively combining the above information on hydrology, ecology, fisheries, and agriculture, the study shows that building new dams is not an efficient way to increase economic growth and reduce poverty in the region. In fact, such efforts are countereffective and, at best, transfer welfare from the Inner Niger Delta to the Upper Niger region (Zwarts *et al.* 2006).

Rather than building more dams in the Upper Niger, the study advises to aim additional efforts at improving the efficiency of the existing infrastructure, as well as of current economic activities in the Inner Niger Delta itself. This approach will also provide greater certainty for the essential ecoregional network functioning of the Inner Delta. Several of these recommendations seem to have been adopted by the Mali government. The attention of economic development within the Inner Niger Delta has increased, as well as the continued efforts to improve the irrigation efficiency in the agriculture sector upstream.

Box A.2. Nature conservation at the cost of local livelihoods?

From a conservation and development perspective a strong statement is provided by Schmidt-Soltau (2002) and Cernea and Schmidt-Soltau (2006) in a description of an impact assessment in a national park project in South-West Cameroon. In 1986 Korup National Park was created, covering an area of 1,259 km^2. It soon became famous; the British Sunday newspaper *The Observer* introduced Korup National Park to the world with a special full-colour supplement titled 'Paradise lost?'

Here the paper could have ended, if the area had been solely inhabited and utilised by mammals, fishes, birds, and insects; but the perception of Africa as a continent of a vast wilderness with abundant freely ranging wild animals waiting for tourists and researchers to enjoy is flawed. In reality, there is no 'no man's land' in Africa. The wilderness is often communal land shared between villages. In the case of Korup National Park the land is home to 1,400 people. Nearly 30,000 individuals from 187 villages are utilising the park and its surrounding area for their livelihood as hunters, gatherers, fisher-folk, and farmers.

An assessment of the impacts of prohibiting any further exploitation of the forest showed that even if the project were to use its entire budget to compensate the traditional owners on an annual basis, the villagers – not considering the impact on their subsistence – would be forced to contribute €31- per person (or 19 percent of their annual cash income) to the conservation of rainforests. Yet, 81 percent of respondents saw the forests as their source of livelihood and therefore supported the idea of forest conservation, but with a desire to be more involved in park management and to be allowed to continue traditional exploitation.

Obviously the clashing interest between strict protection and sustainable use of nature conservation areas requires an analysis from both a biophysical and a socioeconomic point of view. Valuation of ecosystem services within an impact assessment framework provides an effective tool.

Additional information on this case can be found in Abdel-Dayem *et al.* (2005), Attia *et al.*, (2005), Ministry of Water Resources and Irrigation (2005), and World Bank (2005).

Aral Sea Wetland Restoration Strategy

Main messages

• Semiquantitative valuation of ecosystem services, expressed in terms of each service delivered (i.e. not monetised) works well for comparison of alternative intervention strategies (high strategic level). Participatory MCA is an effective tool, capable of dealing with limited level of detail in data.
• Monetisation of services in a CBA works well at project level when discussing concrete and well-defined investments within the framework of the selected overall strategy.
• Construction of an 'ecosystem services–values' table provides a good visualisation of the variety of services, and the table is a good communication tool.
• Valuation of ecosystem services leads to better, more sustainability oriented decisions.

Introduction to the case

In the early 1960s, the Government of the former Soviet Union decided to intensify and expand its irrigation activities in Central Asia. The irrigation water was taken from the Amudarya and the Syrdarya, the two main rivers contributing water to the Aral Sea. The result of this large-scale intensification of water use for irrigation has been shrinking of the Aral Sea, desiccation of large areas around the Aral Sea, and increasing salination of its waters. The Aral Sea today is practically devoid of higher forms of life because of its salinity. Other environmental effects concern the reduced availability of (flood)water in the deltas of the Amudarya and Syrdarya rivers, considerable loss of biodiversity, loss of vegetation and fisheries, the occurrence of salt and dust-laden winds, and deteriorating health conditions because of salination of groundwater. In 1995, about 10 percent of the original wetlands remained in the deltas, largely maintained by a mix of incidental floodwaters and saline drainage water flowing into constructed water reservoirs.

Context of the case study: the planning process

In 1992, after the collapse of the Soviet Union, five Central Asian States: Kazakhstan, Uzbekistan, Kyrgyzstan, Turkmenistan, and Tajikistan decided to tackle what had become known as the 'Aral Sea crisis'. They signed an agreement for cooperation on management, utilization, and protection of water resources in the Aral Sea catchment area. The international community offered help, coordinated by the World Bank in the Aral Sea Programme (ASP). To manage the ASP, the Interstate Council on the Aral Sea (ICAS) was set up.

The Aral Sea Programme is based on seven decisions made by the Heads of State of the five Republics. The Aral Sea Wetland Restoration Project (ASWRP), which is the subject of this case description, was created to answer to Decision No. 4, which states:

To undertake research work and to decide upon the existing engineering options, to prepare projects and to create artificially watered landscape ecosystems in the deltas of the Amudarya and Syrdarya rivers and on the exposed Aral Sea beds. Furthermore to undertake the required melioration work in order to restore the original environmental landscape in the above-mentioned areas.

The geographical area of study is the Amudarya Delta south of the Aral Sea, in the semiautonomous republic of Karakalpakstan, Uzbekistan. The Dutch Government, through a World Bank Trust Fund, provided funding for the study.

The problems of desiccation, desertification, loss of productive resources, and decline in living conditions have been documented extensively by local nongovernmental organisations (NGOs) and the Uzbek authorities. Highly contradictory reports on living conditions in the Amu Darya Delta and a wide array of sometimes unrealistic solutions developed by different institutes have made the ICAS request a comprehensive study of all existing information and the development of a coherent strategy for the delta.

The ICAS, in consultation with the World Bank, had drafted the T.o.R. for the study. In 1995, a consortium of Dutch consultants, in close consultation with local Uzbek institutes, developed a coherent strategy for the restoration of the Amu Darya Delta that was broadly accepted by local stakeholders and government authorities and an investment programme of priority pilot projects. One pilot project, the restoration of the Sudoche wetlands, was designed in detail, which in 2007 was successfully implemented.

Assessment context

The main objective of the study was to bring a halt to and, if possible mitigate, the deteriorating environmental conditions and the detrimental effects on the local population in the Amu Darya Delta by advocating wetlands restoration. The objective links human well-being directly to environmental conditions. Even although ecosystem services were not mentioned as such, the project description very obviously referred to the concept. In their work plan, the consultants took the ecosystem services of a dynamic seminatural wetland system as the point of departure, and used valuation of these services as a means to structure the decision-making process on a future development strategy for the delta.

The study has all of the characteristics of an SEA integrated into a strategy development process. It started with a baseline study on major environmental, hydrological, and socioeconomic issues in the region. The strategy development process was based on the development and comparison of alternative strategies, in a participatory manner, making use of local knowledge, aimed at providing relevant social, economic, and environmental information for decision making on the future development of the Amu Darya Delta. A large number of local experts/scientists summarised available scientific information, existing plans, and ongoing activities in the area. This information proved to be extremely valuable.[1]

The possibilities to restore wetlands in the Amu Darya Delta depend on the future availability of water which cannot be managed by interventions in the delta. Three scenarios of future river discharge were considered: availability of water in the delta will (i) decrease due to further wasteful irrigation practises; (ii) will increase due to a successful Aral Sea programme; and (iii) will not change. For the development of strategies the first option was considered to be unworkable, because less water would render any restoration effort useless. Within these scenario boundaries, five alternative strategies were developed which differed in the surface area of wetlands to be restored, the amount of water allocated to each watershed, and in mixed or separate use of river discharge, and (saline) drainage water from the irrigation schemes.

---

[1] Significant inputs have been provided by the Central Asian Scientific Research Institute for Irrigation (SPA SANIIRI), the Design Institutes Uzgipromeliovodkhoz and Vodproekt, the State Committee for Nature Protection (Goskompriroda), Karakalpakvodhoz based in Nukus in Karakalpakstan, and the Karakalpak Branch of the Uzbek Academy of Science in Nukus.

Table A.6 *Most important ecosystem services of the Amu Darya delta ecosystems.*

Lakes and marshes	Floodplains	Drylands (groundwater)
Maintenance of groundwaterlevel counteracting desertification (prevention of dust transport by winds).		
Maintenance of biological diversity (medicinal herbs, genetic resources, etc.)		
Fish reproduction and growth		
	Quality and regeneration capacity of pastures (livestock)	
Water supply for agri/aquaculture		
Reed production for construction/processing		
Hunting for musk rats (and water fowl/other animals)		
	Wood and liquorice production	
	Protection of infrastructure	

Ecosystem services and valuation

Three main ecosystems were identified in the Amu Darya Delta, providing key ecosystem services: permanent lakes and marshes, seasonally flooded plains, and drylands with groundwater at 2–5 metres supporting dense vegetation (Table A.6). The larger part of the delta nowadays consists of degraded steppes that no longer function as a part of the delta ecosystem and do not provide any relevant services. The upstream half of the delta has been converted to irrigated land.

The Amu Darya Delta ecosystem services were first determined qualitatively and later on quantified where possible, based on information from local scientists and a socioeconomic survey. Services were assessed for three situations:

(1) The former natural state when 90 percent of the delta could become flooded during summer floods;
(2) The present state, leaving only 10 percent of the original wetland area, mainly artificially maintained; and
(3) Restoration potential with the presently available quantity of water.

Social, economic, and ecological values derived from wetland ecosystem services were quantified in semiquantitative terms. The values referred to are (estimates of) numbers of beneficiaries, jobs, or production levels of various land use forms. For the pilot project a number of services were monetised in a financial and economic cost–benefit analysis. Local scientists from the Nukus Academy of Sciences, government agencies of the autonomous region of Karakalpakstan, and representatives from the delta population provided input.

Maintenance of biodiversity was supported by various legal instruments: Five mammal and eight fish species were listed in the 'Uzbekistan Red Book' of

Table A.7 *Simplified ecosystem services – values matrix for Amy Darya wetlands.*

Wetland services	Social values	Economic values	Ecological values
Recharge of groundwater	Fundamental function for the maintenance of all other ecological processes		
Prevention of dust/salt transport by wind	living conditions/ health	Protection of irrigation schemes	
Maintenance of biological diversity.		genetic reservoirs (wild ancestors/ medicinal)	Many red listed/ threatened species.
Fish spawning/ nursing		fisheries and canning plant	survival aquatic organisms.
Pastures		cattle raising	
Reedlands		processing industry	
Water supply		agriculture, aquaculture	
Muskrat, waterfowl	Local hunting (meat /skins)	Fur & meat industry	
Liquorice production and other wood resources	Fire and construction wood for local use.	Liquorice roots for export. Dried plants for fodder.	

threatened species – some considered extinct. Approximately 13 of 22 threatened bird species in Uzbekistan occur in the delta.

With the available information an ecosystem services–values matrix was constructed to provide insight in the multifunctional character of the natural environment in relation to human activities (Table A.7).

In order to perform a multicriteria analysis, a decision hierarchy for evaluating alternative water management strategies for the delta was constructed, based on the valuation of ecosystem services (Figure A.2). A decision tree was constructed during a workshop with all involved international and local experts, reflecting the outcome of intense debate. Components were based on the values of wetlands for society, which were divided into the three main groups: (i) living conditions, (ii) local economy, and (iii) ecology. Criteria were, where necessary, further divided into subcriteria. For example, resource productivity was subdivided into livestock,

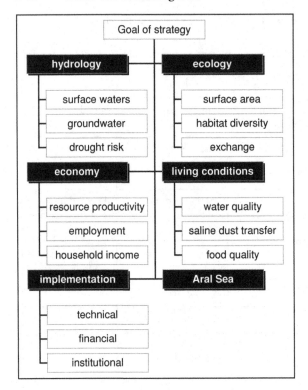

Figure A.2 Decision tree for strategy selection.

fisheries, reed industry, liquorice industry, and muskrat (fur) industry. Because water is such an overwhelmingly important aspect, hydrology was considered to be a fourth main component taking into account neglected values or those of possible future importance (e.g. tourism).

Two other important components for the choice of a strategy were defined: (i) the implementation feasibility of the proposed strategy, and (ii) the existence of the Aral Sea. (Any management strategy for the delta has its impact on the quality of the Aral Sea itself.).

The performance of each criterion under the five strategies was compared and ranked (e.g. for surface water strategy 1 performs better than strategy 4, 4 performing better than 3, etc.). This ranking was done during a workshop in which all local and external experts were present, with additional input from local special-interest organisations. The outcome of the exercise was broadly acknowledged.

The final weighing of the main components was done in a workshop with high-level regional and national decision makers who had to determine whether, for example, the economic component should carry more weight than the ecological component. The outcome of this weighing is an overall ranking of the different strategies, reflecting the input of expert and local stakeholders in the criteria and the decision makers' input in the relative importance of the components. This intense process created a sense of shared responsibility for the outcome between local stakeholders and local and national authorities.

The final step in the project was to define and prioritise a series of concrete pilot projects, based on the chosen strategy. A programme design, including cost–benefit analysis, was included.

Valuation of ecosystem services in the CBA of the pilot programme

The monetary benefits of the pilot programme have been calculated on the basis of 'with minus without' the project (incremental benefits). The calculated incremental benefits of the pilot project are based on the main product categories (or provisioning services) of the project area, which are muskrat, ducks, cattle, liquorice, reeds, fisheries, and aquaculture. The benefits have been calculated on the basis of off-farm prices only. No downstream activities that may create added value of products (e.g. production and sale of canned fish) have been included. Other additional benefits not included are the reduced costs for repairs of civil works (restored wetlands control erratic floods), reduced production losses, and the value of constructions for added safety and health of the local population.

This resulted in an internal rate of return (IRR) of 10.9 percent versus 10.2 percent for the financial versus economic CBA. The conclusion was that the programme was acceptable, especially in view of the many additional benefits and downstream effects. The project may, for example, give rise to additional investments, such as fish farming, rehabilitation of a fish cannery, and increased small-scale agricultural activities for vegetables and fruit trees. These activities are not necessary for the pilot project, and their benefits have not been accounted for in the financial and economic analysis.

The Sudoche Lake Rehabilitation Project was funded by GEF. A preliminary environmental appraisal revealed that further EIA was not required for this project. A rapid five-day appraisal of the impact four years after completion, revealed that (Karimsakov (2006) through De Schutter, personal communication):

• Incomes of both poorest and richest households have increased;
• The numbers of cattle have increased;
• Production of hay for own use and selling on regional market has increased;
• Cutting of reeds and selling of reed-fiber mats (boards) has increased;
• Fish consumption has increased up to 15 kg a week per family;
• Population of muskrats increased.

According to the Ministry of Agriculture and Water Resources of the Republic of Karakalpakstan, fish catches in Lake Sudoche have increased from 20 tonnes before restoration in the year 2002 to more than 110 tonnes in 2005, 2 years after the restoration. In all visited villages, proof of improved well-being of the local population is visible (new motorcycles, boats, fishing nets, satellite antennas, new buildings, as well as herds of sheep and goats). Maybe the most convincing argument of all: the number of young families has increased. Expansion of restoration efforts further downstream is considered possible with the available amount of water. A more-detailed quantitative study would be interesting, but it is not foreseen.

Decision making

With the outcome of the final weighing for the selection of the best strategy, high-level decision makers were shown to be very well aware of the ecological disaster taking place in the delta. Ecological values received highest rankings, followed by socioeconomic values and implementation feasibility, rating equally important. From a strategic perspective it was apparent for all, that restoration of the ecology of the area had to receive first priority because 'everything else depends on the health of the environment', as pointed out by one representative. The Aral Sea rated lowest, indicating that people, for the time being, have given up on the Aral Sea itself in favour of the delta area.

The services–values matrix was a helpful tool as it provided immediate insight in the social, economic, and ecological consequences of interventions. Presenting the matrix for the former, the present and possible future restored situation proved to be a very strong tool to convince decision makers of the values of wetlands. It proved to them that restoration of (natural) wetland services might be a better option than the continued construction of water retention and irrigation works. The former, with its focus on one service only, that is, water supply for irrigation, denies that other services exist. Moreover, multifunctional wetlands can cope with the dynamics of the delta system and stops further land degradation.

The presence of many threatened animal species provided important arguments for the donor to invest in the pilot project, although the main arguments to start the pilot project were of a socioeconomic character.

Valuation of ecosystem services was instrumental in changing the course of development from technocratic and unsustainable interventions towards the restoration of natural processes, which are much better capable of creating added value to inhabitants under the dynamic conditions of the water-stressed delta. The process followed created a strong coalition of local stakeholders and authorities, resulting in necessary pressure to convince national government and the donor community to invest in a pilot project.

Box A.3. Comprehensive Everglades Restoration Program – Florida, USA

The Greater Everglades ecosystem covers more than 69,000 km^2 and is a mosaic of interrelated terrestrial, freshwater and marine systems. Changes in land use and hydrology reduced the spatial extent of the Everglades wetland system to less than 50 percent of its original area by 1990 and dramatically altered the natural flow of water. Agricultural drainage waters caused eutrophication. In December 2000, the U.S. Congress approved the US$7.8 billion Comprehensive Everglades Restoration Program (CERP). Fifty-two engineering projects with associated land purchases were proposed. The initially calculated restoration costs included land acquisition (US$93 million), construction (US$218.3 million), operations and maintenance (US$81 million), and monitoring (US$10 million) expenses in an average year during the 50-year planning period.

Annual use benefits amounted to US$29.2 million resulting from additional water supplies to agricultural and municipal water users in South Florida. In the initial studies no estimates were provided for recreational or nonuse benefits. If the usual criterion that total benefits equal or exceed total costs is applied, the unquantified recreational and nonuse benefits would need to be around US$370 million per year. Milon and Hodges (2000) and Milon and Scrogin (2006) point out that public policy should not be based on a presumption of very large nonuse benefits for each and every ecosystem restoration project. They refer to a 1998 interview survey among 500 households where respondents were asked to select between alternative plans that differed in the extent of ecosystem restoration and dollar costs the household would pay through increased utility taxes, restrictions on household water use, and reductions in farmland acreage in South Florida. This resulted in a willingness to pay ranging from US$54 million to US$355 million annually.

This type of survey shows significant nonuse benefits accruing to Floridians from Everglades restoration, and Floridians express a willingness to pay a significant part of the estimated costs. Whether these benefits extend to people who live outside Florida and whether non-Floridians are willing to make a financial commitment to Everglades restoration remains to be determined. Englehardt (1998) provides a similar argument by turning the reasoning around. Billions of dollars have been contributed voluntarily to pay for the protection of natural areas. This success suggests that society values environmental benefits in preference to other consumption, and that remaining natural areas can be protected in public policy decisions by evaluating ecological benefits explicitly.

SEA boundary conditions

Duration of the strategy development, including all preparatory studies and participatory process, was 12 months. The total cost of the studies amounted to US$1 million. The investment cost for the proposed programme of projects was US$20 million. The Sudoche pilot project was implemented at an approximate cost of US$4 million.

The level of detail was sufficient for a MCA exercise. By focussing on multiple values of ecosystem services instead of translating services directly into monetary values, it became apparent for local stakeholders as well as government representatives that ecological values, expressed in their own terms, received highest ranking. Discussing values expressed in their own terms, and more importantly, recognising stakeholders for each ecosystem service, did not distract the discussion to aggregated figures on money.

In a later stage, when discussing concrete investments, cost–benefit analysis was the proper tool to provide sufficient and convincing arguments that the investments are justified. First-hand accounts from the region give the impression that the decisions turned out to be good ones.

Case based on the author's personal files and the following additional sources: Euroconsult and the Wetland Group (1996), de Schutter (2002), and Dukhovny and de Schutter (2003).

Strategic Catchment Assessment in uMhlathuze municipality, South Africa

Main messages

- Identification and valuation of ecosystem services can inform a local spatial planning process on development constraints and opportunities.
- Monetisation of ecosystem services puts environmental considerations on the decision makers' agenda.

Introduction to the case

The towns of Richards Bay and Empangeni are situated approximately 200 km north of Durban, Kwazulu-Natal, overlooking the Mhlathuze Estuary. Richards Bay is the closest port to Johannesburg, South Africa's economic centre. In 2002, Richards Bay and Empangeni as well as the surrounding rural and tribal areas merged to form the City of uMhlathuze, with 300,000 inhabitants, covering 796 km^2. Unemployment is high (41%). However, economic activity in tribal areas, such as production for own use, arts and crafts, and informal sales, are generally disregarded (uMhlathuze Municipality, 2004). The tribal population creates their own informal employment, thus highlighting the importance of an environment providing free ecosystem services to sustain their livelihoods.

Industry has consistently shown the highest growth rate in the country. With the natural environment already 75 percent transformed, it is evident that conflict between the environment and development will continue to grow in uMhlathuze, unless proper planning takes place. Biodiversity issues in the City of uMhlathuze have lead to various conflict situations during the past couple of years. The classic 'development' versus 'conservation' situation exists, with the local municipality mostly in favour of development as a result of the poor socioeconomic climate that exists in Kwazulu-Natal. The area has, however, been identified as a biodiversity hotspot, and in order to alleviate the conflict and time delays that arise during Environmental Impact Assessments, the uMhlathuze Municipality opted to undertake a Strategic Catchment Assessment.

Context of the case study: the planning process

Environmental sustainability and quality of life are becoming major points of focus for politicians and officials at the local level involved in development planning. A combination of growing community awareness and new legislation is the key driver behind this new focus. The uMhlathuze Municipality has the task to enable sustainable development, which inevitably leads to conflict between environmentalists and developers during EIA procedures because of two key reasons:

- Few workable processes are in place to guide planners towards sustainable development; and
- Very little environmental information is available to inform planning decisions.

Critics are largely arguing that the uMhlathuze Municipality has no 'plan' for the management of its natural biodiversity assets and therefore every piece of untransformed land that is proposed for land conversion has to be rigorously challenged during EIA processes. At this moment, the Municipality has no means or criteria to judge the role or usefulness of any particular land parcel in terms of its use for sustainable development or conservation. This lack of direction gives critics ample scope for litigation and legal challenges. Therefore, in order to ensure sustainable land use planning and decision making, the City of uMhlathuze appointed FutureWorks as consultants, who developed a catchment-based process for assessing, incorporating, and monitoring environmental sustainability into strategic planning.

All municipalities in South Africa are required by the Municipal Systems Act (Act 32 of 2000) to undertake an Integrated Development Planning (IDP) process to which SEA can add value, by providing a practical guide to integrating the concept of sustainability into the planning process. The Performance Management Regulations of this Act states that the Spatial Development Framework, reflected in the IDP, must 'contain a strategic assessment of the environmental impact of the spatial development framework'. In terms of the 'White Paper on Spatial Planning and Land Use Management', each Municipality must compile a spatial development framework of which one of the components must be an SEA.

Biodiversity Management Plans, as well as the Invasive Species Monitoring, Control and Eradication Plan, must form part of the Municipality's Integrated Development Plan (IDP). The Integrated Development Plan is therefore a powerful sustainable development tool for Local Authorities through which biodiversity management and planning can be encouraged and linked to existing planning procedures and processes.

Assessment context

The Strategic Catchment Assessment aims to plug key information gaps such that Municipal Planners and Land Managers will have a strategic decision-making tool. The question may be asked: why this project focussed on providing a tool for urban planners and not for environmental practitioners? Planning integrates the social and economic development needs of an area with the environmental resources available to it. Formerly, urban planning has focussed primarily on the finance, skills, and infrastructure available for development. However, this planning focus now has to expand to include the environment as a priority. The Strategic Catchment Assessment (SCA) focussed on evaluating the environmental sustainability status only; it did not assess social and economic issues in the area.

The SCA followed a four-step approach:

(1) For reasons of transparency and to encourage cooperation a Catchment Forum Group was formed consisting of local specialists as well as interested parties,

20 persons in all. Feedback meetings ensured continued stakeholder interaction and decision making.

(2) Hydrological units were defined that contain both the surface and subsurface drainage systems of specific land areas, and ecosystem services were defined in a landscape assessment.

(3) A status quo assessment of the catchment units provided information on the current environmental sustainability of the catchment areas.

(4) Strategic land use planning and management interventions were developed in response to the observations from the present status of each catchment unit. This information should be used to proactively inform strategic and sectoral planning.

The balance between supply of, and demand for, environmental goods and services in each Catchment Unit is determined based on a key set of environmental goods and services demanded by people in the catchment. Each catchment was then rated *red*, *orange*, or *green*. *Green catchments* are in good condition and currently developed within environmentally sustainable limits. They are generally environmental opportunity areas under proper management and proactive action. *Orange catchments* are in moderate condition and are a nearing unsustainable state. These catchments are being stressed by current land use, and environmental quality is declining. A combination of remedial, management and proactive action is required. *Red catchments* are in poor condition and already unsustainable. These catchments are under stress and the environmental quality has already declined significantly. Remedial and management action is required.

Ecosystem services and valuation

Catchments have been shown to be effective environmental entities for assessing the synergistic impacts of urban development and for integrating the environment into urban planning. The Strategic Catchment Assessment Process accounted for the balance between supply of environmental goods and services by the natural environment and the demand for these goods and services by people. These ecosystem services are currently used free of charge.

The Strategic Catchment Assessment revealed that:

- Two of the eight catchment units are rated *red*. The use and demand for environmental services have largely exceeded supply, and remedial measures are needed.
- Five catchments are rated *orange*. The use of environmental services has affected the ability of the natural environment to provide good quality and a high volume of environmental services. In some cases remedial action is required, but for all these areas future development must proceed with caution.
- One catchment is rated *green*. This catchment is a high-opportunity zone for sustainable development, maximising the benefits provided by high environmental service supply.

It is estimated that, in uMhlathuze, the overall value of the ecosystems supplied is approximately R1.7 billion per annum. Nutrient cycling and waste management, water supply, water regulation, and flood and drought management are some of the

Table A.8 *Annual value of individual ecosystem services and of services per ecosystem.*

Value of ecosystem services (Annual value in R millions)		Value of services per ecosystem (Annual value in R millions)	
Atmosphere regulation – CO_2, etc	23,39	Dams and lakes	162.54
Climate regulation – urban heat sinks	0.00	Floodplains – disturbed	32.54
Flood and drought management	244.11	Floodplains – undisturbed	27.42
Water regulation – timing, rate	137.39	Forest – coastal	34.12
Water supply – volume	297.92	Forest – dunes	37.36
Erosion control	16.10	Forest – riparian and swamp	29.62
Soil formation	0.65	Grasslands – primary	9.37
Nutrient cycling	714.90	Grasslands – utility	0.06
Waste treatment – assimilation and dilution	137.74	Grasslands – secondary	4.62
Pollination – legume and fruit crops	1.53	Rivers and streams	49.47
Disease and pest control	9.74	Sandy beaches and foredunes	1.67
Refugia – wildlife and fish nursery	15.90	Thicket – alien plants	3.53
Food production	30.18	Thicket	3.90
Raw materials – housing, medicinals, craft	20.90	Wetlands – estuarine	433.47
Genetic resources – chemicals	2.33	Wetlands	570.89
Recreation	37.73	Savanna/woodlands	9.52
Cultural	67.20	Nearshore ocean	347.62
Total annual value	1,757.71	Total annual value	1,757.72

most highly valued services. If the above results are taken into consideration, it is clear that the value of ecosystems in uMhlathuze is being eroded by unsustainable practises. If the Municipality wants to ensure the continuation of free service delivery by the environment, it would have to put in place management actions (FutureWorks, 2004).

Table A.8 presents the annual value of each of the key ecosystem services supplied by the natural assets of the uMhlathuze Municipality. Because different habitats deliver these services in different combinations, it is important to understand the total value of these habitats. It is clear that water-related habitats generate some of the greatest values in terms of service delivery. Wetlands have a particularly high value, relating to the high costs of trying to replace a vital but finite resource.

The Status Quo Report, prepared for the uMhlathuze Municipality, is presented in four poster-like pages:

- Page 1 – Pictorial Catchment View showing photographs of typical features.
- Page 2 – General Catchment Information: summary of the Sustainability Status Quo including different land covers, catchment population, levels of engineering services, and key environmental services and their value; positive and negative environmental aspects of the catchment.
- Page 3 – Environmental Sustainability Status Quo contains colour coded indicator information for the catchment: '*Red:*' '*Orange:*' and '*Green:*' When comparing different Catchment Units, this page is very useful.
- Page 4 – Implications and Interventions/Guidelines: provides the implications for land use planning and management, including key environmental opportunities and constraints and legal and other implications for current development scenarios.

Decision making

Instead of identifying and declaring conservation-worthy areas a 'no-go', the study stresses the ecosystem services that the environment provides free of charge to this Municipality. The experience has been positive. Politicians reacted negatively to the term 'biodiversity' but more positively once they realised that environmental services have an economic value.

The land cover mapping, produced for the SCA, provides the relevant information that could be used to identify sensitive habitats and linkages between ecosystems that need to be maintained. The Municipality embarked upon a process to negotiate these areas in an effort to identify (i) sensitive ecosystems that should be conserved, (ii) linkages between ecosystems, and (iii) areas that could be developed without impacting on the area's ability to provide environmental services. More importantly, (iv) it would identify the management actions that need to be implemented in the area in order to ensure not only the survival for key biodiversity assets but also the sustainable use of biodiversity resources to benefit all residents of uMhlathuze.

One of the first strategic responses to the assessment was to review the Industrial Expansion Strategy with respect to:

- Destruction of the nationally significant natural habitats;
- Downstream impacts on a regional fisheries resource;
- The Municipality's primary tourism and recreation zone;
- Human and environmental health risks; and
- Water supply constraints.

An Environmental Services Management Policy and Plan has been established with the aim to include provincial conservation targets into local biodiversity planning; to resolve conflict between 'conservation' and 'development' parties and to form a partnership; to alleviate delays during EIA's as a result of biodiversity concerns; to identify sensitive areas upfront in planning and to avoid impacts; to define functional spatial management units for management to optimise the delivery of environmental services; and to develop management plans to secure these services (Jordan, 2006).

This resulted in Environmental Services Management Plan Zones distinguishing different planning zones:

* 'Red' – Conservation Zone (Level 1)
* 'Green' – Open Space Buffer or Linkage Zone (Level 2)
* 'Clear' – Development Zone (Level 3)

The considered way forward is the integration of Environmental Services Zones into the Spatial Development Framework (SDF) and Land Use Management System (LUMS) processes and the establishment of Nature Reserves by means of proclamation for core areas.

Two main lessons indicated by the municipality staff in charge of the process:

(1) The use of ecosystems services and focus on the value of these services for society was of key importance to convince local councils that biodiversity conservation makes economic sense;
(2) Planners are in the best position to influence sustainable development, so educate them.

Box A.4. Environmental Services Management Plan – Durban, Republic of South Africa

The planning of Durban's social and economic uplift agenda relies to a large degree on the most appropriate usage of the city's open spaces. These open spaces provide important services, such as water for sanitation and drinking and raw materials for building, and protect citizens from natural disasters such as floods. Open spaces also enhance Durban's stature and attractiveness as a lifestyle city and a popular tourist destination. Visitors to the city are very often impressed by the abundant greenery and well-maintained open spaces that are available for leisure, recreation, and tourism.

To this end, the eThekwini Municipality has prepared an Environmental Services Management Plan (EESMP) that aims to protect and enhance the value of a network of open spaces throughout the city precincts, and which will ensure the continued supply of environmental services for the benefit of all residents. Various types of open spaces and ecosystems provide varying mixes of environmental goods and services; for example, wetlands are worth around R200,000 per hectare per annum while forests have a value of around R21,000 per hectare per annum. Research in the field is ongoing, but currently available figures are widely accepted as a useful guide and tool for providing 'order of magnitude' estimates of the value of open space to humanity. It has been estimated that the total replacement value of the environmental goods and services supplied by the 2002 open space system is R3.1 billion per annum. It is noteworthy that this excludes the value of the role of open space in the tourism industry of Durban which itself was estimated to be worth R3.3 billion in 2001 (about €330 million).

The Plan was developed from a review and refinement of the potential Durban Metropolitan Open Space System (D'MOSS). Central to the Plan is a system of linked open spaces, which deliver a range of environmental services to the citizens. River catchments were used as logical units to assess the supply of, and demand for, environmental services. There is undeveloped land throughout the Unicity, all of which supplies environmental services. Of this undeveloped land, 62,000 hectares is the critical or minimum amount, identified in the plan, that is required to sustain the supply of environmental services. By including only critical areas in the system to ensure the supply of environmental services, about 18 percent less land is required than in the 1999 D'MOSS Framework Plan.

In March 2003 the EESMP was approved by the eThekwini Municipal Council as their official policy for the planning and management of the city's open space system. This has paved the way for active implementation of the open space plan.

(Case based on eThekwini Municipality (2002).)

Case based on Van Der Watern *et al.* (2004).

Wareham Managed Realignment in the UK

Main messages

- Absolute monetary values of ecosystem services are difficult to establish when relying on metadata or transfer of data from other areas. Local data collection is needed but it is laborious.
- When dealing with different alternatives, relative difference in values provides a good basis for comparison.
- Sensitivity analysis is an important tool to avoid the risk of major errors and to focus efforts for further research on most relevant issues.

Introduction to the case

Near the village of Wareham (Dorsett, UK), 400 hectares of grazing land and 26 properties are protected by 20 km of tidal flood banks. These banks are in poor condition and are nearing the end of their design life. The responsible authority (Environmental Agency – EA) recently started a process to decide what action should be taken to manage flood safety.

Context of the case study: the planning process

Flood and Coastal defence (sea level rise): Since 2002, the national policy on Flood and Coastal Erosion Risk Management has been based on the '*Making Space for Water*'. This is a strategic approach focussing on sustainable management of flood risks in an environmentally, economically, and socially sound manner. On the local level this approach is elaborated in Local Shoreline Management Plans (SMPs).

All SMPs are subject to public consultation and take into account wider social, environmental, and economic objectives.

Under the *Habitats Regulation* compensation for the loss of habitats is required. One of the consequences from climate change is loss of intertidal habitats (salt marshes). These can be created in Wareham when the managed realignment options are practised (see below).

For any project, plan, or change in present flood risk management a policy appraisal is required, paying attention to environmental costs and benefits. Because several impacts on ecosystem services will need to be taken into consideration, there is a wider scope for the use of economic valuation techniques as part of the appraisal process.

Assessment context

An earlier policy appraisal, focussing on the promising alternative of managed realignment, did not apply economic values to any environmental or nonmarket impacts. That is why in this case study economic values are applied to ecosystem service changes under different scenarios.

Scenario's include the following:

(1) *Doing nothing/baseline*: This is not a viable option but used as baseline;
(2) *Do the minimum*: A continuation of the present level of maintenance that delays failure;
(3) *Improvement*: Maintaining appropriate standards of defence in the face of sea level rise;
(4) *Managed realignment (unconstrained)*: Removing all tidal flood banks. Some secondary defences will be maintained in order to protect a small number of key habitats that cannot be replaced elsewhere.
(5) *Managed realignment (constrained)*: Removal of tidal flood banks in some sections, with secondary defences for selected habitats and maintenance of flood banks in other sectors.

The study is conducted as case material in the handbook Economic Valuation of Environmental Effects that is being developed for the EA. The focus lies on how valuations of ecosystem services in one place (existing studies) can be used in other situations.

Ecosystem services and valuation

Key ecosystem services identified where:

- *Supporting services*: nutrient storage (regulating services: water purification); soil formation, and primary production;
- *Provision services*: loss of (marginal) grazing land, fisheries changes, and nursery function;
- *Regulating services*; carbon storage (climate regulation), erosion regulation is captured in the scope of the appraisal; and

- *Cultural services*: recreation and tourism (fishing, navigation, bird watching, shooting, walking, local business), aesthetic values (warning for double counting), and cultural heritage values;

Initial identification of ecosystem services and key costs (flood damage, loss of land, new habitat creation, recreation, archaeology, and local economy) where discussed in a workshop. Key stakeholders provided comments and more detailed information. Quantitative valuation in the case study is based on data transfer from other studies in similar (wetland) circumstances:

- Agricultural land is 65 percent of the agricultural market value;
- Recreational services are quantified but rarely in monetary terms (travel cost, stated preference). Here entrance fees are used.
- Other services (fisheries, carbon storage, nutrient cycling, navigation) are not included in appraisals because they are not translated in monetary terms.
- Focus lies on the economic valuation of habitat changes, based on existing valuation studies (metaanalysis) of wetland habitats. The main methodological problem was that it is not always clear which services (fisheries, water supply, etc.) are included in the analysis of the wetland type value. Metaanalysis is not suited to the valuation of specific services, because functions are used to provide a general 'habitat' value. It may be possible that some or all of the other services identified are already included in the general habitat value estimates. Besides, values are given in absolute figures while alternative use (of the wetland type) is always possible (other types of tourism or fisheries). Concerning habitat valuation in relation with compensation, valuation is different, because compensation is legally required; than valuation is irrelevant. Last, values are given for a certain year (say 2006) and cannot be used directly in, say, 2012.
- The above reasoning is also true for recreation values; they may be included in the general wetland ecosystem services benefits.

So, there is substantial uncertainty about most elements of the case study and the appropriate monetary values to apply to these due to:

- Scientific evidence gaps on a. o. timing of defence failures, sea level rise, types of habitat created, sediment availability, carbon and nutrient sequestration in different habitat types, fish population, and species/people using different habitat types;
- Economic/human behaviour evidence gaps on a. o. river-based tourism, trips across flood banks, recreational boating and recreation in new habitats;
- Valuation evidence gaps on archaeological values, recreational values, nutrient sequestration values and dealing with uncertainties;
- Methodological uncertainties: see above.

Results of the valuation can be divided into: (a) estimates of the absolute value of the ecosystem services and (b) estimates of the differences in value between the baseline and each of the other options.

When valuing the ecosystem services, the *improve* option appears to be by far the least beneficial option as well in (a) as in (b). The absolute as well as the relative value of ecosystem services is the highest in option *MR unconstrained*, with *do nothing* and *MR constrained* second and third. Additional sensitivity analysis demonstrates that the absolute value of the ecosystem services is highly uncertain and it suggests that the *improve* option is a poor performer under a wide range of assumptions.

Habitat valuing makes the difference between *do nothing* and *MR*: the increasing area of reed beds in *do nothing* is valued higher than the grazing lands that are protected in *MR* variant. The relative value of a reed bed thus is a key issue to be examined further. A relatively small adaptation in values can change the balance in favour of another option.

The main barriers for incorporation of impacts on the value of ecosystem services into project appraisal are:

- The high level of uncertainty regarding data and economic value estimates (scientific data: habitats created, exact timing of defence failures, sea level rise, sediment availability etc; data on economic/human behaviour: tourism, boating, etc.). Via an audit system uncertainties should be documented so that other parties can always verify results;
- The complexity involved in applying benefits from one situation to another and the implications for those conducting appraisals (methodological issues).

The potential for valuation to provide useful input to policymaking, discussions, option formulation and prioritisation is recognised. However, valuations should be robust. In practise this is difficult because of the barriers mentioned. Introduction of standard values, clear guidance, templates, and training will help.

Decision making

On the balance it seems highly likely that the favoured *MR* options will be significantly better in Net Present Value terms. However, these results are preliminary and highlight a number of key uncertainties: recreation, local economy, fisheries, and so forth. A next stage will focus on how to distinguish between various MR options. It is a staged process, and ecosystem service valuation can be helpful in planning and prioritising stages. The case study does not go into detail on the influence on decision making, because it is still too early in the process. However, the authors of the case study argue that inclusion of ecosystem services in the appraisal provides evidence that, in the context of climate change and biodiversity loss, some schemes have a higher cost–benefit ratio. This makes expenditure of public money more efficient.

Case based on Eftec, Economics for the Environment Consultancy Ltd. (2007).

Climate policies and the *Stern Review*

Main messages

- The most far-reaching policy changes for improving the functioning of ecosystem services can be achieved by making the Treasury the champion of the economic valuation study. They have both the authority and the means to follow up on the recommendations.
- Boundary conditions, such as timing, communication, and ownership, are more important in terms of generating societal impact than the quality of the study.

Introduction to the case

The global climate is an overriding condition for the functioning of all ecosystems and the ecosystem services that they provide. Conversely, ecosystems themselves play an important role in climate stabilisation and regulation. The intricate relations between climate and ecosystems work at a global level; from a climate perspective the globe is functioning as one large ecosystem. Changes in the global climate lead to fundamental changes throughout the world's ecosystems and therefore also affect the economic sectors that depend on these ecosystems. The *Stern Review* is one of the best-known assessments to estimate the economic impact of climate change.

The *Stern Review on the Economics of Climate Change* was released on 30 October 2006 by economist Lord Stern of Brentford. It discusses the effect of climate change and global warming on the world economy. Although the economic side of climate change has received more attention, this report is the largest and most widely discussed report. A team of economists at HM Treasury prepared the review, while independent academics were involved as consultants. The *Stern Review* emphasised the need for urgent action to combat and mitigate climate change.

Context of the case study: the planning process

The Chancellor of the Exchequer, Gordon Brown, announced on 19 July 2005 that he had asked Sir Nicholas Stern to lead a major review of the economics of climate change. This in order to better understand the nature of the economic challenges and to understand how these challenges can be met both in the UK and globally. The fierce political debate preceding the request from Gordon Brown concentrated around several issues:

- *Lack of political consensus on climate change in the UK*: This motivated the former UK Chancellor of the Exchequer, Nigel Lawson, to write a letter to *The Times* criticising this political divide.
- *Lack of knowledge on the economics of climate change*: The House of Lords Economics Affairs Select Committee already undertook an inquiry into the economics of climate change in 2005. The committee recommended the UK Government to make greater efforts to assess the costs and benefits of climate change mitigation and adaptation.
- *UK's divide on the position regarding the Kyoto Protocol and the Intergovernmental Panel on Climate Change (IPCC)*: This divide was illustrated by a debate between two eminent scientists. Michael Grubb, Chief Economist of the Carbon Trust, said that Kyoto's targets helped to achieve the actions asked for by the House of Lords report. David Pearce, senior advisor to the House of Lords committee, criticised IPCC for a lack of input of economists in climate policy-making. He recommended HM Treasury to take a more active role. The House of Lords report pointed at the mismatch between the costs and benefits of climate policy as estimated by independent academics and as assumed by politicians.

Ecosystem services and valuation

The *Stern Review* starts by explaining the possible climate change scenarios. If climate change goes unchecked average temperatures could rise by 5°C from preindustrial levels. Even although this affects all countries, the poorest countries will suffer earliest and most. Global warming will result in more serious floods affecting many millions of people. By the year 2050, rising sea levels may result in 200 million permanently displaced due to rising sea levels, heavier floods, and drought. Global warming is likely to also seriously affect global food production and could result in the extinction of 15–40 percent of species.

Since the industrial revolution the level of greenhouse gases in the atmosphere has risen from 280 parts per million (ppm) CO_2 equivalent (CO_2e) to a current level of 430 ppm CO_2e. Anything higher than 450–550 ppm CO_2 would substantially increase the risk of very harmful impacts. Lower values might not be feasible as this would impose very high adjustment costs in the near term.

The main message of the *Stern Report* is that what is presently being done has a limited effect on the climate over the next 40 or 50 years. However, what we do in the coming 10–20 years may have a profound effect on the climate in the second half of this century. The benefits of strong, early action considerably outweigh the costs. Climate change could cost the world at least 5 percent of GDP each year, or even up to 20 percent according to more dramatic predictions. Emission reduction costs could be limited to about 1 percent of global GDP. The damage of each tonne of CO_2 emitted amounts to at least US$85. Yet, emission reduction costs less than US$25 a tonne. A less carbon intensive world could eventually save the economy some US$2.5 trillion a year.

Stern characterises climate change as 'the greatest and widest-ranging market failure ever seen' (Stern, 2006). According to the review several policy element are needed to correct this market failure:

• *Carbon pricing*: taxation, emissions trading or regulation, resulting in higher charges for carbon-intensive goods to show the full social costs of people's actions. The aim should be a global carbon price across countries and sectors. Emissions trading schemes, like that operating across the EU, should be expanded and linked.
• *Policy on technology*: stimulate large-scale development and use of low-carbon and high-efficiency products. Energy research and development support should at least be doubled; support for low-carbon technologies should be five times higher. Introduction of international product standards.
• *International action*. First, integration of climate change into development policy, and with an increased support for development assistance by the industrialised countries. Second, international funding to (i) explore the best ways to stop deforestation (urgent action required); (ii) improve regional information on climate change impacts; (iii) develop crop varieties more resilient to drought and flood.

The *Stern Review* has been under heavy criticism by some economists, for the discount rate he used in his calculations, for not considering costs past 2200, and for the argument that significantly slowing climate change requires deep emission cuts everywhere (Tol and Yohe, 2006; Nordhaus, 2007). A comprehensive overview of the concerns is provided in Table A.9. These critiques come at no surprise since

Table A.9 *Reasons for concern over damage estimates in the Stern Review (source: Yohe and Tol, 2007).*

Source of concern	Reason for concern
No new literature and no new models supporting damage estimates	Damage estimates are three standard deviations higher than the mean of earlier peer-reviewed estimates
Impacts of climate change	*Water*: does not address adaptation
	Sea level rise: does not address adaptation
	Food: ignores growth
	Health: ignores growth
	Refugees: uses most pessimistic scenarios
	Catastrophic risk: double-counts its sources
Very low discount rate employed in damage estimates	Future impacts weigh heavily
	High residuals past 2200
	Leads to inefficient investment
Mitigation cost estimates truncated at 2050	Mitigation must continue past 2050
No justification of the 550 parts-per million target	Lower target implied
	Damages metric not comparable

earlier studies showed a much less severe impact of climate change than the cost estimates published in the *Stern Review*.

Box A.5.   Cost of Policy Inaction on Biodiversity

Impressed by the impact of the *Stern Review on the Economics of Climate Change*, the European Commission requested a consortium of researchers to conduct a similar study for the economic impact of biodiversity loss. The study was commissioned in late 2007 and the first results were presented on 29 May at the 9th Meeting of the Conference of the Parties (COP9) to the Convention on Biological Diversity in Bonn.

One of the main components of the economic review is the assessment of the Cost of Policy Inaction (COPI) with regard to the conservation of biodiversity. The COPI research was carried out by a consortium of institutes led by Alterra in the Netherlands. The COPI study starts off by stating that the 2010 targets for biodiversity will not be reached under continuation of the present biodiversity policies. It demonstrates that the absence of additional biodiversity policies come at a considerable price. These costs result from the fact that natural systems will no longer be able to supply valuable services, such as carbon storage in forests and the supply of sufficient amounts of clean freshwater. A first, very rough estimate shows this loss to be around 7 percent of the Gross World Product (GWP) by 2050.

Some examples of annual losses (between brackets in billions of €) related to policy inaction for a number of ecosystem services: food, fiber, fuel (– €192 billion), air quality maintenance (– €2,019 billion), soil quality maintenance (– €1,856 billion), climate regulation (– €9,093 billion), water regulation and purification (– €782 billion), culture and recreation (– €303 billion), resulting in a total economic value for all calculated ecosystem services of – €13,861 billion. (Case based on Bratt and Ten Brink (2008).)

Decision making

The *Stern Review* attracted more attention than any other economic valuation study in history. Influential people from all over the world were inspired by the *Stern Review* to stress the urgency of immediate action.

A range of *industrial stakeholders* responded positively to the *Stern Review*, considering the climate challenge as a business opportunity rather than a threat to their industry. For example, fourteen of UK's leading companies who met in the Prince of Wales' Corporate Leaders Group on Climate Change, expressed a desire to discuss how Britain could obtain 'first mover advantage' in a 'massive new global market'. Obviously, more conservative responses were also made by, for example, UK and EU industries. These industries fear their competitiveness will certainly suffer if countries, such as the United States, China, or India, would not make decisive climate commitments.

A number of esteemed *scientists* embraced the *Stern Review*, stressing the need to act now rather then later, thereby advising governments worldwide to follow the recommendations made by the *Stern Review*. These included a number of economic Nobel Prize winners, such as Robert M. Solow (Nobel Prize economist 1987), James Mirrlees (1996), Amartya Sen (1998), and Joseph Stiglitz (2001). The latter stated that the *Stern Review* 'makes clear that the question is not whether we can afford to act, but whether we can afford not to act. To be sure, there are uncertainties, but what it makes clear is that the downside uncertainties –aggravated by the complex dynamics of long delays, complex interactions, and strong nonlinearities – make a compelling case for action'. (HM Treasury, 2006). Although, the scientific community was certainly not unanimous about the underlying assumptions and calculations, the majority supported Stern's conclusion that the expected benefits of tackling climate change far outweigh the expected costs.

The most significant impact of the *Stern Review* was seen in the *policy* arena. A number of governments responded by announcing expansion of their climate policies. British Prime Minister, Tony Blair, stated that the Review demonstrated 'overwhelming' scientific evidence of global warming and its 'disastrous' consequences if the world failed to act. A number of measures would follow soon after this recognition (see the next section). Australian Prime Minister, John Howard, announced AU$60 million in projects to help cut greenhouse-gas emissions. Much of this funding was directed at the nonrenewable coal industry. The European Commission also strongly acknowledged the *Stern Review* by stating the need to act now.

Since its publication, the team of the *Stern Review* widely discussed the results of the *Review* with policymakers, academics, and business leaders across the world, in particular, in the EU, China, India, Japan, Indonesia, Africa, the United States, Canada, and Australia. Being struck by the progress on developing policy to reduce energy intensity as well as overall emissions, Stern published a working paper in which he reflected on these positive policy changes (Stern, 2007a, 2007b). Among others, he reports the following progress in climate change policies from around the world:

- In the *UK*, the Climate Change Bill was introduced in Parliament on 14 November 2007 and completed its passage through the House of Lords on 31 March 2008. It will shortly go to the House of Commons for consideration. The aim is to receive Royal Assent by summer 2008. The Climate Change Bill contains provisions that will set a legally binding target for reducing carbon dioxide emission in England and Wales by at least 26 percent by 2020 and at least 60 per cent by 2050, compared to 1990 levels.[2] The proposed Scottish Bill is expected to have even more stringent targets for the reduction of emissions.
- The EU took a more stringent position on carbon trading, sending a strong signal on the role of carbon markets at the centre of the EU's strategy to deliver deeper emissions cuts. The member states also agreed on a new independent EU commitment to reduce greenhouse gases by at least 20 percent by 2020, and pledging to go further up to 30 percent compared to 1990 levels by 2020, as part of an international agreement by other developed economies. Moreover, the EU has agreed mandatory targets on renewable sources in energy use, and the phasing out of the traditional light bulbs.
- *China* is beginning to implement energy efficiency measures, which involve energy efficiency and changes to taxation of vehicle sales. Also, a new tax on energy-intensive products for export has been introduced.
- In *India*, the Integrated Energy Policy under the 11th Five Year Plan is being taken forward – including changes to energy subsidies, plans for more efficient coal-fired power plants, and further development of innovative new technologies for renewable energy.
- Willingness to act in *Japan* is mainly shown during debates between government, industry, and civil society on the challenges of designing further domestic and international action. Rapid technological progress is made in plug-in hybrid vehicles and solar technology. Japan's responsibility in climate issues in trade and foreign investment, particularly with China and India, is also recognised.

[2] In more detail, the Bill involves the following aspects (www.parliament.uk):

- Requires the Government to publish five yearly carbon budgets as from 2008;
- Creates a Committee on Climate Change;
- Requires the Committee on Climate Change to advise the Government on the levels of carbon budgets to be set, the balance between domestic emissions reductions and the use of carbon credits, and whether the 2050 target should be increased;
- Places a duty on the Government to assess the risk to the UK from the impacts of climate change;
- Provides powers to establish trading schemes for the purpose of limiting greenhouse gas;
- Confers powers to create waste reduction pilot schemes;
- Amends the provisions of the Energy Act 2004 on renewable transport fuel obligations.

- In *Africa*, climate change has risen sharply up the agenda. By making climate change one of the key themes for the Summit in January 2007, African leaders have become increasingly aware of the vulnerability of their countries, as well as the opportunities for adaptation, sustainable land management and low-carbon development.

- Also in the *United States*, there have been some significant initiatives to reduce dependency on fossil fuels, and some states, cities and businesses have set objectives to limit greenhouse-gas emissions. For example, California has committed to making a 25-percent reduction in emissions compared to 1990 levels by 2020, and 80 percent reductions by 2050. Regional emissions trading schemes, covering most of the North Eastern states of the United States, are developing rapidly. At the national level, the government presented plans to improve efficiency, reduce emissions and improve energy security particularly in the transport sector.

Box A.6. Carbon trading in the Iwokrama Forest, Guyana

The Iwokrama Forest in central Guyana has nearly one million acres (371,000 hectares) of mostly primary rainforest. It is managed by the Iwokrama International Centre for Rainforest Conservation and Development with the aim to show how tropical forests can be conserved and sustainably used to provide ecological, social, and economic benefits to local, national, and international communities.

By integrating human needs and values into business development and conservation strategies, partnerships with local communities are established by Iwokrama so they can assist in forest management and get direct benefits through joint business development such as ecotourism and sustainable forestry.

In late March 2008, the Centre announced a deal with a British-led international investment company with the aim to secure the future of the Iwokrama forest, while opening up the way for financial markets to play a key role in safeguarding the fate of this forest in Guyana (WBCSD 2008; Howden, 2007). The director of Canopy Capital, who sealed the deal with the Iwokrama rainforest, said: 'How can it be that Google's services are worth billions but those from all the world's rainforests amount to nothing?' The deal is the first serious attempt to pay for the ecosystem services provided by rainforests, such as carbon storage and other ecosystem services. Canopy Capital expects to sell the carbon storage and other rights at a profit within 18 months.

In late 2002, an economic valuation study was conducted in which a first estimate was made of the Total Economic Value of the Iwokrama Forest (Van Beukering and van Heeren, 2002). Sustainable management of the forest was estimated to generate higher economic returns in the order of US$15–31 million over a 30-year period, and proved to benefit especially the local communities. The study also emphasises the need to invest in the early stages of management in order to generate higher and more sustainable benefit flows over the longer term. The deal with Canopy Capital seems to follow this route.

SEA boundary conditions

The *Stern Review* was conducted in a period of slightly more than a year. The Review was announced in July 2005 and was ultimately published on 30 October 2006. It was prepared by a team of economists at HM Treasury. Independent academics acted as consultants only. The team was lead by Lord Stern, at that moment Head of the Government Economic Service and former World Bank Chief Economist.

The Stern study is a review of existing studies on the impact of climate change. The data used in the study are therefore not original. Stern based his study largely on the 2001 IPCC and said that the more up to date information from the latest report gives stronger warnings than the 2001 report. In an April 2008 statement Stern said that the 2007 IPCC report vindicated his findings. Critics of Stern blame him for cherry picking, selecting the scenarios and assumptions that fit best to the message he wants to get across.

Additional information on this case can be found in Byatt *et al.* (2006), Carter *et al.* (2006), Dasagupta (2006), Dietz *et al.* (2007a, 2007b), Hamid *et al.* (2007), Henderson (2007), Stern and Taylor (2007), Tol and Yohe (2007), and Yohe and Tol (2007).

Natural gas extraction in the Wadden Sea, the Netherlands

Main messages

- Economic valuation increases the transparency of complex systems. By explicitly highlighting the crucial uncertainties of certain economic activities, environmental conditionality for continuation of projects can be defined in the approval procedure.
- Economic valuation does not necessarily prevent actual implementation of projects that impact ecosystem services, but it may affect the design of the intervention such that costs and benefits are traded off in a rational manner.

Introduction to the case

The Dutch Wadden Sea is a shallow, semienclosed part of the North Sea, mainly consisting of tidal mud flats, sand flats, sea gullies and salt marshes. The area is bordered by a series of dune barrier islands, the 'Wadden islands'. The Wadden Sea stretches along the North Sea coast from Den Helder in the Netherlands up to Esbjerg in Denmark and is the largest tidal wetland area in Europe. Most of the sea and the uninhabited islands are National Nature Reserve, which is regulated by the Nature Conservation Law and a spatial planning act (PKB). The area is owned by the State and managed by the Ministry of Nature management, Agriculture and Food quality (LNV) and Ministry of Public Works, Transport, and Water Management (V&W).

The entire area constitutes approximately 250,000 hectares; the nature reserve is ca. 150,000 hectares. The Wadden Sea is of international importance being a nursery of marine life, a resting, moulting and feeding area for several millions of migratory birds, and a habitat for thousands of birds, seals, and many other species. The area has been selected for European protection as part of the Natura 2000 Network (EUCC, 2008).

The Wadden Sea is not only a region of ecological importance but provides many economic benefits as well. The region, especially the Wadden islands, is a key recreational area for the Netherlands and Germany. Other important activities include, for example, fisheries, military practises, wind energy, and gas exploitation. Amongst these activities, probably because of its extractive nature, gas exploitation has been a key issue in research and policy debate over the past years.

An estimated 200 billion cubic metres of gas are located below the Wadden Sea distributed over several small fields. In 1969, the 'Nederlandse Aardoliemaatschappij B.V.' (NAM), Mobil and Elf Petroland received a concession with respect to gas exploitation from the Wadden Sea. Due to increasing environmental concern, the parties agreed to a moratorium on drilling from 1984 to 1993. In the mid-1990s, the NAM started several test drills from three locations on the mainland. Six gas fields were found and NAM intended to start exploitation again (NAM, 2006).

These plans resulted in public debate and research efforts on the effects of gas exploitation on the Wadden Sea. Van Wetten *et al.* (1999) in their economic valuation study indicated that negative impacts from gas exploitation on ecology and societal losses could amount to 7 to 32 billion guilders (€3 billion to €15 billion). In 1999, the government rejected the plans of the NAM and exploitation was cancelled. However, an advisory board ('Committee Meijer') was appointed to further investigate the actual consequences of exploitation. The committee concluded that negative effects on ecology were very limited and gas exploitation should be allowed under strict regulations. As a result, the government approved the plans of the NAM and gas exploitation has started in 2007. Before the actual licences were appointed, the NAM performed an Environmental Impact Assessment (EIA) that further underlined the conclusion of the 'Committee Meijer': there are no ecological reasons to prohibit gas exploitation.

Context of the case study: the planning process

Nature management in the area is determined by the Key Spatial Planning Decision (PKB Waddenzee), a national planning instrument to combine economic development with environmental protection of the area. Through the various PKB's the government promotes sustainable development by controlling the extent of fisheries, gas exploitation, recreation, tourism, and military activities. The PKB is binding upon all national, provincial, and municipal authorities. The third draft PKB determining the future of the area for the next ten years is still under discussion (EUCC, 2008).

One of the key issues in the policy debate is the exploitation of gas from the Wadden Sea. Gas exploitation from the Wadden Sea is an important contributor to the Dutch economy, providing a yearly benefit of approximately €5 billion per year, which is substantial part of the total revenues of fossil fuels in the Netherlands. In addition, by reducing the dependency on fuel imports from the Middle East,

Dutch natural gas plays an important role in the European Union as well (Ministry of Economic Affairs, 2004).

Besides economic benefits, gas exploitation may also have an impact on the environment. Gas exploitation can, for instance, cause subsidence of the sea floor, which would affect the area's tidal mud flats, sand flats, sea gullies, and salt marshes, including its flora and fauna (NAM, 2006). As the Wadden Sea is an internationally important wetland from nature conservation point of view, the proposed gas exploitation raised fierce public debate.

The 'Soil Subsidence Study Wadden Sea' commissioned by NAM (1998) was one of the first studies to determine the effects of soil subsidence from gas exploitation in the Wadden Sea. The study determined the abiotic and biotic changes in the Wadden Sea and concluded that the effects of gas exploitation on ecology would be very limited.

In response to this study van Wetten *et al.* (1999) published the report 'The Dark Side of Wadden Gas'. They argued that NAM (1998) did not take into consideration the effects on ecosystem services, such as water regulating, drinking water supply, tourism, and so forth. In addition, van Wetten *et al.* (1999) point out that the economic value of these ecosystem services has been underestimated and not properly defined in previous studies. Therefore, they conducted an economic valuation study of the Wadden Sea, including a Cost–Benefit Analysis (CBA) of gas exploitation. The Total Economic Value (TEV) of the Wadden Sea should than be taken into account in decision making instead of the sectoral approach taken before.

Assessment context

In 2005 the NCEA advised several times on the Wadden Sea and its immediate surroundings. The Dutch cabinet's latest viewpoint is that the exploration of gas from fields beneath the Wadden Sea and the Lauwers Lake should in principle be possible, on condition that it remains within natural boundaries. The NCEA's advice on scoping guidelines for the EIA report concentrated on the effects of subsidence on the Wadden system and on the water system of the Lauwers Lake. Because important nature conservation values exist in both areas it will be necessary – in addition to geological, hydrological, and morphological changes – to offer a good insight into the effects on the natural environment. The cabinet suggests a 'hand-on-the-tap' approach: exploration can only take place within the natural boundaries. If these boundaries are exceeded, the production will be reduced or even stopped (a clear example of application of the precautionary principle). Commissioned by the government, the NCEA was involved in the realisation of this approach.

Practically at the same time as the advice on gas exploration, the government started an SEA procedure for alterations to the national spatial plan of the Wadden Sea. As a result of the SEA Scoping memorandum, the NCEA concluded that the issue at stake is how the suggested nature assessment should be interpreted: is it a matter of appropriate assessment or is it – due to (still) lacking conservation targets for qualifying species and habitats – just an indicative picture of the effects on the natural environment? The NCEA advised to offer a deeper insight into the intervention–effect relationship, for every activity and every qualifying habitat and/or species: that is, which mechanisms will be affected, what is the magnitude of the effect, on which scale, how long will the impact last and how long will it take to recover?

Box A.7. Private sector and the Pagbilao mangrove forest in the Philippines

Mangrove forests and swamps are rapidly declining in many parts of the world. This has resulted in the loss of important ecosystem services, including forest products, tidal wave control, breeding ground for fish, and so forth (Spaninks and Van Beukering, 1997). One of the major threats to mangroves in the Philippines is the rapidly increasing aquaculture industry.

The Pagbilao mangrove forest is one of the last remaining mangrove forests on Southern Luzon. Even before World War II, the area was a favourite to poachers who gather the 'bakawan', which is a good material for charcoal. Already in 1975, the 145 hectare area was declared as the Pagbilao Mangrove Experimental Forest by virtue of Bureau of Forest Development (BFD) Administrative Order No. 7 (s. 1975). This declaration provided the necessary protection (and funding) and further poaching was prevented. In the 1980s and 1990s, aquaculture was the main driver of further conversion of the Pagbilao forest.

In 1996, an economic valuation study was conducted with the aim to demonstrate the importance of the Pagbilao mangrove forest. Comparisons were made between the total value of conservation of mangroves and the economic value generated by alternative uses, such as aquaculture and forestry. In addition to economic value, equity and sustainability objectives were taken into account and analyzed according to the perspective of different types of decision makers involved. Combining economic valuation and cost–benefit analysis, the study concludes that if economic efficiency is maximised, conversion to aquaculture is the preferred alternative. However, if equity and sustainability objectives are included in a multicriteria setting, more sustainable use is the preferred alternative (Janssen and Padilla, 1999; Gilbert and Janssen, 1998).

For decades, government funding was insufficient to properly manage this wetland. Partly inspired by the results of the valuation study, the private sector, such as the coal-fired plant owner Mirant Philippines, came in and joined in the effort. The project was dubbed 'Carbon Sink Initiative'; it helped in rehabilitating forests that were able to absorb air pollutants such as carbon. Moreover, new investments were made to revitalise the wetland experimental functions. The mangrove forest is nowadays claimed to be a living proof of successful rehabilitation.

Ecosystem services and valuation

Van Wetten *et al.* (1999), following De Groot (1992), define four ecological functions: (i) regulation functions, (ii) habitat functions, (iii) information functions, and (iv) production functions. For each of these functions, the ecosystem services of the Wadden Sea are selected and valued. Bequest and existence values were not

Table A.10 *Total Economic Value of the Wadden Sea in million euros per year (source: van Wetten et al., 1999).*

Ecosystem function	Valuation method	Value (million €)
Regulation functions		
– CO_2 storage	Benefit transfer	35
– Flood protection	Damage cost avoided	213
– Protection against salt spray	Shadow price	4
– Strategic drinking water supply	Shadow price (replacement costs)	353
– Seawater purification	Benefit transfer	649
– Pest control potatoes	Shadow price	737
– Natural accession land	Shadow price (replacement costs)	0.45
Habitat functions		
– Refuge nature	Shadow price (investments by public bodies)	266
– Breeding ground mussel	Market price	154
– Breeding ground plaice and sole	Market price	803
– Breeding ground shrimp	Market price	92
Information functions		
– Tourism and recreation	Market price	771
Production functions		
– Production mussels	Market price	231
– Production cockles	Market price	21
– Production lugworm	Market price	0.90
– Production shrimp	Market price	83
Total Economic Value (TEV)		**4,416**

included in the study. Using several valuation techniques, van Wetten *et al.* (1999) estimated the TEV at approximately €4.4 billion (see Table A.10).

In contrast to NAM (1998), van Wetten *et al.* (1999) state that gas exploitation can lead to considerable negative effects and damage to the Wadden Sea ecosystem. They estimate the damage to ecosystem functions that depend on sand flats, sea gullies, salt marshes, beaches, and dunes are one-third of their estimated value (these include purification of seawater, refuge nature, breeding ground, recreation, and tourism and production). Estimations of damage to other functions are based on the effects described by NAM (1998). The total costs, in case serious effects occur as a result of gas exploitation, are approximately €1.1 billion (see Table A.11).

The economic benefits of gas exploitation on the Wadden Sea were estimated based on three different scenarios, each with a different time period in which exploitation takes place. Applying a 4-percent discount rate under the baseline

Table A.11 *Costs of gas exploitation in the Wadden Sea (in million € per year) (source: van Wetten et al., 1999).*

Ecosystem function	Costs (million €/yr)
Flood protection	3
Strategic drinking water supply	47
Seawater purification	216
Damage from water to agriculture/houses	10
Loss of land	28
Refuge nature	89
Breeding ground	350
Tourism and recreation	257
Production	112
Total costs	**1,111**

scenario (realization period 2011–2025), van Wetten *et al.* (1999) estimate the aggregated (present) value of the benefits to be between €3 to €18 billion over the same period.

Finally, these results were applied in a CBA for three different scenarios. The least harmful scenario (no damage in years 1–5, 50-percent damage in years 6–10, 100-percent damage in years 11–50), results in a societal loss of €3 to €15 billion as a result of gas exploitation from the Wadden Sea. Van Wetten *et al.* (1999) admit that these results are based on limited scientific knowledge of several of the presented values and include many uncertainties. Therefore, the results might involve double counting and overestimation of costs and benefits. On the other hand, intrinsic values were not included, and contingent valuation studies to estimate such values were lacking.

Decision making

In December 1999, the government eventually decided not to give permission for gas exploitation. However, uncertainties and discussions about the effects of gas exploitation continued. In 2003, the government appointed the 'Committee Meijer', an advising committee, to give an integral advice on the Wadden Sea. The committee published their findings in a report: 'Space for Wadden' (Meijer *et al.*, 2004). They concluded that there are no ecological reasons to prohibit exploitation. The main reason for this conclusion is that the dynamic system of the Wadden Sea compensates for soil subsidence. Due to natural dynamics and the supply of sand and mud from the North Sea, the effects of soil subsidence resulting from gas exploitation will be balanced by increased sedimentation and soil accretion. Gas could be exploited without negative consequences. The committee therefore recommended that gas exploitation from the Wadden Sea should take place under strict regulations.

After the advice of the 'Commissie Meijer' the government in 2004 approved gas exploitation from the Wadden Sea. In 2006, NAM published the Environmental Impact Assessment. The main conclusion of the EIA is that gas exploitation does not have negative consequences for the environment. An independent committee confirmed this conclusion and the government issued the licenses. Gas has been extracted from two new gas fields at Moddergat (province of Friesland) since February 2007. From 2008 and 2009, gas will also be exploited from other locations in the province of Groningen. Exploitation will continue for 35 years. In total, gas exploitation involves six gas fields, containing about 25 billion cubic metres of gas (NAM, 2008).

Although the economic valuation studies did not halt the project, it increased the awareness of policy makers about the potential economic losses of ecosystem services and thus affected the design of the gas exploitation infrastructure. Gas exploitation can only take place from outside the boundaries of the Wadden Sea, with oil pipes entering the gas reserves in a sideward direction.

Moreover, clear conditions were set with regard to possible unforeseen environmental impact that may occur in the future. The Dutch cabinet introduced a 'hand-on-the-tap' approach: exploration can only take place within the natural boundaries. If these boundaries are exceeded, the production will be reduced or even stopped. Commissioned by the government, the Netherlands Commission for Environmental Assessment (NCEA) was involved in the realisation of this approach.

Additional information on this case can be found in CBS (2008) and NCEA (2006).

Sustainable financing of marine parks in the Antilles

Messages

• Contingent valuation as a means to value ecosystem services can be effectively applied, in this case leading to implementation of measures guaranteeing better management of national parks and financial sustainability of the management operations.

Introduction to the case

The ecosystems of the Netherlands Antilles,[3] with their coral reefs, humid elfin forests, and semidesert scrublands, not only contain the richest biodiversity in the Kingdom of the Netherlands, but also represent an irreplaceable tourism resource – the most important source of income for the islands. The marine ecosystems along the coasts (i.e. coral reefs, seagrass beds, mangroves) are also essential for healthy fisheries, and both the marine and terrestrial ecosystems provide a buffer against erosion and hurricane damage. According to a recent report by the World

---

[3] The Netherlands Antilles, previously known as the Netherlands West Indies or Dutch Antilles/West Indies, is part of the Lesser Antilles and consists of two groups of islands in the Caribbean Sea: Curaçao and Bonaire, just off the Venezuelan coast, and Sint Eustatius, Saba and Sint Maarten, located southeast of the Virgin Islands.

Resources Institute, the tourism and ecological value of all the coral reefs of the Antilles is estimated at US$24–144 million per year (WRI, 2004).

Well-managed nature reserves are the cornerstone of nature policy and both marine and terrestrial nature parks have been established, or are in an advanced stage of establishment, on all the islands. Good management requires funds for infrastructure, personnel, maintenance, education, and public information, but funding for the recurrent annual operating costs of the nature parks has been plagued by instability and deficits. This is caused by dependency on one-time project subsidies, limited ad hoc financial assistance from local government authorities and fluctuating revenues from tourism (Spergel, 2005).

The experience in the region to date demonstrates that self-financing is a viable option for many of the region's protected areas, particularly those that attract large numbers of visitors. Several protected areas, such as the Bonaire and Saba Marine Parks, now have effective revenue generation strategies, and as a result are among the best managed in the region (Geoghegan, 1998). Economic valuation studies helped to establish these systems of sustainable financing.

The Bonaire Marine Park (BMP) study (Dixon *et al.*, 1993a) played an important role in the establishment of the financial systems underlying the marine parks in the Antilles. It is a representative case study for the region and explicitly combines analysis of ecological and economic factors. The BMP study took place during the early 1990s, a period in which the management of the Park was being revised and improvements were necessary because of serious concerns about the lack of formal management, an increase in diver activity and the consequences of coastal development in general. The results indicate that proper management can yield both protection and development benefits, but questions of ecosystem carrying capacity and national retention of revenues raise important issues for longer-term sustainability.

Context of the case study: the planning process

Throughout the last two decades in the Caribbean as in the rest of the world, there has been a rapid increase in the number of declared protected areas. A 1992 survey identified 175 protected areas in the insular Caribbean, and that number is likely to have increased in the intervening years. However, only a very small percentage of these declared protected areas exist in actual fact. Most are paper parks in which no management occurs. The motivation to establish protected areas is often based on the perception that such areas enhance a country's competitiveness in the tourism sector. However, while the political will to establish protected areas may be strong, the will to budget for their management has shown itself to be very weak, in the face of urgent national priorities and continuous fiscal crisis (Geoghegan, 1998).

In recent years there have been increasing calls to transform paper parks into managed protected areas, and to establish new protected areas to tap the ecotourism market and to provide a measure of protection against development pressures, particularly in the coastal zone. Given the limited ability of most governments in the region to meet the costs of management, alternative sources of revenue are being explored. Pressure to establish self-financing protected areas is also coming from international development and lending agencies, which often bear the capital start-up costs of protected areas and want to assure that their investments are secure. The implementation of mechanisms for financial sustainability has become a routine conditionality of loans and grants for protected areas. Governments in the region are therefore looking at alternatives to government revenues for financing protected areas (Geoghegan, 1998).

Park development in the Netherlands Antilles has not followed the more traditional sequence of events whereby enabling legislation is passed first, followed by the designation of one or more protected areas. De facto establishment and even management of the parks were initiated prior to their legislation. Management of protected areas by NGO's is also rather unique in the Caribbean region. STINAPA (National Parks Foundation) and its sister NGOs on the various islands are managing protected areas irrespective of land tenure. The Bonaire Marine Park is managed under contract by STINAPA Bonaire, but in the case of the marine parks in Curacao and Saba management agreements with Governments are nonexisting or less formal. Nevertheless the partnership between Government and NGOs has served the protected area system well (Van't Hof, 1991).

After its establishment in the early 1980s, the failure to introduce a visitor fee system in 1981 created serious financial difficulties for the Bonaire Marine Park (BMP). Eventually, with no staff or funding, the Park became a 'paper park'; management and control of access were left to the dive operators.

In 1990, the Island Government of Bonaire commissioned an evaluation of the situation, which resulted in the following major recommendations:

- Introduce a visitor fee system;
- Introduce a licensing system for commercial water sports operators; and
- Create a new institutional structure for BMP, including representation from the tourism industry.

On the basis of these recommendations the Dutch Government approved funding and technical assistance for the revitalization of BMP for a period of three years: US$125,000 for operational costs and capital expenditure plus US$28,000 in technical assistance was allocated for the first year, and US$250,000 was reserved for subsequent years. Allocation of funding in the second and third years of the project would be subject to approval of annual budgets. One condition to the grant was the requirement that a visitor fee be introduced, which would eliminate the need for further financial assistance beyond 1993.

The Park was re-established and revenues were being generated by the introduction of an annual admission fee of US$10 per diver to help pay expenses. In 1992 the fees (called 'admission tickets') raised over US$170,000, which was enough to cover salaries, operating costs and capital depreciation. Revenues were also produced by sales of souvenirs and books and from donations (Dixon *et al.*, 1993a).

Nowadays, Bonaire has one of the most sustainable marine parks in financial terms, in the world. Table A.12 provides an overview of the relative contributions from various financial sources of the nature parks in the Netherlands Antilles, as well as the level of self-sufficiency.

Table A.12 *Relative contributions from various financial sources and the degree to which these cover the basic requirements for two nature parks per island (one park on St Maarten) (source: Spergel, 2005).*

Island	Island government (2002–2003)	Other grants (2002–2003)	Self-generated revenue (2002–2003)	Available budget as percentage of basic requirements (2002)	
Saba	17%	30%	53%	100%	40%
St Eustatius	21%	51%	28%	100%	17%
St Maarten	17%	78%	5%	100%	21%
Bonaire	6%	4%	90%	100%	78%
Curaçao	26%	5%	69%	100%	59%

Ecosystem services and valuation

Bonaire, a crescent shape island with an area of 288 km^2, is located in the Caribbean Sea approximately 100 km north of the coast of Venezuela. The reefs around Bonaire form a narrow fringing reef, which starts practically at the shoreline and extends to a maximum of 300 m offshore. Approximately 55 other species of coral can be found on the reefs. The whole area is protected as part of the Bonaire National Marine Park (since 1999). The site is of international importance being designated as a RAMSAR site (STINAPA, 2008).

To evaluate the success of the BMP in providing protection to the marine ecosystem, first, a visitors' survey was conducted of among SCUBA divers to obtain their perceptions of the condition of the Park and their rating of selected parameters in comparison to other Caribbean areas or to the condition of BMP in the past. These questions helped to assess the environmental carrying capacity of the BMP from a diver's perspective. Second, photo analysis was carried out to analyse coral cover and species diversity.

The majority of the divers interviewed rated the condition of the reefs high and the overall condition of the reefs in Bonaire better than or equal to any other destination they visited. The results of the photo analysis indicated that increased diver use was having an adverse impact on the coral reefs, that is, the extent of coral cover has decreased significantly at the most-frequented dive sites. The results of the analysis suggest that there may be a critical level of 4,000 to 6,000 dives per year above which impact becomes significant. Based on the number of available dive sites, Dixon *et al.* (1993a) estimated the 'annual carrying capacity' at 190,000 to 200,000 dives per year (the average visiting diver makes 10 or 11 dives during his or her stay on Bonaire). Annual use was already more than 180,000 in 1992, so the threshold was likely to be reached with an expected loss of reef biodiversity.

The quantification of the costs and benefits derived from the ecosystem services provided by the BMP was based on the assumption that Bonaire is attractive because its unique resources are protected. Aided by its protected status, a significant privately operated sector is successfully marketing Bonaire as a tourist destination. However, if

Table A.13 *Revenues and costs associated with the Bonaire Marine Park (source: Dixon et al., 1993).*

Revenues	US$ million
Direct Revenue	
Diver fees (1992)	0.19
Indirect (private sector) Revenues (gross)	
Hotels (rooms/meals)	10.4
Dive operation (including retail sales)	4.8
Restaurants, souvenirs, car rentals, misc. services	4.7
Local air transport	3.3
Subtotal	23.2
Costs	
Costs of Protection	
Direct costs – Establishment, initial operation, rehabilitation	0.52
- Annual recurring costs	0.15
Indirect costs and opportunity costs	?

protection of the marine ecosystem is not maintained, much of Bonaire's attraction would be lost, and along with it the associated revenues currently accruing to the private and public sectors.

The main categories of benefits included in the economic analysis are gross revenues to the private sector and BMP user fees. Of the primary uses of the waters contained in the Park, only revenues from dive-based tourism are considered, as the other uses of BMP waters are less dependent on the protection offered by the Park. Land-based supporting activities to dive-tourism include hotels, restaurants, souvenir sales, and car rental. Table A.13 lists the main revenues and costs, including divers' fees, associated with Bonaire Marine Park. In 1992, diver and other direct use fees, the one source of 'direct' revenues from use of BMP, summed up to about US$190,000. This amount is very small in comparison to other park-related gross revenues.

Given the controversy surrounding the institution of a user fee system, a contingent valuation survey was conducted in late 1991 to get inference of visitor's general perception of and willingness to pay user fees for the BMP. An overwhelming 92 percent agreed that the user fee system is reasonable and would be willing to pay the proposed rate of US$10 per diver per year.

Approximately 80 percent of those surveyed said that they would be willing to pay at least US$20 per diver per year, 48 percent would be willing to pay at least US$30 per diver per year, and 16 percent would be willing to pay US$50 per diver per year, yielding an average value for WTP of US$27.40 (excluding the 8 percent who were not willing to pay a fee).

Clearly the average willingness-to-pay exceeded the relatively modest US$10 fee instituted in 1992. The difference between what people would be willing to pay for a good or service and what they actually pay is known as consumers' surplus. At the current rate of dive visitation (an estimated 18,700 divers in 1992) admission fees and estimated consumer surplus total US$512,000 per year, of which US$325,000 is the consumer surplus.

Decision making

Bonaire and its marine park are representative of the issues facing many marine protected areas in the Caribbean. The case of Bonaire illustrates the difficult trade-offs that exist in combining economic and ecological goals. Its marine ecology is rich, protected, but threatened. In late 2005 another long-held ambition became reality: several employees of STINAPA, including the Director, the Managers, and the Chief Rangers, acquired police-type ticketing and law enforcement powers.

On 31 March 2005, the 1991 legislation covering Marine Park usage fees was changed with the inauguration of the Nature Fee. With the introduction of this legislation all the users of the Bonaire National Marine Park, not solely the divers, pay a user's fee. The most significant changes include:

- Marine Park tags also admit entrance to Washington/Slagbaai National Park.
- The price of Marine Park tags for SCUBA divers changed to US$25.00 for a year pass or US$10 for a day pass.
- Swimmers, board sailers, and all other users of the Marine Park are now required to have to pay US$10 for a year pass.

Recently, it was decided that tag receipts go directly to STINAPA and are used entirely for the management of Bonaire's National Parks (STINAPA, 2008).

The model of sustainable financing as developed in Bonaire is referred to in the Caribbean, and even receives worldwide attention. Although it is hard to prove this, the Bonaire model is often regarded in policy discussions on marine funding as the classical example of how to develop a system of payments for environmental services for marine ecosystem systems.

Box A.8.   Marine recreation operators join forces with conservationists in Hawaii

Hawaii's coastal waters are blessed with miles of exquisite coral reefs. More than 25 percent of the islands' marine life is found nowhere else on Earth. Because government budgets for marine protection are not sufficient, particularly in Hawaii where government funding for marine management is among the lowest in the nation, private action and private money are essential in helping to ensure that the Hawaiian reefs are protected.

Recognising the need to support the state's work, dive and snorkel operators and local conservation organisations joined forces in 2005 to raise money for marine conservation on the Big Island and Maui. Through an innovative new programme called the Reef Fund, dive and snorkel operators solicit voluntary donations from their clients to fund high-priority marine protection programmes on their islands, such as the repair and installation of mooring buoys, the protection of nesting and resting beaches for rare and endangered sea turtles and monk seals, and the establishment of local education and outreach programmes to protect marine resources.

Originally, the private industry feared the negative impact for their business of requesting their customers to contribute to conservation. A 2003 survey done by Cesar and Van Beukering for TNC and the State Division of Aquatic Resources, however, indicated that 80 percent of those surveyed were willing to pay at least US$5 per snorkel per dive day for marine resource protection programmes if the funds went to a private or nonprofit institution and were not managed by a government agency (Van Beukering and Cesar, 2004; Van Beukering *et al.*, 2004).

While the majority of other fee-based marine protection funds around the world are mandated by the local or national governments, Hawaii's fund is voluntary. On Maui, the Reef Fund is coordinated by the local nonprofit Hawaii Wildlife Fund (Maui Reef Fund, wildhawaii.org/). On the Big Island, the fund is managed by the Waimea-based nonprofit Malama Kai (Big Island Reef Fund, www.malama-kai.org/). Donations collected by marine recreation operators are pooled into a collective fund on each island and managed by the nonprofit organisation, which is advised by a committee of operators, conservationists, scientists, and other stakeholders. The advisory committees decide how the funds will be spent on their islands.

Additional information on this case can be found in Parsons and Thru (in press).

Payments for Environmental Services in Costa Rica

Main messages

- Payment for ecosystem services (PES) has played a major role in changing Costa Rican destructive and rapid deforestation into sustainable management, with tangible and convincing results.
- PES facilitates market processes between individual landowners and the world carbon market.
- PES can ease the existing inequity in distribution of costs and benefits, when benefits of ecosystem services accrue to the global community, while the opportunity cost of not converting a forest lies with local landowners.
- Through the explicit quantification of the societal demand and supply for ecosystem services, economic valuation can play an important role in the emergence of PES.

Introduction to the case

The Costa Rica Payments for Environmental Services (PES)[4] programme was initiated in 1997, becoming one of the first country-wide PES programmes in the

---

[4] Pagos por Servicios Ambientales (PSA) in Spanish.

world, and the first to adopt the terminology of environmental services and PES. Since its inception, it has become a point of reference for environmental authorities and practitioners around the world, as well as becoming one of the pillars of Costa Rica's image as a 'green' country that is a model for sustainable development.

The programme was fostered by the 1996 changes in the Forest Law that created the legal framework to pay landowners for the provision of four types of ecosystem services: (i) carbon sequestration, (ii) watershed protection, (iii) scenic beauty, and (iv) maintenance of biodiversity. Originally, Costa Rica's government expected a large influx of funds through the sale of carbon sequestration and biodiversity prospecting by pharmaceutical companies, but neither materialised and, while there are still high expectations for the former, the latter has been almost abandoned. Nevertheless, the Costa Rican government pressed ahead, earmarking for PES a 3.5-percent tax on fuels and putting in place the programme management agency, the Fondo Nacional de Financiamiento Forestal (FONAFIFO) (FAO, 2007).

Since its inception in 1997, a large area of privately owned land has participated in the project. As of October 2005, approximately 250,000 hectares where under contract: 95 percent of them for protection of natural forests and 4 percent for reforestation activities. By 2005, public and private water users were paying some US$500,000 a year for watershed conservation through forest protection (FAO, 2007).

Context of the case study: the planning process

In the last two decades, Costa Rica transformed from one of the most rapidly deforesting countries in the world to one of the foremost pioneers in reforestation, forest management, and forest protection. The predominant vision of development and economic growth in Costa Rica has been linked until recently with agro-export production, which has affected legal and institutional frameworks, economic policies, and land use decisions (De Camino et al., 2000).

In 1950, forests covered more than one-half of Costa Rica; by 1995, forest cover had declined to 25 percent of the national territory. Costa Rica had one of the highest deforestation rates in the world in the 1980s (Ortiz and Kellenberg, 2002). In the past years, however, the deforestation rate has fallen dramatically due to a remarkable set of institutional innovations in Costa Rican forestry in the mid 1990s (Chomitz et al., 1998).

Box A.9. Drinking water for New York City from restoration of the Catskill Watershed

In New York City, where the quality of drinking water had fallen below standards required by the U.S. Environmental Protection Agency (EPA), authorities opted to restore the polluted Catskill Watershed that had previously provided the city with the ecosystem service of water purification. Once the input of sewage and pesticides to the watershed area was reduced, natural abiotic processes, such as soil adsorption and filtration of chemicals, together with biotic recycling via root systems and soil microorganisms, improved water quality to levels that met government standards. The cost of this investment in natural capital was estimated between US$1–1.5 billion, which contrasted dramatically with the estimated US$6–8 billion cost of constructing a water filtration plant plus the US$300 million annual running costs. (Case based on Chichilnisky and Heal (1998).)

Intent of Costa Rica's PES programme

The PES programme was established in Costa Rica in 1997. This programme evolved in two phases (Sánchez-Azofeifa, 2007). The *first phase* (1997–2000) coincided with a significant drop in the national rate of deforestation (1997–2000), relative to the 1986–1997 time period and the high rates of forest clearing that occurred from the 1960s to the early 1980s. Recently, there has been a net increase in forest cover, mostly due to land abandonment. The *second phase* of the PES programme relates to the implementation of the Ecomarkets project (2001–today) and involves a comprehensive microtargeting scheme and the provision of new ecosystem services (e.g., drinking water) that were not part of the first phase.

Costa Rica established a PES programme within a framework created by three laws. The 1995 Environment Law 7554 mandates a 'balanced and ecologically driven environment' for all. The 1996 Forestry Law 7575 mandates 'rational use' of all natural resources and prohibits landcover change in forests. Finally, the 1998 Biodiversity Law promotes the conservation and 'rational use' of biodiversity resources.

In the first phase, payments were designed to address relevant forest conservation failures from a legal and institutional standpoint. Forest landowners were compensated by the PES programme if they planted or maintained natural forest on their land and recognised carbon sequestration, watershed protection, scenic beauty, and biodiversity as important services. These four services were not measured at once on a piece of land. A set of services with identical value was assumed to be provided by each enrolled parcel. The first phase had a 'first come, first served' policy; parcel size does not play a role in enrollment. Farm size, however, as well as human capital, and household economic did play a role in participation in the programme. Large landowners were disproportionately represented at national and regional levels.

The PES programme has to compete with other land use returns. Average PES returns varied from US$22 to US$42 per hectare per year before fencing, tree planting, and certification costs. Cattle ranching is the main competing land use, showing returns from US$8 to US$125, depending on land type, location and ranching practises. One measure of cattle-ranching returns is the cost of pasture renting. In Cordillera Central pasture rental ranges from US$20 to US$30 per hectare per year (Sánchez-Azofeifa *et al.*, 2007).

Implementation of PES

Table A.14 shows three types of contracts under the first phase of the Costa Rican PES programme:

• *Forest conservation contracts*: landowners are required to protect existing (primary or secondary) forest for five years. Land–cover change is not allowed. Payments for

Table A.14 *Distribution of payments per contract type (source: Ortiz and Kellenberg, 2002).*

Contract type	Total Payment (US$)*	Distribution by year				
		1	2	3	4	5
Forest Conservation Easements (C)	210	20%	20%	20%	20%	20%
Sustainable Forest Management	327	50%	20%	10%	10%	10%
Reforestation	537	50%	20%	15%	10%	5%

US$1 = 346 colones

forest management contracts amounted to US$210 per hectare over a five-year period;

- *Sustainable forest management contracts* (briefly interrupted in 2000): landowners with a 'sustainable logging plan' received compensation for low-intensity logging and keeping forest services intact. Contracts lasted 15 years, but payments were received during the first five years, amounting to US$327 per hectare.
- *Reforestation contracts*: owners are required to plant trees on agricultural/abandoned land and to maintain the plantation for 15 years. Contracts lasted 15 years, with most payment in the first two years amounting to US$537 per hectare.

A PES contract remains with the property if it is sold. The greenhouse-gas–mitigation potential of a parcel is transferred from owner to the national government. Costa Rica can sell abatement units on an international market. Registration for individuals is limited to a maximum of 300 hectares per year and a minimum of 2 hectares per year. Indigenous groups, however, may register up to 600 hectares per year. Coalitions acting through local NGOs have no limit. NGOs can enhance participation by playing an intermediary role between smallholders and authorities. The PES programme is administered by FONAFIFO, a public forestry-financing agency created under Forestry Law 7575 in 1996. Inspection responsibilities, however, lie with the Sistema Nacional de Areas de Conservacion (SINAC) and with the Ministerio del Ambiente y Energía (MINAE).

Funding sources for PES scheme

The most important source of funding for the original PES programme was a 15-percent consumer tax on fossil fuels established under the 1996 Forestry Law. FONAFIFO should receive one-third of the revenue, but the Ministry of Finance rarely delivered that amount. In 2001 the legislature adopted the Ley de Simplificación y Eficiencia Tributaria, which assigns 3.5 percent of the tax revenue directly to the PES programme. In theory this provided less money; in practise actual transfers from the Ministry of Finance were increased. The PES programme received an average of US$6.4 million per year.

The PES programme also receives funding from voluntary contracts with private hydroelectric producers. FONAFIFO is being reimbursed by these producers for payments given to individual landowners, for example, those upstream in a watershed. Such private agreements generated some US$100,000 to finance about 2,400 hectarers of PES contracts. When fully implemented these agreements are expected to grow to about US$600,000 annually, covering close to 18,000 hectares.

Carbon trading was expected to provide significant funding. However, no significant market for carbon abatement has emerged. Only Norway has bought 200 million tons of carbon sequestration for US$2 million in 1997.

A World Bank loan and a Global Environmental Facility (GEF) grant of US$8 million contributed to a programme called *Ecomercados* (the name of the second phase of the PES programme, which started in the year 2000). US$5 million was used for conservation contracts along the Mesoamerican Biological Corridor. The other US$3 million was used for capacity development purposes of relevant organisation such as FONAFIFO, SINAC, and MINAE (Sánchez-Azofeifa *et al.*, 2007).

Total Distribution of PES Contracts

In the first phase of the PES programme around 300,000 hectares of primary, secondary, or planted forest received funding, with a mean project size of approximately 102 hectares. The largest project was 4,025 hectares. The stated size limits were not fully enforced; 202 projects were over the 300-hectare maximum and 60 contained less than the 2-hectare minimum. From 1997 to 2000, the number of participants entering the programme decreased, probably because funds were not delivered as expected. Payments for conservation alone were larger than the sum of the payments made for reforestation and forest management), but conservation contracts had the lowest payments per unit area. Reforestation and management contracts generally held steady over the years, whereas conservation payments fell (e.g. more than US$20 million in 1997; almost US$12 million in 1999; and less than US$4 million in 2001) (Sánchez-Azofeifa *et al.*, 2007).

Ecosystem services and valuation

In 1993, the World Bank prepared the Forest Sector Review for Costa Rica. The review was the first attempt by the Bank to calculate the total economic value of Costa Rican forests. Table A.15 shows the economic values of various forest activities according to the Bank's study. Twenty-eight percent of the value corresponds to market values (especially of wood) and 72 percent to nonmarket values. In the most pessimistic (from the Costa Rican perspective) distribution of benefits, 66 percent of the environmental services of forests are enjoyed by the global community and only 34 percent by Costa Rica. The cumulative annual rent is US$208 million, of which US$137 million is enjoyed by the global community without compensation for Costa Rican farmers, and US$71 million is received by Costa Rica. Although the study has several weaknesses, such as the inclusion of primary forests only and the exaggerated value of carbon sequestration, the study highlight some important points: the value of environmental services is high, the global community receives the major benefits of these services, and owners of the resources that provide these services are not compensated for their full value (De Camino *et al.*, 2000).

In 1994, the World Bank issued another study of the value of primary forest in Costa Rica (see Table A.16). The study arrives at a rent of US$102–US$214 hectares

Table A.15 *Total Economic Values of Costa Rican Forests according to the World Bank, 1993 (source: De Camino et al., 2000).*

Product or service	Total value (US$M)	Value per hectare (US$)	Value per hectare per year (US$ at an 8% discount rate)	Base
Carbon sequestration	1,098	845	68	1.3 mil hectares
Sustainable logging	403	620	50	0.65 mil hectares
Existence and option value	383	295	24	1.3 mil hectares
Ecotourism	272	209	16	1.3 mil hectares
Hydroelectric power	36	207	17	0.18 mil hectares
Pharmaceutical	3	2.3	0.2	1.3 mil hectares
Urban and rural water	59	47	3.8	1.3 mil hectares
Total	2,254	2,225	179	
Total market	403	620	50	28%
Total nonmarket	1,851	1,605	128	72%
Total Costa Rica	664	1,001	81	34%
Total World	1,612	1,224	98	66%

per year without considering the value of wood, and US$170–US$282 hectares per year by considering the value of wood. Although the values assigned by each study differ, both studies support the importance of payment for environmental services (De Camino *et al.*, 2000).

In 1996, MINAE commissioned the Costa Rican Tropical Science Center to conduct a study to obtain a scientific basis for assigning a value to environmental services. The Costa Rican Tropical Science Center recommended payments for all four environmental services recognised under the PES programme. The study distinguished between primary and secondary forests, departing from the assumption that secondary forests provide fewer environmental services than natural forests (Table A.17). However, the study did not reveal the criteria that are used to distinguish between primary and secondary forests, or how compensation should be calculated for reforestation, forest management, forest conservation, or agroforestry systems.

On February 26, 1997, MINAE specified PES amounts. The World Bank and CCT studies suggested fixing a quantity per hectare and year or a single payment for one full rotation or cutting cycle. Instead, MINAE fixed a payment for environmental services for a period of five years and as a percentage of the costs of establishing and managing different kinds of forests. This amount is intended as a lump-sum

Table A.16 *Environmental Values of Primary Forests (De Camino et al., 2000).*

Types of benefits	Average value (US$ per hectare)	
- Urban water supply	2.3–4.6	
- Loss of hydroelectric productivity	10.0–20.0	
- Protection of agricultural lands	0.25–2.0	
- Flood control	4.0–9.0	
Subtotal hydrological benefits		16.6–35.6
Carbon sequestration		60.0–120.0
Ecotourism (recreation or nonconsumptive value)		12.6–25.1
Future pharmaceuticals (optional value)		0.15
Transfer of fund (existing and optional values)		12.8–32.0
Total		102.2–213.7
Net present value (6%)		1,277.5–2,671.3

Table A.17 *Findings of the TSC Study on Recommended Compensatory Payments in US$ per year per hectare (De Camino et al., 2000).*

Environmental service type	Primary forest	Secondary forest
Carbon sequestration	38	29.3
Waer conservation	5	2.5
Biodiversity	10	7.5
Natural beauty	5	2.5
Total	**58**	**41.8**

compensation for all environmental services. This decision was made to avoid disrupting forest management.

Despite these attempts to estimate the economic value of ecosystem services provided by the Costa Rican forests, both the Forestry Law (1996) and Biodiversity Law (1998) do not define the type of financial instrument nor the monetary amount that should be paid. Therefore, it is essential that FONAFIFO has solid scientific information as input in the negotiations of voluntary agreements, which form an important contribution to the payment mechanism.

One example of scientific support is the study by Reyes *et al.* (2002). Based on replacement and maintenance cost, they estimated a range of values for the ecological services provided by forests in several watersheds. These values range from US$100 hectares per year (Peñas Blancas watershed) to US$176 hectares per year

(Pejibaye watershed). This implies that if forest cover is preferred in relation to the provision of hydrological services and is to be guaranteed in the long term, the landowners would have to receive at least US$100 per hectare per year in terms of additional income in order to protect forest cover or commit themselves to reforestation activities.

Additional information on this case can be found in Kishor and Constantino (1993), World Bank (1993), and WWF (2007).

Water transfer in Spain

Main messages

- When it is obvious that significant impacts on ecosystem services can be expected from a plan, ignoring such impacts may lead to opposition and ultimately the cancellation of the plan. Not studying (the impacts on) ecosystem services and their respective ecological, social, and economic values thus can have serious repercussions.
- The Ebro Delta combines multiple ecosystem services. One important service is its role in maintaining internationally important biological diversity. This has resulted in a protected status of parts of the delta. Ignoring this important aspect has contributed greatly to the failure of the water transfer plan to get approval.

Introduction to the case

The Spanish National Hydrological Plan (SNHP) was passed into law by the Spanish Parliament in July 2001. The SNHP identifies an elaborate programme of infrastructure development and management to assure constant water supply all over Spain. This plan, with a projected capital cost of €4.2 billion, consists of two main parts: (1) transfer of 1,050 cubic hectometres (hm^3) of water per year from the basin of the Ebro River to other river basins in the north, southeast, and south of Spain; as well as (2) a block of 889 public water works affecting other Spanish river basins.

The chief SNHP objective is to transfer water from the Ebro Basin to four other river basins in the east of Spain. The project contains two large transfer projects from the lower Ebro River: the Northern Transfer, planned to transfer 189 hm^3 to the metropolitan area of Barcelona for urban uses; and the Southern Transfer, planned to transfer 861 hm^3 to the Levante region and southeast Spain. Almost 70 percent of this transfer would be used for agricultural purposes, with the remaining 30 percent being for urban uses (WWF, 2006).

These water transfers would lead to serious impacts on the Ebro River, which is of high economic and environmental importance for several reasons: The Ebro river basin counts close to 3 million inhabitants, with almost 50,000 living in the Ebro Delta. It has been estimated that economic activities associated with the ecosystems of the Ebro Delta produce an annual turnover of €120 million from fisheries, aquaculture, agriculture and tourism (Day *et al.*, 2006).

Designated as a Natura 2000 zone and Ramsar site, the Ebro Delta is the third most important wetland in Spain with a significant importance at a European level.

Water transfer would lead to the deterioration of the Ebro Delta ecosystem. The area is the second most important SPA (Special Protection Area) in Spain after the Doñana National Park.

New dams will also need to be constructed in the high Pyrenees Mountains to regulate the water flow of the Ebro, which will lead to additional serious environmental and social impacts (WWF, 2006).

Context of the case study: the planning process

The SNHP approved by Congress in 2001 set out the government vision on how it intends to regulate, manage and plan the water resources and all their related uses within Spain. As one step leading to the approval of the SNHP, in 1999, the Water Law (1985) was modified to adapt it to the purposes and needs of the SNHP. The Plan claimed to comply with the requirements of the European Water Framework Directive in terms of sustainable water use, environmental protection, reduction of pollution through efficient water planning, use of economic analyses and instruments, approval and action programmes and cost recovery principles (Tortajada, 2006). However, extensive analyses indicated that the Plan was not compatible with the Water Directive, mainly in economic and environmental terms (see, for example, Albiac *et al.*, 2006; Albiac *et al.*, 2003; Biswas and Tortajada, 2003; Embid, 2003; Garrido, 2003; Getches, 2003; Hanemann, 2003; Howitt, 2003).

Two regions in the basin from which water was to be transferred, Aragón and Cataluña, strongly opposed the Plan. Aragón argued that the National Hydrological Plan could not be justified on economic, environmental nor on social grounds. Furthermore, implementation of proper demand management practise in the water-importing regions would make the water transfer unnecessary. Numerous analyses indicated that environmental and economic sustainability principles were mostly ignored. The lack of a social assessment was further reason to question the plan. The plan simply stated that water transfer would not have any impacts on the economic activities in the donor basin. It would neither have negative consequences on population distribution within the donor basin.

According to environmental studies, the implementation of the plan would aggravate the ecological problems already occurring in the area downstream from the diversion point, especially in the Ebro Delta and estuary (see, for example, Ibáñez and Prat, 2003; Arrojo Agudo, 2001). A major criticism was the approval of the National Hydrological Plan by the national government before an environmental impact assessment was carried out. The plan did not consider any of the impacts that such a large water transfer would have on biodiversity, wetlands, ecological flow, and expected changes in land use in the Ebro Delta. Neither did it take into account the impacts on human activities, such as fisheries and rice production.

Cost–benefit estimates of the plan and its strategic environmental assessment were considered inaccurate or completely lacking (Hanemann, 2003). An example is provided by the revenues per cubic metre of exported water. The revenues were expected to compensate negative impacts in the Ebro river basin resulting from the transfer. This revenue was not based on any economic analyses, but was an administrative charge. The compensation was insufficient to mitigate the expected adverse impacts of the water transfer in the exporting region. (Tortajada, 2006).

Assessment context

For the construction of infrastructure funding was needed from the European Commission. This provided a platform for the Government of Aragon and several environmental groups to submit a formal complaint to the European Commission. This resulted in hearing in the European Parliament, and a Seminar on the plan and its expected impact in October 2003. The European Union organised this seminar to promote dialogue between the Governments of Spain, Aragon, and the environmental groups. Following the outcomes of discussions and several technical studies from within the European Commission, it was recommended not to provide financial support for the National Hydrological Plan. However, before the European Commission could take a final decision, the 2004 elections in Spain resulted in the change of the ruling political party and the cancellation of the 2001 National Hydrological Plan (Tortajada, 2006).

Spain's newly elected socialist government enacted the Law 11/2005 of 22 June, paving the way for a new water policy called the Programa AGUA, Actuaciones para la Gestión y la Utilización del Agua, 'actions for management and use of water'. The AGUA programme targets Mediterranean Spain as a priority case for action. The law, Actuaciones en el Litoral Mediterraneo RDL 2/2004, forecasts an additional 1,063hm^3 of water and is estimated to cost €3.8 billion. Ten provinces are included (from Girona in the northeast to Malaga in the southwest) and these provinces are within the catchment area of five hydrographic zones. Twenty-one desalination facilities are planned for six provinces on the Spanish Mediterranean coast to supplement their water needs.

Controversially, AGUA challenges the geographical ideology of the river basin: whereas the National Hydrological Plan sought to balance basin deficits in the Mediterranean region by transferring water from distant basins, AGUA's vision extends the geographical boundaries of the river basin itself beyond its coastal limits to tap the marine waters and littoral saline aquifers of the Mediterranean coast. Spain has over 1,500 km of coastline and numerous coastal aquifers with brackish groundwater, which can be desalinated. Furthermore, unlike the Ebro transfer, supplies of desalinated water can be predicted independently of climate changes and drought. In theory at least, the opportunities to supply desalinated water to recipient basins are limitless. Sustaining a basin's freshwater needs is simply a matter of financial investment to pay for the facilities' construction and running costs. Conceptually, therefore, resources are traded – as long as the 'value-added' to water (through goods and services produced) exceeds the cost of freshwater provision, the system can be considered economically sustainable. The Spanish government believes these balances can be achieved because the revenues from the service (including tourism) and agriculture sectors offset the financial cost of desalinating water (Downward and Taylor, 2007).

Ecosystem services and valuation

The Ebro Delta is an important wetland in the western Mediterranean, with economic and ecological values. It is an important bird habitat, being the second most important special protection area for birds (SPA) in Spain. Some 8,000 hectares of the delta have been designated as a natural park in 1986. The international importance of the natural values of the Ebro Delta has been widely recognised. In 1984, the delta was declared an area of special interest for conservation of halophytic vegetation by the Council of Europe. It has also been recognised as an area of European importance for conservation of aquatic vegetation. In 1993, it was included in the list of Ramsar areas and is part of the Natura 2000 network. Wetland area has been steadily reduced from approximately 250 km^2 in 1900 to 80 km^2 in 1990 due to conversion to agriculture and other uses (Day *et al.*, 2006).

Of a total of 330 species of birds observed in the delta, over one hundred species breed in the area. A total of 55 species are endangered or migratory species included in Annex I of the EU Birds Directive. The Ebro Delta has international importance for breeding for at least 24 migratory species and 13 wintering species. The fish fauna in the delta is also very rich, with 55 species observed, including six endemic species of the western Mediterranean coast. The last viable world population of a globally threatened freshwater mussel occurs in the lower river. There are 18 habitats included in the 92/43/EEC Directive for the Conservation of Natural Habitats and Wild Flora and Fauna, from which two are of priority conservation and eight are locally endangered.

Agriculture, fisheries, aquaculture and tourism are economic activities depending on the delta. The total annual economic value is about €120 million. Agriculture accounts for a gross economic benefit of about €60 million, tourism about €30 million, fisheries about €20 million, and aquaculture about €10 million. Rice cultivation is the main human activity in the delta. Rice fields cover about 60 percent of the delta, playing a crucial role in the economy as well as the ecology of the delta. Rice production is about 120,000 tonnes per year, the third most important of the European Union. An extensive irrigation system delivers fresh water from the Ebro River to the rice fields. Apart from economic productivity, rice fields provide many more important ecosystem services, such as wintering area of migratory birds, the prevention of saltwater intrusion, and denitrification in soils.

The Ebro River influences some of the most important fishing areas in the western Mediterranean, with an average catch of about 6,000 tonnes per year. The importance of aquaculture is increasing, focussed on the cultivation of about 3,000 tonnes of mussels and oysters per year in Fangar and Alfacs bays. Since the creation of the natural park, tourism has increased substantially. The estimated number of visitors is over half a million people per year (Day *et al.*, 2006).

The elements of the SNHP that include water transfers from the Ebro have caused great controversy, especially because of the different perspectives and uncertainty in defining the environmental and socioeconomic impacts in the donor and receiving basins (Alcácer-Santos). In the SNHP plan various economic arguments are used in support of supply-oriented water management, through construction of reservoirs and water transfer. Other alternatives such as demand management and water conservation are left in the shade. A study carried out by the University of Zaragoza for WWF shows that real costs of the SNHP were highly underestimated, in fact the SNHP made a negative contribution to economy of €3.5 billion (Arrojo *et al.*, 2002).

Decision making

In this case study, it was not so much the presence of economic valuation studies, but the lack of proper estimates of the real costs and benefits, that influenced decision making with regard to the plan. The analytical approaches used to formulate such a complex plan were generally considered to be inadequate. Among critics there was agreement that additional economic evaluation studies on the impacts of the water transfer were needed. For example, economic analysis of the long-term elasticity of water demand for urban and industrial uses in the project area was lacking. Similarly, a study was lacking to measure the willingness of the farmers to pay for water; such willingness was needed to ensure financial feasibility of the planned sale of imported water. Another study deemed necessary would estimate the present and future cost of water supply.

It was generally believed that cost recovery in combination with water transfer would lead to substantially increased water prices for the urban and industrial consumers. This would significantly reduce the total water requirements. Cost recovery for the agricultural sector would also mean that rates for agricultural water would go up substantially. This would also lead to reduced water requirements as marginal agriculture would disappear. Consequently, transferred water would not be economically interesting for many farmers as costs would be higher than the marginal value of water. Simply stated, returns on crops would be insufficient to pay for the transferred water.

The main rationale for the water transfer project was the need for water in the coastal areas of the south. This need was, however, fundamentally questionable. Forecasts of future water demand were likely to be significantly less if demand management practises, such as full cost recovery, proper levels of water tariffs, more efficient water management in the urban, industrial and agricultural sectors, treatment and reuse of wastewaters, and so forth, were considered. Furthermore, available cost-effective alternative options were ignored. For example, the provision of desalinated seawater or groundwater along the coastal areas could be provided more economically. Moreover, through desalination water would be available in about 2–4 years. The water transfer project would take at least 10 years.

From the EU perspective, this plan was also unlikely to be accepted and funded. The National Hydrological Plan did not comply with the following Community texts: (i) Treaty of the European Community in the content and numbering arising from the 1997 Treaty of Amsterdam; (ii) European Parliament and Council Directive 2000/60/EC of 23rd October 2000, establishing a Community framework of action in the field of water policy; (iii) Council Directive 79/409/EEC, of 2nd April 1979, regarding the conservation of wild birds; and (iv) Council Directive 92/43/EEC, of 21st May 1992, regarding the conservation of natural habitats and wild fauna and flora.

A full-fledged SEA at plan level could have avoided the total tearing down of the plan. A proper integrated assessment of economic, social, and environmental consequences, including the analysis of alternatives based on water demand management would have shown in an early stage the nonviability of the plan in its original form.

Additional information on this case can be found in IUCN (2004).

Compensation payments by Exxon Valdez

Main messages

- Economic valuation of ecosystem services provides acceptable clues for legal procedures and fines.
- The ecosystem service of maintenance of biodiversity can be monetised as a bequest value by using a stated preference methodology. The valuation similarly provided result accepted in legal procedures.
- This case set an example for liability claims for damage inflicted upon biodiversity.

Introduction to the case

Around midnight, on 24 March 1989, the *Exxon Valdez* ran aground on Bligh Reef near the coast of Alaska. The oil tanker was carrying 1,264,155 barrels of oil. Approximately 257,000 barrels, or 38,800 tonnes, were spilled. More than four summer seasons and US$2.1 billion (Exxon's account) were spent before the cleaning effort was called off. Various methods were used to remove oil from the beaches. Not all beaches were cleaned; some beaches remain oiled today. The cleanup effort included some 10,000 workers, 1,000 boats and roughly 100 aircraft. These became known as Exxon's 'army, navy and air force'. In spite of these enormous efforts, many believe that storms and wave action were more effective in cleaning the beaches. The spill region has more than 15,000 km of shoreline.

The Exxon Valdez oil spill (EVOS) is one of the largest ever in the United States. It is widely considered as the environmentally most damaging oil spill worldwide. The timing of the spill, the remote and spectacular location, and the abundance of wildlife in the region made it an environmental disaster well beyond the scope of other spills. Partly because it is also the most publicised and studied environmental tragedy in history, the disaster can be considered to be extremely influential in changing policies. For example, much has been accomplished over the years to prevent another Exxon Valdez–type accident.

Assessment context

Following the EVOS, the State of Alaska and the United States acted as trustees to protect and assess damage to the environment. Immediately after the EVOS, the Trustees began a series of studies – the Natural Resource Damage Assessment – to determine the effects of the oil spill on the environment, both its resources and services (e.g. marine and terrestrial mammals, birds, fish and shellfish, archaeological resources, and subsistence). These documents describe the studies necessary to determine the extent and magnitude of injury to natural resources of Prince William Sound and the adjacent Gulf of Alaska, including several economic valuation studies.

The studies to assess injury were designed to support: (i) the development of restoration plans to promote the long-term recovery of natural resources, and (ii) the determination of damages to be claimed for the loss of services of the natural resources. These documents were the start of a process of research and consultations, which eventually resulted in a settlement agreement and the development and implementation of a restoration plan for the entire affected region.

Ecosystem services and valuation

The various impact studies contributed to the establishment of a draft Restoration Plan in 1993. The draft Restoration Plan was analysed in the final Environmental Impact Statement (FEIS), comparing the potential environmental impacts of the draft Restoration Plan, as the Proposed Action 5, and four other alternatives.

The alternatives included:

- No action, normal agency management would occur, but no restoration actions would be funded from by the Trustees;
- Habitat Protection, habitat acquisition and protection actions would be the only restoration actions pursued;
- Limited Restoration, a mix of habitat protection, monitoring and research, and general restoration actions would be implemented for the most severely damaged resources and services;
- Restoration, habitat protection, monitoring and research, and general restoration actions would be implemented for all damaged resources and services; and
- Proposed Action (Draft Restoration Plan) uses all three restoration categories to restore damaged resources and services, but places greater emphasis on monitoring and research than any other alternative. While emphasizing habitat protection; general restoration actions would be used primarily for resources that were still not recovering.

Economic valuation

Following the research and planning process, several federal studies were proposed to assess the economic value of injury to natural resources associated with the EVOS. These would cover eight major areas: (i) commercial fishing, (ii) public land values, (iii) recreation, (iv) subsistence, (v) intrinsic values, (vi) research programmes, (vii) archaeological resources, and (viii) petroleum price impacts.

Ultimately, four ecosystem services were actually valued in economic terms:

- Replacement costs of birds and mammals (Brown, 1992);
- Recreational and sports fishing losses (Carson and Haneman, 1992; Mills, 1992);
- Tourism industry (McDowell Group, 1992); and
- Contingent valuation (CV) of lost passive use values (Carson et al., 1992).

Replacement costs of birds and mammals

The study estimates values based on the costs of relocation, replacement and rehabilitation for some of the shorebirds, seabirds, and the marine and terrestrial mammals that may have suffered injury or were destroyed in the oil spill. A likely range of costs is estimated and a best estimate is selected (see Table A.18).

Table A.18 *Range of estimates for replacement costs of mammals and birds (in 1989 US$).*

	Lowest	Highest
Marine mammals	700 (Harbour seal)	300,000 (Killer Whale)
Terrestrial mammals	125–250 (White tailed Deer)	300–500 (Brown Bear)
Seabirds and eagles	167 (Gull)	22,000 (Eagle)

Box A.10. Administrative penalties for damage to coral reefs in Hawaii

Throughout the years, the United States has enacted several laws that enable trustees to recover damages for injuries to resources under certain circumstances. Funds recovered via Natural Resource Damage Assessments (NRDAs) are commonly used to pay for restoration of the injured resources. The original procedure of NRDAs was that trustees assess the damage, determine the amount of physical restoration that is necessary, and seek the cost of restoration from the responsible party.

Economic valuation is not traditionally used in standard NRDA. It is the cost of the restoration that matters to the trustees, not the value of the resources injured. However, this approach to damage assessment is gradually changing. More and more, trustees have pursued both NRDA damages and civil penalties for the same incident. For example, supported by economic valuation studies, the Florida Keys National Marine Sanctuary introduced a schedule of escalating fines for injury to living coral based on the area of impact.

A similar process of integrating economic valuation in setting penalties is ongoing in Hawaii. Hawaii's coral reef ecosystems provide a wide range of services to coastal populations, such as fisheries, tourism, biodiversity, and natural protection. These same coral reefs are under constant pressure from damaging activities, such as anchoring, ship grounding, and coastal development. Penalizing such damage not only discourages potential violators from exercising pressure on the reef but also provides the wardens with means for better management.

Because of the absence of a workable system of penalties for coral reef damage in Hawaii many violators were not punished for doing damage to the reef, despite the fact that the damage to the reef was well-documented and the violators were identified. It was hard to determine a reasonable penalty because of the way the present law is written. Recently, the bill H.B.3176 was proposed to allow the Department of Land and Natural Resources (DLNR) the authority to impose a fine for large-scale reef damage (State of Hawaii, 2008). The fine of up to US$5,000 per square metre is consistent with laws in other states and with the value of the reef. H.B.3176 addresses the urgent needs to have natural resource laws that are complete, clear, and enforceable, and providing appropriate opportunities for administrative enforcement.

In setting the level of the penalties, ample use was made of an economic valuation study for coral reefs in Hawaii (Cesar and Van Beukering, 2004). Without even attempting to measure their intrinsic value, this study shows that coral reefs, if properly managed, contribute enormously to the welfare of Hawaii through a variety of quantifiable benefits. The net benefits of recreational, amenity, biodiversity, fishery, and education spillover values are estimated at US$360 million a year for Hawaii's economy, and the overall asset value of the state of Hawaii's 1,660 km^2 of potential reef area in the main Hawaiian Islands is estimated at nearly US$10 billion. Converted to value per square metre, the economic value can be as high as US$2,600.

Additional information on this case can be found in Brown (1992).

Recreational and sports fishing losses

Those planning economic studies to assess damages from lost recreational uses identified sport fishing as the recreational activity with the most potential for rigorous evaluation of the spill's impact. The study indicated the impacts on sport fishing industry through the analysis of annual survey results that has been conducted since 1977 by the Sport Fish Division of the Alaska Department of Fish and Game among anglers who sport fished in Alaska. During a five-year period from 1984 to 1988, the estimated number of anglers who fished the area where oil was spilled in 1989 increased continuously. However, these increasing trends changed after the EVOS. The results of the study indicate that in the oil spill area, the estimated number of anglers decreased 13 percent from 120,160 in 1988 to 104,739 in 1989, the number of household trips decreased 15 percent from 270,956 to 230,520, the number of days fished decreased 6 percent from 312,521 to 294,598, and the number of fish harvested decreased 10 percent from 352.630 to 318,981.

Another study determined a range of monetary values for recreational fish losses after the oil spill (Carson and Haneman, 1992). A lower-bound estimate was found by considering the reduction in fishing days between 1988 and 1989 in the immediate spill area (i.e. 17,923 days), ignoring whether households participated in the oil spill clean-up, and valuing lost days at an average value of US$204 per day. This calculation yields a lower bound estimate of US$3.6 million. An upper bound was found by considering the lost days for 1989 (i.e. 127,527) and 1990 (i.e. 40,669) in the South central area based on a prediction from a simple trend regression equation using the pre-1989 data coupled with a higher value of US$300 per day. This calculation yields an upper-bound estimate of US$50.5 million.

Tourism industry

Two research techniques were utilised. The first reviewed all existing data which were accessible and which might indicate impacts of the oil spill on the 1989 visitor season. The second technique included executive interviews of two major groups: (i) tourist-affected businesses and (2) relevant government agencies and organisations.

Overall, the EVOS had major effects on Alaskan tourism industry. Some examples of the identified impacts include:

- The negative effects of the spill directly felt by visitors were as follows: visitor spending decreased with 8 percent in South central Alaska and 35 percent in Southwest Alaska from previous summer spending, the two major affected areas. The net result was a loss of US$19 million in visitor spending.
- A potential loss of 9,400 visitors was determined for the summer of 1989, representing US$5.5 million in in-state expenditures.

Contingent Valuation of lost passive use values
The CV study was designed to measure the loss of passive use values arising from damage to natural resources caused by the oil spill. Respondents were told that if no action is taken over the next ten years another oil spill would almost certainly cause damages to Prince William Sound comparable to those of the Exxon Valdez spill. Respondents were then asked their willingness to pay for a realistic programme that would prevent with certainty the damages, which would be caused by such a spill. The median household willingness to pay for the spill prevention plan was found to be US$31. Multiplying this number by an adjusted number of U.S. households resulted in a damage estimate of US$2.8 billion.

Decision making

On 8 October 1991, Exxon agreed to pay the United States and the State of Alaska US$900 million over ten years to restore the damaged resources by the spill, and the reduced or lost services (human uses) they provide. Exxon was fined US$150 million, the largest fine ever imposed for an environmental crime. The court forgave US$125 million of that fine in recognition of Exxon's cooperation in cleaning up the spill and paying certain private claims.

The various impact studies made an important contribution to the settlement and the adaptation of the Restoration Plan and related policies. In 1989, 72 studies were being carried out in 10 categories of natural resources and related services. Research has been continuing on the effects of residual oil in the ecosystem and on the natural recovery process ever since. The Trustee Council adopted the Restoration Plan for the civil settlement funds in 1994 after an extensive public process. More than 2,000 people participated in the meetings or sent in written comments.

A major lesson of this disaster was that the spill prevention and response capability in Prince William Sound was fundamentally inadequate. Debate continues over whether a spill the size of the Exxon Valdez disaster can be contained and removed once it's on the water. But there is little doubt that today the ability of industry and government to respond is considerably strengthened from what it was in 1989 (EVOSTC, 2008).

The Exxon Valdez case was also crucial for the further development and acceptance of economic valuation in environmental policy making. After the Exxon Valdez oil spill, contingent valuation studies gained new prominence in the natural resource damage assessment process. It was in this context that NOAA convened a blue-ribbon panel, chaired by two Nobel laureates (i.e. Kenneth Arrow and Robert Solow), to explore whether or not contingent valuation studies were reliable enough to measure total value (direct plus passive use) for the natural resource damage assessment process. To some extent, the panel's recommendations shaped the development of the method, use of the results of stated preference studies by Federal agencies, and the direction of research in the area since 1992.

The NOAA panel concluded that stated preference studies could provide estimates reliable enough to be the starting point of a judicial process of damage assessment, including lost passive-use values (58 Federal Register 460, January 15, 1993). Moreover, the panel gave several specific and fairly stringent recommendations on how stated preference studies should be designed and administered to ensure reliability and validity (Arrow *et al.* 1993).

References

Abdel-Dayem, S., Hoevenaars, J., Mollinga, P. P., Scheumann, W., Slootweg, R. and van Steenbergen, F. (2004). Reclaiming Drainage. Toward an Integrated Approach. IBRD Agriculture and Rural Development Department, Report No. 1. http://siteresources.worldbank.org/INTARD/Resources/Drainage_final.pdf. (Summary published in *Irrigation and Drainage Systems*, **19**, 71–87.)

Abdel-Dayem S., Hoevenaars J., Mollinga P. P., Scheumann W., Slootweg R. and van Steenbergen F. (2005). Agricultural drainage – towards an integrated approach. *Irrigation and Drainage Management*, **19**, 71–87.

Abelson, Ph. H. (1991). Resources of plant germplasm. *Science*, **253**, 833.

Abelson, Ph. H. (1997). Editorial. Evolution of higher education. *Science*, **277**, 747.

Adams, W. M. (2004). *Against Extinction: The Story of Conservation*. London: Earthscan.

Adams, W. M., Aveling, R., Brockington, D., Dickson, B., Elliott. J., Hutton, J., Roy, D., Viva, B. and Wolmer, W. (2004). Biodiversity conservation and the eradication of poverty. *Science*, **306**, 1146–9.

ADB (2006). Summary Environmental Impact Assessment Project Number: 36052. Pakistan: North-West Frontier Province Road Development Sector and Sub regional Connectivity Project Peshawar–Torkham Subproject. Asian Development Bank, Manila.

Agarwal, B. (1992). The gender and environment debate: lessons from India. *Feminist Studies*, **18**, 119–58.

Agrawala, S. (1998). Structural and process history of the intergovernmental panel on climate change. *Climatic Change*, **39**, 621–42.

Ahmad, N. and Lahiri-dutt, K. (2006). Concerns in coal mining displacement and rehabilitation in India engendering mining communities: examining the missing gender. *Technology and Development*, **10**, 313.

Albiac, J., Uche, J., Valero, A., Serra, L., Meyer, A. and Tapia, J. (2003). The economic unsustainability of the Spanish National Hydrological Plan. *Water Resources Development*, **19**, 437–58.

Albiac, J., Hanemann, M., Calatrava, J., Uche, J. and Tapia, J. (in press). The rise and fall of the Ebro water transfer. *Natural Resources Journal*.

André, P., Enserink, B., Connor, D. and Croal, P. (2006). *Public Participation*. International best practise principles. *IAIA Special Publications Series No. 4*. www.iaia.org/Non_Members/pubs_Ref_Material/pubs_ref_materia_index.htm.

Androulakis, I. and Karakassis, I. (2006). Evaluation of the EIA system performance in Greece, using quality indicators. *Environmental Impact Assessment Review*, **26**, 242–56.

Anonymous. (1997). Business and Biodiversity: A Guide for the Private Sector. The World Conservation Union (IUCN) and the World Business Council for Sustainable Development (WBCSD).

Anonymous. (1999). *Climate Action: The Atlantic Forest in Brazil*. Arlington, VA: Nature Conservancy (TNC).

Anonymous. (2002a). Biodiversity Resources. *A Companium Volume to Business and Biodiversity: The Handbook for Cooperate Action*. Earthwatch Institute (Europe), International Union for Conservation of Nature and Natural Resources, World Business Council for Sustainable Development. (ISBN 2-940240-28-0.)

Anonymous. (2002b). *Business and Biodiversity: The Handbook for Corporate Action*. Earthwatch Institute (Europe), International Union for Conservation of Nature and Natural Resources, World Business Council for Sustainable Development. (ISBN 2-940240-28-0.)

Anonymous. (2004). *Planning Policy Statement 22: Renewable Energy*. Norwich, United Kingdom: Published for the Office of the Deputy Prime Minister, under licence from the Controller of Her Majesty's Stationery Office.

Anonymous. (2008a). 'Pascua-Lama Mine' in Keystone Mining Post. http://keystoneminingpost.com/Mining/Geology/South/Tin.aspx.

Anonymous. (2008b). Sugar for Biofuel to Displace Kenya's Tana Delta Wildlife. Environment News Service (ENS). www.ens-newswire.com/ens/jun2008/2008-06-26-03.asp.

Armstrong, A. (2004). A Black-headed Dwarf Chameleon – translocation project in Durban, *Palmnut Post*, **7**, 8–10.

Arrojo A. P. (2001) (in Spanish). *The debate on the National Hydrological Plan*. Fundación Nueva Cultura del Agua, Bakeaz, Bilbao.

Arrojo, P., Míguelez, E. and Atwi, M. (2002). *Análisis y valoración socioeconómica de los trasvases del Ebro previstos en el Plan Hidrológico Nacional Español*. Elaborada por WWF/Adena por Fundación Nueva Cultura del Agua. Departamento de Análisis Económico de la Universidad de Zaragoza, Zaragoza.

Arrow, K., Solow, R, Portney, P.R., Leamer, E. E., Radner, R. and Schuman, H. (1993). Report of the NOAA Panel on contingent valuation. *Federal Register*, **58**, 4601–14.

Arts, J. (1998). EIA Follow-up. *On the Role of Ex-Post Evaluation in Environmental Impact Assessment*. Groningen: Geo Press.

Arts, J., Caldwell, P. and Morrison-Saunders, A. (2001). Environmental impact assessment follow-up: good practice and future directions – findings from a workshop at the IAIA 2000 Conference. *Impact Assessment and Project Appraisal*, **19**, 175–85.

Ash, N. and Jenkins, M. (2007). Biodiversity and Poverty Reduction: The Importance of Biodiversity for Ecosystem Services. Cambridge: UNEP-WCMC, 219c Huntingdon Rd, CB3 0DL 01223 277314. www.unep-wcmc.org.

Atkinson, S. F., Bhatia, S., Schoolmaster, F. S. and Waller, W. T. (2000). Treatment of biodiversity impacts in a sample of US environmental impact statements. *Impact Assessment and Project Appraisal*, **18**, 271–82.

Attia F., Fahmi H., Gambarelli G., Hoevenaars J., Slootweg R. and Abdel-Dayem S. (2005). *West Delta Water Conservation and Irrigation Rehabilitation Project. Drainframe Analysis. Main Report.* Internal report. Arab Republic of Egypt: Ministry of Irrigation and Water Resources and Washington, DC: World Bank.

Balfors, B. and Mörtberg, U. (2002). Landscape ecological assessment: evaluating ecological effects of urbanisation scenarios. *Proceedings of the 21st Annual IAIA Conference*, The Hague, the Netherlands, June 2002.

Balfors, B., Mörtberg, U., Gontier, M. and Brokking, P. (2005). Impacts of region-wide urban development on biodiversity in strategic environmental assessment. *Journal of Environmental Assessment Policy and Management*, **7**, 229–46.

Beattie, A. J. (1991). Biodiversity and bioresources – the forgotten connection. *Search*, **22**, 59–61.

Beisner, B. E., Haydon, D. T. and Cuddington, K. (2003). Alternative stable states in ecology. *Frontiers in Ecology and the Environment*, **1**, 376–82.

Benett, E. L. and Robinson, J. G. (2000). *Hunting of Wildlife in Tropical Forests, Implications for Biodiversity and Forest Peoples.* Environment Department Papers, 76. Washington, DC: World Bank.

Bengtsson, J. (1998). Which species? What kind of diversity? Which ecosystem function? Some problems in studies of relations between biodiversity and ecosystem function. *Applied Soil Ecology*, **10**, 191–9.

Bertand, N. (2002). *Biodiversity Resources.* A companion volume to *Business and Biodiversity: The Handbook for Corporate Action*, IUCN. www.iucn.org.

Bhaskar, R. (1979). *The Possibility of Naturalism. A Philosophical Critique of the Contemporary Human Sciences.* Brighton: Harvester.

Bina, O. (2007). A critical review of the dominant lines of argumentation on the need for strategic environmental assessment. *Environmental Impact Assessment Review*, **27**, 585–606.

Bishop, J., Kapila, S., Hicks, F., Mitchell, P. and Vorhies, F. (2008). Building Biodiversity Business. Shell Int. Ltd. and IUCN. www.iucn.org/dbtw-wpd/edocs/2008-002.pdf.

Biswas, A. K., and Tortajada, C. (2003). An assessment of the Spanish National Hydrological Plan. *Water Resources Development*, **19**, 377–97.

Blaikie, P. and Jeanrenaud, S. (1997). Biodiversity and human welfare. In *Social Change and Conservation: Environmental Politics and Impacts of National Parks and Protected Areas*, eds. K. B. Ghimire and M. P. Pimbert. London: Earthscan, pp. 46–70.

Bower, D. J. (1989). Genetic resources worldwide. *TIBTECH*, **7**, 111–16.

Braat, L. and Ten Brink, P. (eds.) (2008). *The Cost of Policy Inaction: The Case of Not Meeting the 2010 Biodiversity Target (Executive Summary).* European Commission, DG Environment. Brussels.

Bräuer, I., Müssner, R., Marsden, K., Oosterhuis, F., Rayment, M., Miller, C. and Dodoková, A. (2006). The Use of Market Incentives to Preserve Biodiversity Final Report. A Project under the Framework Contract for Economic Analysis. ENV.G.1/FRA/2004/0081.

Bridgewater, P. and Arico, S. (2002). Conserving and managing biodiversity sustainably: the roles of science and society. *Natural Resources Forum*, **26**, 1–4.

Briffett, C., Obbard, J. P. and Mackee J. (2003). Toward SEA for the developing nations of Asia. *Environmental Impact Assessment Review*, **23**, 171–96.

Brockelman, W. Y. (1989). Priorities for biodiversity research. *Journal of the Science Society of Thailand*, **15**, 231–5.

Brockington, D. (2003). Injustice and conservation – is "local support" necessary for sustainable protected areas? *Policy Matters*, **12**, 22–30.

Brooks, T. M., Mittermeier, R. A., Mittermeier, C. G., da Fonseca, G. A. B., Rylands, A. B., Konstant, W. R., Flick, P., Pilgrim, J., Oldfield, S., Magin, G. and Hilton-Taylor, C. (2002). Habitat loss and extinction in the hotspots of biodiversity. *Conservation Biology*, **16**, 909–23.

Brouwer, H. and Van Den Tempel, R. (2006) (in Dutch). Chapter 4: Effectenanalyse (Analysis of effects). In *Praktijkboek Flora- en Faunawet* [Putting the flora and fauna law in practise]. The Hague, the Netherlands: SDU.

Brown, G., Jr. (1992). Replacement Costs of Birds and Mammals. Distributed by the State of Alaska Attorney General's Office.

Brownlie, S. (2005a). Systematic Conservation Planning, Land Use Planning and SEA in South Africa. Presentation at the International Experience and Perspectives in SEA conference, Prague 26–30 September 2005. International Association for Impact Assessment.

Brownlie, S., De Villiers, C., Driver, A., Job, N. and Von Hase, A. (2005b). Systematic conservation planning in the Cape Floristic Region and Succulent Karoo, South Africa: enabling sound spatial planning improved impact assessment. *Journal of Environmental Assessment Policy and Management*, **7**, 201–28.

Brownlie, S., Walmsley, B. and Tarr, P. (2006). Guidance Document on Biodiversity, Impact Assessment and Decision Making in Southern Africa. As part of IAIA's Capacity Building in Biodiversity and Impact Assessment project. The Southern African Institute for Environmental assessment in association with De Villiers Brownlie Associates. www.saiea.com/cbbia/html/guidance/main.html.

Bröring, U. and Wiegleb, G. (2005). Assessing biodiversity in SEA. In *Implementing Strategic Environmental Assessment*, eds. M. Schmidt, E. João and E. Albrecht. Berlin: Springer-Verlag.

BSR. (2005). Environmental Markets: Opportunities and Risks for Business. Business for Social Responsibility (BSR). www.bsr.org.

Bunting, A. H. (1990). The pleasures of diversity. *Biological Journal of the Linnean Society*, **39**, 79–87.

Burrows, L. (2004). United Kingdom: Integration of Biodiversity Issues into SEA: Somerset Country Council. Case study compiled for the drafting of CBD guidelines on Biodiversity in SEA. Somerset County Council, UK. www.eia.nl/ncea/pdfs/sea/casestudies/7_uk_sea_transport_somerset_county.pdf.

Byatt, I., Castles, I., Goklany, I. M., Henderson, D., Lawson, N., McKitrick, R., Morris, J., Peacock, A., Robinson, C. and Skidelsky, R. (2006). The *Stern Review*: a dual critique. Part II – economic aspects. *World Economics*, **7**, 199–229.

Byron, H. (2000). Biodiversity Impact. *Biodiversity and Environmental Impact Assessment: A Good Practise Guide for Road Schemes*. RSPB, WWF-UK, English Nature and Wildlife Trusts, Sandy, UK.

Byron, H. J., Treweek, J. R., Sheate, W. R. and Thompson, S. (2000). Road developments in the UK: an analysis of ecological assessment in environmental impact statements produced between 1993 and 1997. *Journal of Environmental Planning and Management*, **43**, 71–97.

Byron, H. and Treweek, J. (eds.). (2005). Special Issue on Strategic Environmental Assessment and Biodiversity. *Journal of Environmental Assessment Planning and Management*, **7**.

Canter, L. W. (1996). *Environmental Impact Assessment*. 2nd edn. New York: McGraw Hill.

Canter, L. W. and Canty, G. A. (1993). Impact significance determination – basic considerations and sequenced approach. *Environmental Impact Assessment Review*, **13**, 275–297.

Capistrano, D., Samper, C., Lee, M. J. and Raudsepp-Hearne, C. (eds.). (2005). *Ecosystems and Human Well-Being: Multiscale Assessments*. Findings of the Sub-global Assessments Working Group of the Millennium Ecosystem Assessment. The Millennium Ecosystem Assessment series v. 4. Washington, DC/Covelo/London: Island Press.

Cardinale, B. J., Srivastava, D. S., Duffy, J. E., Wright, J. P., Downing, A. L., Sankaran, M. and Jouseau, C. (2006). Effects of biodiversity on the functioning of trophic groups and ecosystems. *Nature*, **443**, 989–92.

Carpenter, S. R., DeFries, R., Dietz, T., Mooney, H. A., Polasky, S., Reid, W. V. and Scholes, R. J. (2006). Millennium Ecosystem Assessment: research needs. *Science*, **314**, 257–8.

Carlile, P. R. (2002). A pragmatic view of knowledge and boundaries: boundary objects in new product development. *Organization Science* **13**, 242–55.

Carson, R. T. and Hanemann, W. M. (1992). A Preliminary Economic Analysis of Recreational Fishing Losses Related to the *Exxon Valdez* Oil Spill. A report to the Attorney General of the State of Alaska.

Carson, R. T., Mitchell, W. R. C., Hanemann, M., Kopp, R. J., Presser, S. and Ruud, P. A. (1992). A Contingent Valuation Study of Lost Passive Use Values Resulting from the *Exxon Valdez* Oil Spill. A report to the Attorney General of the State of Alaska.

Carter, R. M., Freitas, C. R. de, Goklany, I. M., Holland, D. and Lindzen, R. S. (2006). The *Stern Review*: a dual critique – Part I: the science. *World Economics*, **7**, 167–98.

Cash, D. W., Clark, W. C., Alcock, F., Dickson, N. M., Eckley, N., Guston, D. H., Jäger, J. and Mitchell R. B. (2003). Knowledge systems for sustainable development. *Proceedings of the National Academy of Sciences of the United States of America*, published online 30 May 2003. www.pnas.org/cgi/content/abstract/1231332100v1.

CBD. (1992) Text of the Convention on Biological Diversity. www.cbd.int/convention/convention.shtml.

CBD. (2000a). Convention on Biological Diversity: Decision V/6 Ecosystem Approach. www.cbd.int/convention/cop-5-dec.shtml?m=COP-05&id=7148&lg=0.

CBD. (2000b). Access to Genetic Resources. Draft Guidelines on Access and Benefit-sharing regarding the Utilization of Genetic Resources. Note by the Executive Secretary. UNEP/CBD/COP/5/INF/21. www.cbd.int/doc/meetings/cop/cop-05/information/cop-05-inf-21-en.pdf.

CBD. (2000c). The Cartagena Protocol on Biosafety. www.cbd.int/biosafety/default.shtml.

CBD. (2001). Workshop on Biological Diversity and Tourism. Item 3 of the provisional agenda held during 4–7 June 2001, Santo Domingo.

CBD. (2002). Decision VI/7 A. Further development of guidelines for incorporating biodiversity-related issues into environmental-impact-assessment legislation or processes and in strategic impact assessment. www.cbd.int/convention/cop-6-dec.shtml?m=COP-06&id=7181&lg=0.

CBD. (2003). Proposals for further development and refinement of the guidelines for incorporating biodiversity-related issues into environmental impact assessment legislation or procedures and in strategic impact assessment: report on ongoing work. Note by the Executive Secretary. (UNEP/CBD/SBSTTA/9/INF/17).

CBD. (2004a). Decision VII/11 Ecosystem Approach. www.cbd.int/convention/cop-7-dec.shtml?m=COP-07&id=7748&lg=0.

CBD. (2004b). Ad Hoc Open-Ended Working Group on AccessaAnd Benefit-Sharing. Strategic Plan: Future Evaluation of Progress. The Need, and Possible Options, for Indicators for Access to Genetic Resources and in Particular for the Fair and Equitable Sharing of Benefits Arising from the Utilization of Genetic Resources. Note by the Executive Secretary. UNEP/CBD/WG-ABS/3/6. www.cbd.int/doc/meetings/abs/abswg-03/official/abswg-03-06-en.doc.

CBD. (2004c). Decision VII/16. Akwé Kon Voluntary Guidelines for the Conduct of Cultural, Environmental and Social Impact Assessment regarding Developments Proposed to Take Place on, or which are Likely to Impact on, Sacred Sites and on Lands and Waters Traditionally Occupied or Used by Indigenous and Local Communities. www.cbd.int/decisions/default.aspx?m=COP-07&id=7753&lg=0#_ftn57.

CBD. (2004d). The Impact of Trade Liberalization on Agricultural Biological Diversity. A Synthesis of Assessment frameworks. Note by the Executive Secretary UNEP/CBD/COP/7/INF/15. www.cbd.int/doc/meetings/cop/cop-07/information/cop-07-inf-15-en.doc.

CBD. (2005). Thinking the Unthinking: A Roundtable Discussion on Biodiversity Offsets. Convention on Biodiversity (CBD). SBSTTA 11, Montreal, 29 November 2005.

CBD. (2006). Decision VIII/28 Impact Assessment: Voluntary Guidelines on Biodiversity-Inclusive Impact Assessment. www.cbd.int/convention/cop-8-dec.shtml?m=COP-08&id=11042&lg=0.

CBS. (2008). *Webmagazine, 21 April 2008*. Central Bureau of Statistics, Netherlands. www.cbs.nl/nl-NL/menu/themas/macro-economie/publicaties/artikelen/archief/2008/2008-2441-wm.htm.

CEAA. (1996). A Guide on Biodiversity and Environmental Assessment. Canadian Environmental Assessment Agency (CEAA), Ministry of Supply and Services.

CEQ. (1993). *Incorporating Biodiversity Considerations into Environmental Impact Analysis under the National Environmental Policy Act*. Washington, DC: Council on Environmental Quality, US Government Printing Office.

Cernea, M. M. (1996). Understanding and preventing impoverishment from displacement: reflections on the state of knowledge. In *Understanding*

Impoverishment: The Consequences of Development Induced Displacement, ed. C. McDowell. Oxford: Berghahn Books.

Cernea, M. M. and McDowell, Ch. (eds.). (2000). *Risks and Reconstruction: Experiences of Resettlers and Refugees*. Washington, DC: World Bank.

Cernea, M. M. and Schmidt-Soltau, K. (2006). Poverty risks and national parks: policy issues in conservation and resettlement. *World Development*, **34**, 1808–30.

CES. (2001). *Impact of Iron Ore Mining on the Flora and Fauna of Kudremukh National Park and Environs: A Rapid Assessment unpublished report*. Bangalore, India: Centre for Ecological Sciences, Indian Institute of Science.

Cesar, H. S. J. and Van Beukering, P. J. H. (2004). Economic valuation of the coral reefs of Hawaii, *Pacific Science*, **58**, 231–42.

Chaker, A., El-Fadl K., Chamas, L. and Hatjian, B. (2006). A review of strategic environmental assessment in 12 selected countries. *Environmental Impact Assessment Review*, **26**, 15–56.

Cherp, A. (2001). EA legislation and practice in Central and Eastern Europe and the former USSR: a comparative analysis. *Environmental Impact Assessment, Review*, **21**, 335–62.

Cherp, A. and Antypas, A. (2003). Dealing with continuous reform: towards adaptive EA policy systems in countries in transition. *Journal of Environmental Assessment Policy and Management*, **5**, 455–76.

Chichilnisky, G. and Heal. G. (1998). Economic returns from the biosphere. *Nature*, **391**, 629–30.

Chivian, E. (ed.). (2002). *Biodiversity: Its Importance to Human Health*. Centre for Health and the Global Environment. Cambridge, MA: Harvard Medical School.

Chomitz, K., Brenes, E. and Constantino, L. (1998). *Financing Environmental Services: The Costa Rican Experience and Its Implications*. Washington, DC: Development Research Group, World Bank.

Christensen, P., Kørnøv, L. and Holm Nielsen, E. (2003). EIA as a regulation: does it work. *Journal of Environmental Planning and Management*, **48**, 393–412.

Commission for Environmental Impact Assessment (1997). Advisory Review of the Hidrovía Paraguay-Paraná Navigation Project. Report No. 029-92. Utrecht, the Netherlands.

Commoner, B. (1971). *The Closing Circle: Nature, Man, and Technology*. New York: Alfred A. Knopf.

Commonwealth of Australia. (1999). Environment Protection and Biodiversity Conservation Act No. 91 of 1999 as amended.

Commonwealth of Australia. (2007). *Leading Practice Sustainable Development Program for the Mining Industry*. (ISBN 0-642-725063.)

Consejo Andino de Ministros DeRelaciones Exteriores. (2002). Decision 523: Estrategia Regional de Biodiversidad para los Países del Trópico Andino.

Consorcio Prime Engenharia / Museo Noel Kempff Mercado / Asociación Potlatch (2004). Evaluación ambiental estratégica y revisión / complementación del eeia del corredor de transporte santa cruz – puerto suárez. Resumen ejecutivo. www.eia.nl/ncea/pdfs/sea/casestudies/5_bolivia_sea_route_santa_cruz.pdf.

Convention on International Trade in Endangered Species of Wild Fauna and Flora. (1994). Criteria for amendment of Appendices I and II. Conf. 9.24 (Rev. CoP14). www.cites.org/eng/res/09/09-24R14.shtml.

Convention on Migratory Species. (2002). Resolution 7.2 *Impact Assessment and Migratory Species*. Available at www.wcmc.org.uk/cms/COP/cop7/proceedings/pdf/en/part_I/Res_Rec/RES_7_02_Impact_Assessment.pdf.

Cooney, R. and Dickson, B. (eds.). (2006). *Biodiversity and Precautionary Principle: Risk and Uncertainty in Conservation and Sustainable Use*. London: Earthscan.

Costanza, R. and Daly, H. E. (1992). Natural capital and sustainable development. *Conservation Biology*, **6**, 37–46.

Cowell, R. (2000). Environmental compensation and the mediation of environmental change: making capital out of Cardiff Bay. *Journal of Environmental Planning and Management*, **43**, 689–710.

Coyne, M. 2004. Wetlands: Bush changes administration policy to 'net gain' of resource. Greenwire, 23 April 2004, *Natural Resources*, **10**, Environment and Energy Publishing LLC.

CSIR. (2000). Strategic Environmental Assessment in South Africa. Guideline Document. Department of Environmental Affairs and Tourism, Pretoria, RSA., www.environment.gov.za.

Culhane, P. J., Friesema, H. P. and Beecher, J. A. (1987). *Forecasts and Environmental Decision Making*. London: Westview Press.

Cuperus, R., Canters, K. J., Udo deHaes, H. A. and Friedman, D. S. (1999). Guidelines for ecological compensation associated with highways. *Biological Conservation*, **90**, 41–51.

Dada, O. J. and Okubokimi, A. (2004). Moving the EMP from the Shelf to the Field in Shell Petroleum Development Company, Nigeria. A paper prepared for the IAIA Annual Conference in Vancouver, Canada.

Dahmer, T. D. and Felley, M. L. (2000). Industrial development as a spur to nature conservation: Black-faced Spoonbills in Taiwan. *Proceedings of the 19th Annual IAIA Conference*, Hong Kong, June 2000.

Dalal-Clayton, B. and Sadler, B. (2005). *Strategic Environmental Assessment, A Sourcebook and Reference Guide to International Experience*. London: Earthscan.

Day, J. W., Maltby, E. and Ibáñez, C. (2006). River basin management and delta sustainability: a commentary on the Ebro Delta and the Spanish National Hydrological Plan. *Ecological Engineering*, **26**, 85–99.

David, T. J. (1993; editor). Towards the Wise Use of Wetlands. Report of the Wise Use Project. Ramsar Convention Bureau, Gland, Switzerland.

Davis, R. K. (1963). Recreation planning as an economic problem. *Natural Resource Journal*, **3**, 239–49.

De Camino, R., Segura, O., Arias, L. G. and Perez, I. (2000). *Costa Rica: Forest Strategy and the Evolution of Land Use*. Washington, DC: Environment Division, World Bank.

Deelstra, Y., Nooteboom, S. G., Kohlmann, H. R., Van Den Berg, J. and Innanen, S. (2003). Using knowledge for decision-making purposes in the context of large projects in the Netherlands. *Environmental Impact Assessment Review*, **23**, 515–653.

de Groot, R. S. (1992). *Functions of Nature: Evaluation of Nature in Environmental Planning, Management and Decision Making*. Groningen: Wolters-Noordhoff.

de Groot, R. S., Stuip, M., Finlayson, M. and Davidson, N. (2006). Valuing wetlands: guidance for valuing the benefits derived from wetland ecosystem services. *Ramsar*

Technical Report No. 3 / CBD Technical Series No. 27. Ramsar Convention Secretariat, Gland, Switzerland.

de Groot, W. T. (1992). *Environmental Science Theory: Concepts and Methods in a One-World, Problem-Oriented Paradigm.* New York: Elsevier.

de Jong, J., Oscarsson, A. and Lundmark, G. (2004). Hur behandlas biologisk mångfald i MKB?, [How biodiversity is treated in EIA?] Centrum för biologisk mångfald, Sveriges lantbruksunivesitet, Uppsala.

de Schutter, J. (2002). Water and Environmental Management Project Component E Monitoring of Construction Works for Rehabilitation of Infrastructure for the Sudoche Wetlands near the Amu Darya Delta. Delft, the Netherlands: Resource Analysis and Tashkent, Uzbekistan: VEP SANIIRI.

De Villiers, C. C., Driver, A., Brownlie S., Clark, B., Day, E. G., Euston-Brown, D. I. W., Helme, N. A., Holmes, P. M., Job, N. and Rebelo, A. B. (2005). *Fynbos Forum Ecosystem Guidelines for Environmental Assessment in the Western Cape.* Fynbos Forum. Kirstinbosch, Cape Town: Conservation Unit, Botanical Society of South Africa.

Department of Environmental Affairs and Development Planning. (2006). Provincial Guideline on Biodiversity Offsets – Draft for Public Comment. Republic of South Africa, Provincial Government of the Western Cape.

DFID, EC, UNDP, and World Bank. (2002). Linking Poverty Reduction and Environmental Management: Policy Challenges and Opportunities. UNDP/DFID/EC/World Bank, p. 80.

Diaz, A., Farglione, J., Chaplin, F. S. and Tilman, D. (2006). Biodiversity loss threatens human well-being. *PloS Biology*, **4**(8), e277. Use link doi.10.1371/journal.pbio.0040277 at http://biology.polsjournals.org.

Dietz, S., Anderson, D., Stern, N., Taylor, C., Zenghelis, D. (2007a). Right for the right reasons: a final rejoinder on the *Stern Review. World Economics*, **8**, 229–58.

Dietz, S., Hope, C., Stern, N., Zenghelis, D. (2007b). Reflections on the *Stern Review* (1): A robust case for strong action to reduce the risks of climate change. *World Economics*, **8**, 121–68.

Dietz, T. and Stern, P. C. (eds.). (2008). *Public Participation in Environmental Assessment and Decision Making. Panel on Public Participation in Environmental Assessment and Decision Making. Committee on the Human Dimensions of Global Change, Division of Behavioral and Social Sciences and Education.* Washington, DC: National Academies Press.

Dixon, J.A., Scura, L.F. and Van't Hof, T. (1993a). Meeting ecological and economic goals: marine parks in the Caribbean. *Ambio*, **22**, 117–25.

Dixon, J. A, Scura, L. F. and Van't Hof, T. (1993b). Ecology and microeconomics as 'Joint Products': the Bonaire Marine Park in the Caribbean. In *Biodiversity Coalition*, eds. Perrings *et al.* Norwell, MA: Kluwer Academic Publishers, pp. 125–7.

Doelle, M. and Sinclair A. J. (2006). Time for a new approach to public participation in EA: promoting cooperation and concensus for sustainability. *Environmental Impact Assessment Review*, **26**, 185–205.

Downward, S. R. and Taylor, R. (2007). An assessment of Spain's Programa AGUA and its implications for sustainable water management in the province of Almeria, southeast Spain. *Journal of Environmental Management*, **82**, 277–89.

Dryzek, J. S. (1993). Policy analysis and planning: from science to argument. In *The Argumentative Turn in Policy Analysis and Planning*, eds. F. Fischer and J. Forester. Durham and London: Duke University Press, pp. 212–32.

Dukhovny, V. A. and de Schutter, J. (2003). *South Priaralie-New Perspectives. Final report of the NATO Science for Peace Project SfP 974357*. Haarlem, the Netherlands and SIC-ICWC, Tashkent, Uzbekistan: Ecotec Resource.

Eaton, P. (1985). Tenure and taboo: customary rights and conservation in the South Pacific. In *Third South Pacific National Parks and Reserves Conference: Conference Report*, vol. **2**., Noumea, New Caledonia: South Pacific Commission, pp. 164–75.

EBI. (2003a). Integrating Biodiversity Conservation into Oil and Gas Development. Energy and Biodiversity Initiative BP, Chevron Texaco, Conservation International, Fauna and Flora International, IUCN, the Nature Conservancy, Shell, Smithsonian Institution, Statoil. www.theebi.org.

EBI. (2003b). Good Practice in the Prevention and Mitigation of Primary and Secondary Biodiversity Impacts. Energy and Biodiversity Initiative BP, Chevron Texaco, Conservation International, Fauna and Flora International, IUCN, the Nature Conservancy, Shell, Smithsonian Institution, Statoil. www.theebi.org.

EBI. (2003c). Biodiversity Indicators for Monitoring Impacts and Conservation Actions. Energy and Biodiversity Initiative BP, Chevron Texaco, Conservation International, Fauna and Flora International, IUCN, the Nature Conservancy, Shell, Smithsonian Institution, Statoil. www.theebi.org.

EBI. (2003d). Opportunities for Benefiting Biodiversity Conservation. Energy and Biodiversity Initiative (EBI) BP, Chevron Texaco, Conservation International, Fauna and Flora International, IUCN, the Nature Conservancy, Shell, Smithsonian Institution, Statoil. www.theebi.org.

EBI. (2004). Integrating Biodiversity into Environmental Management Systems. The Energy and Biodiversity Initiative. BP, Chevron Texaco, Conservation International, Fauna and Flora International, IUCN, the Nature Conservancy, Shell, Smithsonian Institution, Statoil. www.theebi.org.

Ecaat, J. (2004). A Review of the Application of Environmental Impact Assessment (EIA) in Uganda. *Report prepared for the United Nations Economic Commission for Africa*.

Edgar, P. W., Griffiths, R. A. and Foster, J. P. (2005). Evaluation of translocation as a tool for mitigating development threats to great crested newts (Triturus cristatus) in England, 1990–2001. *Biological Conservation*, **122**, 45–52.

Eftec, Economics for the Environment Consultancy Ltd. (2007). *Policy Appraisal and the Environment: An introduction to the Valuation of Ecosystem Services*. Department for Environment, Food, and Rural Affairs (DEFRA), UK.

Ehrlich, P. R. and Wilson, E. O. (1991). Biodiversity studies: science and policy. *Science*, **253**, 758–62.

Ehrlich, P. R. (2003). Citation from an article in *New York Times*, 1 July.

Ehrlich, P. R. (2008). Key issues for attention from ecological economists. *Environment and Development Economics*, **13**, 1–20.

Ekstrom, J. (2005). Biodiversity Offsets. Presentation in the Cambridge Conservation Forum (CCF) Seminar, Bird Life International.

El Fadl, K. and El-Fadel, M. (2004). Comparative assessment of EIA systems in MENA countries: challenges and prospects. *Environmental Impact Assessment Review*, **24**, 553–93.

Embid, A. (2003). The transfer from the Ebro Basin to the Mediterranean basins as a decision of the 2001 National Hydrological Plan: the main problems posed. *Water Resources Development*, **19**, 399–411.

Emerton, L. and Muramira, E. (1999). Uganda Biodiversity: Economic Assessment. IUCN World Conservation Union. Unpublished Paper Prepared with the National Environment Management Authority as part of the Uganda National Biodiversity Strategy and Action Plan.

Energy and Biodiversity Initiative. (2004). Integrating Biodiversity into Environmental and Social Impact Assessment Processes. BP, ChevronTexaco, Conservation International, Fauna & Flora International, IUCN, The nature Conservancy, Shell, Smithsonian Institute, Statoil, www.theebi.org.

Englehardt, J. D. (1998). Ecological and economic risk analysis of Everglades: phase I. restoration alternatives. *Risk Analysis*, **6**, 755–71.

Environmental Protection Department. (1997). Environmental Measures on Airport Core Projects, Case I, North Lantau Expressway Project. Hong Kong: Government of Hong Kong.

E&P Forum. (1986). *E&P Forum view on Environmental Impact Assessment (EIA)*. Report No. 2.40/135. London: E&P Forum.

EPA. (1999). Considering Ecological Processes in Environmental Impact Assessments. United States Environmental Protection Agency (EPA). http://es.epa.gov/oeca/ofa/ecol99.html.

EPA. (2006). Environmental Offsets: Position Paper 9. Environmental Protection Authority (EPA), Govt. of Western Australia.

Espinoza, G. and Alzina, V. (ed.). (2001). *Review of Environmental Impact Assessment in Selected Countries of Latin America and the Caribbean: Methodology, Results And Trends*. Washington, DC: Inter-American Development Bank. Centre for Development Studies.

Espinoza, G. and Alzina, V. (2001). Review of Environmental Impact Assessment in Selected Countries of Latin America and the Caribbean; Methodology, Results and Trends. Washington, DC: Inter-American Development Bank. Centre for development studies.

eThekwini Municipality. (2002). *Durban environmental services management plan*. www.durban.gov.za/durban/services/departments/environment/documents_reports/desmp/.

EUCC – The Coastal Union. (2008). www.coastalguide.to/dutch_waddensea/.

Euroconsult and the Wetland Group. (1996). Aral Sea Wetland Restoration Project. Main Report. The executive committee of the Interstate Council for Addressing the Aral Sea Crises / IBRD.

European Commission. (1994). Environmental Impact Assessment Review Checklist European Commission (DGXI), Brussels.

European Commission. (2000). '*Managing Natura 2000 Sites*'. *The Provisions of Article 6 of the 'Habitats' Directive 92/43/EEC*. (ISBN 92–828-9048–1.) Luxembourg: Office for Official Publications of the European Communities.

European Commission. (2006). Handbook for Trade Sustainability Impact Assessment. http://trade.ec.europa.eu/doclib/docs/2006/march/tradoc_127974.pdf.

European Council. (1985). Council Directive 85/337/EEC of 27 June 1985 on the assessment of the effects of certain public and private projects on the environment.

European Council. (1992) Council Directive 92/43/EEC of 21 May 1992 on the conservation of natural habitats and of wild fauna and flora ("Habitat Directive"). http://eur-lex.europa.eu/LexUriServ/LexUriServ. do?uri=CELEX:31992L0043:EN:NOT.

European Council. (1997). Council Directive 97/11/EC of 3 March 1997 amending Directive 85/337/EEC on the assessment of the effects of certain public and private projects on the environment.

European Council. (2001). Directive 2001/42/EC of the European Parliament and of the Council of 27 June 2001 on the assessment of the effects of certain plans and programmes on the environment. Official Journal L 197, 21/07/ 2001, 0030 – 0037. http://ec.europa.eu/environment/eia/sea-legalcontext. htm#legal.

European Council. (1985). 'Council Directive 85/337/EEC of 27 June 1985 on the assessment of the effects of certain public and private projects on the environment'.

European Council. (1997). Council Directive 97/11/EC of 3 March 1997 amending Directive 85/337/EEC on the assessment of the effects of certain public and private projects on the environment. http://eur-lex.europa. eu/LexUriServ/LexUriServ.do?uri=CELEX:31997L0011:EN:NOT.

EVOSTC (Exxon Valdez Oil Spill Trustee Council). (2008). www.evostc.state.ak. us/.

Fahrig, L. (1997). Relative effects of habitat loss and fragmentation on population extinction. *Journal of Wildlife Management*, **61**, 603–10.

FAO. (2007). *The Global Environmental Facility and Payments for Ecosystem Services. A Review of Current Initiatives and Recommendations for Future PES support by GEF and FAO programmes.* By P. Gutman and S. Davidson, WWF Macroeconomic for Sustainable Development Program Office. Report commissioned by FAO for Payments for Ecosystem Services from Agricultural Landscapes-PESAL project, PESAL Papers Series No.1, Rome.

Fischer T. and Gazzola P. (2006). SEA effectiveness criteria – equally valid in all countries? The case of Italy. *Environmental Impact Assessment Review*, **26**, 396–409.

Fish, S., Snashall, D. and Streater, J. (2004). *Offsetting Environmental Impacts to Facilitate Mining.* Australia: Rio Tinto Coal Australia and ERM Australia Pty Ltd.

Folgarait, P. J. (1998). Ant biodiversity and its relationship to ecosystem functioning: a review. *Biodiversity and Conservation*, **7**, 1221–44.

Forman, R. T. T. (2000). Estimate of the area affected ecologically by the road system in the United States. *Conservation Biology*, **14**, 31–5.

Fox, J. and Nino-Murcia, A. (2005). Status of species conservation banking in the United States. *Conservation Biology*, **19**, 996–1007.

Freeman, A. M., III. (2003). *The Measurement of Environmental and Resource Values.* Washington, DC: Resources for the Future (RFF).

Frost, L., Reich, M. R. and Fujisaki, T. (2002). A partnership for Ivermectin: social worlds and boundary objects. In *Public Private Partnerships for Public Health*, ed. M. R. Reich. Cambridge, MA: Harvard University Press, pp. 87–113.

FutureWorks. (2004). *uMhlathuze Strategic Catchment Assessment: A Tool for Sustainable Land Use Management and Planning.* Report prepared for the uMhlathuze Municipality. Ministry of Water Resources and Irrigation. (2005). *Integrated Irrigation Improvement and Management Project.* Project Appraisal Document.

Gabriel, D., Roschewitz, I., Tscharntke, T. and Thies, C. (2006). Beta diversity at different spatial scales: plant communities in organic and conventional agriculture. *Ecological Applications*, **16**, 2011–21.

Garrido, A. (2003). An economic appraisal of the Spanish National Hydrological Plan. *Water Resources Development*, **19**, 459–80.

GEF. (2005). *Achieving the Millennium Development Goals – A GEF Progress Report.* Washington, DC: Global Environment Facility.

Geneletti, D. (2002). *Ecological Evaluation for Environmental Impact Assessment.* Netherlands Geographical Studies 301. Utrecht, the Netherlands: KNAG.

Geneletti, D. (2006). Some common shortcomings in the treatment of impacts of linear infrastructure on natural habitats. *Environmental Impact Assessment Review*, **26**, 257–67.

Geneletti, D. (2008). Incorporating biodiversity assets in spatial planning: methodological proposal and development of a planning support system. *Landscape and Urban Planning*, **84**, 252–65.

Geoghegan, T. (1998). *Financing Protected Area Management: Experiences from the Caribbean.* CANARI Technical Report No. 272, p. 17.

George, C. (2004). Integrated Impact Assessment of International Trade Policy and Agreements: the European Union's Sustainability Impact Assessments of Proposed WTO Agreements on Agriculture and Forest Products. Case study compiled for the drafting of CBD guidelines on Biodiversity in SEA. Institute for Development Policy and Management, University of Manchester, UK.

George, C. and Kirkpatrick, C. (2003). Sustainability Impact Assessment of World Trade Negotiations: Current Practice and Lessons for Further Development. Impact Assessment Research Centre Working Paper Series, Paper No: 2. Institute for Development Policy and Management (IDPM), University of Manchester.

Getches, D. (2003). Spain's Ebro River transfers: test case for water policy in the European Union. *Water Resources Development*, **19**, 501–12.

Ghazoul, J. (2007). Recognising the complexities of ecosystem management and the ecosystem services concept. *Gaia*, **16**, 215–21.

Gibson, R. B. (2002). From Wreck Cove to Voisey's Bay: the evolution of federal environmental assessment in Canada. *Impact Assessment and Project Appraisal*, **20**, 151–9.

Gilbert, A. J. and Janssen, R. (1998). Use of environmental functions to communicate the values of a mangrove ecosystem under different management regimes. *Ecological Economics*, **25**, 323–46.

Glazewski, J. and Paterson, A. R. (2005). Biodiversity, genetic modification and the law. In *Environmental Law in South Africa*. 2nd edn., ed. J. Glazewski. London: Lexis Nexis Butterworths, pp. 255–92.

Goldschmidt, T. (1998). *Darwin's Dreampond: Drama in Lake Victoria*. Cambridge, MA: MIT Press.

Gontier, M., Balfors, B. and Mörtberg, U. (2003). Prediction Tools for Biodiversity in EIA. Abstract for the 22nd Annual IAIA Conference, Marrakech, Morocco, June 2003.

Gontier, M., Balfors B. and Mörtberg, U. (2005). Biodiversity in environmental assessment – current practice and tools for prediction. *Environmental Impact Assessment Review*, **26**, 268–86.

Goodland, R. (2003). Policy Options: How to Ensure World Bank Group-Supported Extractive Industries Reduce Poverty and Promote Sustainable Development. Washington, DC. Extractive Industries Review (IFC), p. 111.

Government of Brazil. (1965). Law 4771: Forestry Code.

Government of Brazil. (2000). Law 9985: Protected Area law.

Gray, I. M. and Edwards-Jones, G. (1999). A review of the quality of environmental impact assessments in the Scottish forestry sector. *Forestry*, **72**, 1–10.

Gunderson, L. H., Carpenter, S. R., Folke, C., Olsson, P. and Peterson, G. (2006). Water RATs (Resilience, Adaptability, and Transformability) in lake and wetland social-ecological systems. *Ecology and Society*, **11**, 16.

Gunningham, N. and Young, M. (2002). Redesigning Environmental Regulation: The Case of Biodiversity Conservation. Copyright 2003, Environmental Law, Alliance Worldwide.

Gutman, P. and Davidson, S. (2007). A Review of Innovative International Financial Mechanisms for Biodiversity Conservation with Special Focus on the International Financing of Developing Countries' Protected Areas. WWF-MPO. www.ecnc.nl/jump/page/721/Financing%20mechanisms.html.

Hamid, L., Stern, N. and Taylor, C. (2007). Reflections on the *Stern Review* (2): a growing international opportunity to move strongly on climate change. *World Economics*, **8**, 1–18.

Hamilton, A. C. (2004). Medicinal plants, conservation and livelihoods. *Biodiversity and Conservation*, **13**, 1477–517.

Hanemann, M. (2003). Appendix C: economics. In *A technical review of the Spanish National Hydrological Plan (Ebro River out-of-basin diversion)*. Fundación Universidad Politécnica de Cartagena, Murcia, pp. 41–51.

Hanemann, W. M. (1994). Valuing the environment through contingent valuation. *Journal of Economic Perspectives*, **8**, 19–43.

Hansen, J. (2004). Defusing the global warming time bomb. *Scientific American*, **290**, 69–77.

Hassan, R., Scholes, R. and Ash, N. (2005). *Ecosystem and Human Well-Being: Current State and Trends: Findings of the Conditions and Trends Working Group of the Millennium Ecosystem Assessment*. The Millennium Ecosystem Assessment series vol. 1. Washington, DC/Covelo/London: Island Press.

Heal, G. M. (2000). *Nature and Market Place: Capturing the Value of Ecosystem Services*. Washington, DC: Island Press, p. 203.

Hein, L., van Koppen, K., de Groot, R. S. and van Ierland, E. C. (2006). Spatial scales, stakeholders and the valuation of ecosystem services. *Ecological Economics*, **57**, 209–28.

Henderson, D. (2007). Governments and climate change issues: the case for rethinking. *World Economics*, **8**, 183–228.

Herrera, R. J. (2007). Strategic environmental assessment: the need to transform the environmental assessment paradigms. *Journal of Environmental Assessment Policy and Management*, **9**, 211–34.

Heuvelhof, E. F. T. and Nauta, C. M. (1996). (M)ERKENNING, Onderzoek naar de doorwerking van m.e.r. Evaluatiecommissie Wet Milieubeheer, VROM, The Hague, the Netherlands.

Hilden, M. (1995). Evaluation of the Significance of Environmental Impacts. EIA Process Strengthening Workshop, 4–7April 1995, Canberra, Australia.

Hilden, Rydevik T. and Bjarnadottir, H. (2007). Context awareness and sensitivity in SEA implementation. *Environmental Impact Assessment Review*, **27**, 666–84.

Hildén, T., Furman, E. and Kaljonen, M. (2004). Views on planning and expectations of SEA: the case of transport planning. *Environmental Impact Assessment Review*, **24**, 519–36.

Hong Kong Environmental Protection Department. (1997). Technical Memorandum on Environmental Impact Assessment Process. Environmental Impact Assessment Ordinance, Cap.*499*, *s.16*. www.epd.gov.hk/eia/legis/index3.htm.

Hooper, D. U., Chapin, F. S., III, Ewel, J. J., Hector, A., Inshausti, P., Lavorel, S., Lawton, J. H., Lodge, D. M., Loreau, M., Naeem, S., Schmid, B., Setälä, H., Symstad, A. J., Vandermeer, J. and Wardle, D. A. (2005). Effects of biodiversity on ecosystem functioning: a consensus of current knowledge. Ecological Society of America Report. *Ecological Monographs*, **75**(1), 3–35.

Howden, D. (2007). Take Over Our Rainforest: Guyana's Extraordinary Offer to Britain to Save One of the World's Most Important Carbon Sinks. www.independent.co.uk/environment/climate-change/take-over-our-rainforest-760211.html.

Howitt, R. (2003). Some economic lessons from past hydrological projects and applications to the Ebro River transfer proposal. *Water Resources Development*, **19**, 471–84.

Howlett, M. and Ramesh, M. (1995). *Studying Public Policy: Policy Cycles and Policy Subsystems*. Toronto: Oxford University Press.

Howlett, M. and Ramesh, M. (1998). Policy subsystem configurations and policy change: operationalizing the postpositivist analysis of the politics of the policy process. *Policy Studies Journal*, **26**(3), 466–81.

Hoyt, E. (1988). *Conserving the Wild Relatives of Crops*. Rome: IBPGR/IUCN/WWF.

Hulme, D. and Murphree, M. (1999). Communities, wildlife and the 'new conservation' in Africa. *Journal of International Development*, **11**, 277–86.

IAIA. (1999). Principles of Environmental Impact Assessment Best Practice. International Association for Impact Assessment, USA / Institute of Environmental Assessment, U.K. www.iaia.org/Non_Members/Activity_Resources/key_resources.htm.

IAIA. (2002). Strategic Environmental Assessment Performance Criteria. Special Publication Series No. 1. www.iaia.org/Non_Members/Pubs_Ref_Material/pubs_ref_material_index.htm.

IAIA. (2005). Biodiversity & Ecology Section. Principles on Biodiversity in Impact Assessment. IAIA Special Publication Series, 3, Fargo, U.S.A.

Ibáñez, C. and Prat, N. (2003). The environmental impact of the Spanish National Hydrological Plan on the lower Ebro River and Delta. *Water Resources Development*, **19**, 485–500.

ICMM. (2003). Position Statement on Mining and Protected Areas. International Council on Mining and Metals (ICMM). www.icmm.com/publications/497ICMMPositionStatementonMiningandProtectedAreas.pdf.

ICMM. (2005a). Biodiversity Offsets: A Proposition Paper. International Council on Mining and Metals (ICMM), London. www.icmm.com/news/720biodiversity_proposition_2005.pdf.

ICMM. (2005b). Biodiversity Offsets: A Briefing Paper for the Mining Industry. International Council on Mining and Metals (ICMM), unpublished.

ICMM. (2006). Good Practice Guidance for Mining and Biodiversity. International Council for Mining and Metals, London, UK. www.icmm.com.

IEEM. (2006). Guidelines for Ecological Impact Assessment in The United Kingdom. Institute of Ecology and Environmental Management, Winchester, Hampshire. www.ieem.org.uk/ecia/EcIA%20Approved%207%20July%2006.pdf.

IFC. (2004). A Guide to Biodiversity for the Private Sector. International Finance Corporation – World Bank Group. www.ifc.org/BiodiversityGuide.

IIED. (2002). Drawers of Water II. London, UK: In collaboration with Community Management and Training Services Ltd. (Kenya), Institute of Resource Assessment of the University of Dares Salaam (Tanzania) and Child Health of Makerere University Medical School (Uganda).

Innanen, S. (2004). Environmental impact assessment in Turkey: capacity building for European Union accession. *Impact Assessment and Project Appraisal*, **22**, 141–51.

IPIECA. (2003). The Oil and Gas industry: Operating in Sensitive Environments. Publication of International Petroleum Industry and Environmental Conservation Association.

IPIECA and OGP. (2005). A Guide to Developing Biodiversity Action Plans for the Oil and Gas Sector. International Petroleum Industry Environmental Conservation Association and the International Association of Oil and Gas Producers. www.ipieca.org/activities/biodiversity/downloads/publications/baps.pdf.

IPIECA and OGP. (2006). Key Biodiversity Questions in the Oil and Gas Lifecycle. International Petroleum Industry Environmental Conservation Association and the International Association of Oil and Gas Producers. www.ipieca.org/activities/biodiversity/downloads/publications/bdwg_lifecycle.pdf.

IPCC. (Intergovernmental Panel on Climate Change), UNEP/WMO. (2001). Third Assessment Report, Working Group III, Mitigation. www.ipcc.ch.

IPTRID. (2005). Toward Integrated Planning of Irrigation and Drainage in Egypt. In support of the Integrated Irrigation Improvement and Management Project (IIIMP). Final Report. Fao, Rome. www.fao.org/docrep/008/a0021e/a0021e00.htm.

IUCN. (1994). Red List of Threatened Animals. Gland, IUCN Switzerland and Cambridge, UK.

IUCN. (2004). Assessment and Provision of Environmental Flows in Mediterranean Watercourses – Basic Concepts, Methodologies and Emerging Practice. Mediterranean Case Study environment flow assessment for the Ebro Delta in Spain – Improving links between wetland and catchement management.

IUCN and E&P Forum. (1993). Oil and Gas Exploration and Production in Arctic and Subarctic Onshore Regions. Guidelines for Environmental Protection. IUCN-The World Conservation Union / E&P Forum – The Oil Industry International Exploration and Production Platform Publication no. 2.55/184, London.

IUCN and ICMM. (2004). Integrating Mining and Biodiversity Conservation. Case studies from around the world. IUCN Gland, Switzerland and ICMM, Cambridge, UK.

IUCN, UNEP, and WWF (1980). World Conservation Strategy. Living Resource Conservation for Sustainable Development. http://data.iucn. org/dbtw-wpd/edocs/WCS-004.pdf.

Jordan, T. (2006). *Integrating Biodiversity Issues into Strategic Environmental Planning.* (Powerpoint presentation available at www.iclei-europe.org/uploads/media/ B2_JORDAN_biodiversity_uMhlathuze.pdf.)

Jalakas, L. (1998). Case Study: Comprehensive Planning of the Island Naissaar, Viimsi Municipality, Estonia. In *Strategic Environmental Assessment in Transitional Countries. Emerging Practices.* eds. B. Sadler, J. Dusik, S. Casey and N. Mikulic, REC.

Janssen R. and Padilla, J. E. (1999). Preservation or conversion? Valuation and evaluation of a Mangrove forest in the Philippines. *Environmental and Resource Economics*, **14**, 1573–2.

Jansson, A. M., Hammer, M., Folke, C. and Costanza, R. (eds.). (1994). *Investing in Natural Capital: The Ecological Economics Approach to Sustainability.* Washington, DC: Island Press.

Jenkins, M., Scherr, S. and Inbar, M. (2004). Markets for biodiversity services: potential roles and challenges. *Environment*, **46**, 32–42.

Johnston, R. and Madison, M. (1997). From landmarks to landscapes: a review of current practices in the transfer of development rights. *Journal of the American Planning Association*, **63**, 365–78

Kamppinen M. and Walls M. (1999). Integrating biodiversity into decision-making. *Biodiversity and Conservation*, **8**, 7–16.

Kareiva, P. and Marvier, M. (2003).Conserving biodiversity coldspots. *American Scientist*, **91**, 344–51.

Kessler, J. J., Rood, T., Tekelenburg, T. and Bakkenes, M. (2007). Biodiversity and socioeconomic impacts of selected agro-commodity. *Journal of Environment and Development*, **16**, 131–60.

Khagram, S., Clark, W. C. and Raad, D. F. (2003). From the environment and human security to sustainable security and development. *Journal of Human Development*, **4**, 289–313.

Kiev Protocol. (2003). Protocol on Strategic Environmental Assessment. Convention on Environmental Impact Assessment (EIA) in a Transboundary Context. www. unece.org/env/eia/sea_protocol.htm.

Kishor, N. and Constantino, L. (1993). Forest Management and Competing Land Uses: An Economic Analysis for Costa Rica. World Bank LATEN Dissemination Note 7. Latin America Technical Department.

Kiss, A. (2004). Making biodiversity conservation a land-use priority. In *Getting Biodiversity Projects to Work: Towards Better Conservation and Development*, eds. T. O. McShane and M. P. Wells. New York: Columbia University Press, pp. 98–123.

Klein, J. T. (1990). *Interdisciplinarity. History, Theory, and Practice*. Detroit: Wayne State University Press.

Klein, J. T. (1996). *Crossing Boundaries, Knowledge, Disciplinarities, and Interdisciplinarities*. Charlottesville/London: University Press of Virginia.

Knegtering, E., Drees, J. M., Geertsema, P., Huitema, H. J. and Schoot Uiterkamp, A. J. M. (2005). Use of animal species data in environmental impact assessments. *Environmental Management*, **36**, 862–71.

Kolhoff, A. and Slootweg, R. (2005). Biodiversity in SEA for spatial plans – experiences from the Netherlands. *Journal of Environmental Assessment Policy and Management* 7, 267–86. www.eia.nl/ncea/pdfs/sea/casestudies/4_nl_sea_for_spatial_plans.pdf.

Kolhoff, A. J., Runhaar, H. A. C., Driessen, P. J. (in press). The contribution of capacities and context to EIA system performance and effectiveness: an analytical framework. *Impact Assessment and Project Appraisal*.

Koning, P. C. and Slootweg, R. (1999). Key Ecological Processes in the Assessment of Impacts on Biologoical Diversity. A Study for the Document Retrieval and Screening System for Environmental Impact Assessment (DR-EIA). Mekon Ecology, Leiden (unpublished document).

Koudstaal, R., Slootweg, R. and Boromthanarat, S. (1994). *Wise Use of Wetlands: A Methodology for the Assessment of Functions and Values of Wetlands (with Pak Phanang as an Example Case)*. Haarlem, the Netherlands: The Wetland Group.

Kørnøv, L. and Thissen, W. (2000). Rationality in decision- and policy-making: implications for strategic environmental assessment. *Impact Assessment Project Appraisal*, **18**, 191–200.

Krishnaswamy, J., Mehta, V. K. (2003). Impact of Iron Ore Mining in Kudremukh on Bhadra River Ecosystem. www.wcsindia.org/sedimentreport.pdf.

Krugman, P. and Wells, R. (2006). *Economics*. Worth Publishers, New York.

Kuiper, G. (1997). Compensation of environmental degradation by highways: a Dutch case study. *European Environment*, **7**, 118–25.

Kumari, K. and King, K. (1997). Paradigm Cases to Illustrate the Application of the Incremental Cost Assessment to Biodiversity. www.gefweb.org/Operational_Policies/Eligibility_Criteria/Incremental_Costs/paradigm.htm.

Lafarge. (2000). Lafarge and the Environment. Paris, France: Environment Department, Lafarge Group. www.lafarg.com.

Landell-Mills, N. and Porras, I. T. (2002). *Silver Bullet or Fools' Gold? A Global Review of Markets for Forest Environmental Services and their Impact on the Poor*. London: IIED Publication.

Lee, N. (2006). Bridging the gap between theory and practise in integrated assessment. *Environmental Impact Assessment Review*, **26**, 57–78.

416 · References

Lee, N. and Colley, R. (1992). Reviewing the Quality of Environmental Statements, Occasional Paper 24. 2nd edn. Manchester: Department of Planning and Landscape, University of Manchester.

Lehmann, A., Leathwick, J. R. and Overton, J. M. (2002). Assessing New Zealand fern diversity from spatial prediction of species assemblages. *Biodiversity Conservation*, **11**, 2217–38.

Leknes, E. (2001). The roles of EIA in the decision-making process. *Environmental Impact Assessment Review*, **21**, 309–34.

LeMaitre, D. C. and Gelderblom, C. M. (1998). Biodiversity Impact Assessment: Putting the Theory in Practise. Paper presented at the 18th Annual Meeting of the International Association for Impact Assessment, Christchurch, New Zealand, 19–24 April 1998.

Ligtvoet, W. and Witte, F. (1991). Perturbation through predator introduction: effect on food web and fish yields in Lake Victoria (East Africa). In *Terrestrial and Aquatic Ecosystems: Perturbations and Recovery*, ed. O. Ravera. Chichester: Ellis Horwood Ltd., pp. 263–68.

Lin, C. Y. O. (2006). Social and gendered implications of dam resettlement on the Orang Asli of Peninsular Malaysia gender. *Technology and Development*, **10**, 77–99.

Lindenmayer, D. B. (1994). Wildlife corridors and the mitigation of logging impacts on fauna in wood production forests in south-eastern Australia: a review. *Wildlife Research*, **21**, 323–40.

Liou, M. and Yu Y. (2004). Development and implementation of strategic environmental assessment in Taiwan. *Environmental Impact Assessment Review*, **24**, 337–50.

Litzinger, R. (2008). Damming the Angry River. Available at: www.gbcc.org.uk/30article3.htm (last accessed July 2008).

Loreau, M., Naeem, S., Inchausti, P., Bengston, J., Grime, J. P., Hector, A., Hooper, D. U., Huston, M.A., Raffaelli, D., Schmid, S., Tilman D. and Wardle D. A. (2001). Biodiversity and ecosystem functioning: current knowledge and future challenges. *Science*, **294**, 804–8.

Löwy, I. (1992). The strength of loose concepts – boundary concepts, federative experimental strategies and disciplinary growth: the case of immunology. *History of Science*, **30**, 371–96.

Kareiva, P. and Marvier, M. (2003). Conserving biodiversity coldspots. Recent calls to direct conservation funding to the world's biodiversity hotspots may be bad investment advice. *American Scientist*, **91**, 344–9.

Mack, R. H., Simberloff, D., Lonsdale, W. M., Evans, H. and Clout, M. (2000). Biotic invasions: causes, epidemiology, global consequences, and control. *Ecological Applications*, **10**, 689–710.

Magurran, A. (2003). *Measuring Biological Diversity*. Oxford, UK: Blackwell Publishing, p. 260.

Mandelik, Y., Dayan, T. and Feitelson, E. (2002). Ecological Impact Assessment in Israel: A review of environmental statements. *Proceedings of the 21st Annual IAIA Conference*, The Hague, the Netherlands.

Mandelik, Y., Dayan, T. and Feitelson, E. (2005a). Planning for biodiversity: the role of ecological impact assessment. *Conservation Biology*, **19**, 1254–61.

Mandelik, Y., Dayan, T. and Feitelson, E. (2005b). Issues and dilemmas in ecological scoping: scientific, procedural and economic perspectives. *Impact Assessment Project Appraisal*, **23**, 55–63.

Mathur, H. M. and Marsden, D. (eds). (1998). *Development Projects and Impoverishment Risks*. Delhi: Oxford University Press.

Maurits la Riviere, J. W. (1989). Threats to the world's water. *Scientific American*, **261**, 80–94.

McDowell Group. (1990). *An Assessment of the Impact of the Exxon Valdez Oil Spill on the Alaska Tourism Industry*. Seattle, Washington: Preston, Thorgrimson, Shidler, Gates, and Ellis.

McKenny, B. (2005). Environmental Offset Policies, Principles, and Methods: A Review of Selected Legislative Frameworks. Biodiversity Neutral Initiative. http://biodiversityneutral.org/EnvironmentalOffsetLegislative Frameworks. pdf.

McKinney, L. D. and Murphy, R. (1996). When biologists and engineers collide: habitat conservation planning in the middle of urbanized development. *Environmental Management*, **20**, 955–61.

McNeely, J. A. (1994). Critical issues in the implementation of the Convention on Biological Diversity. In *Widening Perspectives on Biodiversity*, eds. A. F. Krattiger, J. A. McNeely, W. H. Lesser, K. R. Miller, Y. St. Hill and R. Senanayake. Gland: IUCN/IAE. pp. 7–10.

McNeely, J. A. (2006). Using economic instruments to overcome obstacles to in situ conservation of biodiversity. *Integrative Zoology*, **1**, 25–31.

Meijer, W., Lodders-Elfferich, P. C., Hermans, L. M. L. H. A. (2004). *Ruimte voor de Wadden*. Eindrapport Adviesgroep Waddenzee Beleid.

Metrick, A. and Weitzman, M. L. (1998). Conflicts and choices in biodiversity preservation. *Journal of Economic Perspectives*, **12**, 21–34.

Meynell, P.-J. (2005). Use of IUCN Red Listing process as a basis for assessing biodiversity threats and impacts in environmental impact assessment. *Impact Assessment and Project Appraisal*, **23**, 65–72.

Millennium Ecosystem Assessment. (2003). *Ecosystems and Human Well-being: A Framework for Assessment*. Washington, DC: Island Press. www. millenniumassessment.org/en/products.aspx.

Millennium Ecosystem Assessment. (2005a). *Ecosystems and Human Wellbeing: Biodiversity Synthesis*. Washington, DC: World Resources Institute.

Millennium Ecosystem Assessment. (2005b). *Synthesis*. Washington, DC: Island Press.

Miller, K. R., Raid, W. V. and Barber, C. V. (1992). The global biodiversity strategy and its significance for sustainable agriculture. In *Biodiversity-Implications for Global Food Security*, eds. M. S. Swaminathan and S. Jena. Madras, India: Macmillan Publications.

Mills, M. J. (1992). Alaska Sport Fishing in the Aftermath of the *Exxon Valdez* Oil Spill. Special Publication No. 92–5, Alaska Department of Fish and Game, Division of Sport Fish, Anchorage, Alaska.

Milon, J. W. and Hodges, A. (2000). Who wants to pay for Everglades restoration? *The Magazine of Food, Farm and Resource Issues*, Summer, 2000. http://findarticles. com/p/articles/mi_m0HIC/is_2_15/ai_66918325.

Milon, J. W. and Scrogin, D. (2006). Latent preferences and valuation of wetland ecosystem restoration. *Ecological Economics*, **56**, 162–75.

Ministry of Economic Affairs. (2004). *Gaswinning in Nederland. Belang en beleid*. EZ publicatienummer: 04EP10. Ministry of Economic Affairs, The Hague. www. sodm.nl/data/documentatie_data/publicatie_data/gasbrief.pdf.

Missrie, M. and Nelson, K. (2005). Direct Payments for Conservation: Lessons from the Monarch Butterfly Conservation Fund. Research Summary/Paper No. 8, College of Natural Resources, University of Minnesota.

Mitchell, R. C. and Carson, R. T. (1989). *Using Surveys to Value Public Goods: The Contingent Valuation Method*. Baltimore: Johns Hopkins University Press.

Moll, P. and Zander, U. (2006). *Managing the Interface. From Knowledge to Action in Global Change and Sustainability Science*. München: Oekom Verlag,

Mollinga, P. P. (2003). *On the Waterfront. Water Distribution, Technology and Agrarian Change on a South Indian Canal Irrigation System*. Wageningen University Water Resources Series. Hyderabad, India: Orient Longman (orig. 1998).

Mollinga, P. P. (2008). The Rational Organisation of Dissent. Interdisciplinarity in the study of natural resources management. ZEF Working Paper, 33. Center for Development Research, Bonn, Germany. www.zef.de/fileadmin/webfiles/ downloads/zef_wp/wp33 Mollinga.pdf.

Morgan, M. S. and Morrison, M. (eds.). (1999). *Models as Mediators. Perspectives on Natural and Social Science*. Cambridge, UK: Cambridge University Press.

Morrison-Saunders, A. and Fischer, T. (2006). What is wrong with EIA and SEA anyway? A sceptic's perspective on sustainability. *Journal of Environmental Assessment Policy and Management*, **8**, 19–39.

Mwalyosi, R. and Hughes, R. (1998). The performance of EIA in Tanzania (unpublished document).

Myers, N. (1996). Environmental services of biodiversity. *Proceedings of the National Academy of Sciences of the United States of America*, **93**, 2764–69.

Myers, N., Mittelmeier, R. A., Mittelmeier, C. G., da Fonseca G. A. B. and Kent, J. (2000). Biodiversity hotspots for conservation priorities. *Nature*, **403**, 853–8.

NAM (Nederlandse Aardolie Maatschappij B.V.). (1998). *Integrale bodemdalingstudie, hoofdrapport en bijbehorende deelrapporten inzake geomorfologie en infrastructuur, kwelders en vogels*. www.NAM.nl.

NAM (Nederlandse Aardolie Maatschappij B.V.). (2006). *Gaswinning onder de Waddenzee*. www.NAM.nl.

Nelson, J. and Prescott, D. (2003). Business and the Millennium Development Goals: A Framework for Action. The Prince of Wales International Business Leaders Forum in collaboration with the United Nations Development Programme.

Netherlands Commission for Environmental Assessment. (not dated). Strategic Environmental Assessment – Views and Experiences. www.eia.nl/nceia/ products/publications.htm.

Netherlands Commission for Environmental Assessment. (2001). *Proposed Conceptual and Procedural Framework for the Integration of Biological Diversity Considerations with National Systems for Impact Assessment*. (ISBN 90–421-0893–2.)

NCEA. (2006). Netherlands Commission for Environmental Assessment: Annual Report 2005. www.commissiemer.nl/mer/commissie/img/jve05-hoofdtekst. pdf.

Nilsson, M. and. Dalkmann, H. (2001). Decision making and strategic environmental assessment. *Journal of Environmental Assessment and Policy Management*, **3**, 305–27.

Nitz, T. and Brown, A. L. (2001). SEA must learn how policy-making works. *Journal of Environmental Assessment Policy and Management*, **3**, 329–42.

Nooteboom, S. (2007). Impact assessment procedures for sustainable development: a complex theory perspective. *Environmental Impact Assessment Review*, **27**, 645–65.

Nordhaus, W. D. (2007). A review of the Stern Review on the economics of climate change. *Journal of Economic Literature*, **45**, 686–702.

Noss, R. F. (1990). Indicators for monitoring biodiversity. *Conservation Biology*, **4**, 355–64.

Novacek, M. J. and Cleland, E. E. (2001). The current biodiversity extinction event: scenarios for mitigation and recovery. *Proceedings of the National Academy of Sciences of the United States of America*, **98**, 5466–70.

NRC. (2001). *Compensating for Wetland Losses under the Clean Water Act. Committee on Mitigating Wetland Losses, Board on Environmental Studies and Toxicology, Water Science and Technology Board, Division on Earth and Life Studies, National Research Council (NRC)*. Washington, DC: National Academy Press.

NSW. (2002). Green Offsets for Sustainable Development: Concept Paper. New South Wales (NSW) Government proposal for public consultation. www.environment.nsw.gov.au/greenoffsets/index.htm.

NSW. (2006a). *Avoiding and Offsetting Biodiversity Loss: Case Studies*. Published by Department of Environment and Conservation, New South Wales (DEC). (ISBN 1-74137-838-9.)

NSW. (2006b). BioBanking – A Biodiversity Offsets and Banking Scheme: Conserving and Restoring Biodiversity in New South Wales (NSW). Department of Environment and Conservation NSW Publications.

NWRP Project. (2000). *National Water Resources Plan for Egypt – Fisheries and Water Resources. NWRP Technical Report No. 6*. WL/Delft Hydraulics.

OECD. (1996). *Environmental Performance in OECD Countries: Progress in the 1990s*. Paris: OECD.

OECD (2003). *Handbook of National Accounting: Integrated Environmental and Economic Accounting 2003*. United Nations, European Commission, International Monetary Fund, Organisation for Economic Co-operation and Development, and the World Bank. http://unstats.un.org/unsd/envaccounting/seea.asp.

OECD. Development Assistance Committee (2006a). Applying Strategic Environmental Assessment. Good practice guidance for development co operation. DAC Guidelines and Reference Series. www.oecd.org/dataoecd/4/21/37353858.pdf.

OECD. (2006b). Policy Brief: Putting Climate Change Adaptation in the Development Mainstream. www.oecd.org/dataoecd/57/55/36324726.pdf.

OED. (1998). *Recent Experiences with Involuntary Resettlement: Thailand-Pak Mun. No. 17541*. Operations Evaluation Department. Washington, DC: World Bank.

Oindo, B. O., Skidmore A. K. and de Salvo, P. (2003). Mapping habitat and biological diversity in the Maasai Mara ecosystem. *International Journal of Remote Sensing*, **24**, 1053–69.

Oliemans, W., Blok, K. and Alamgir Chowduhury, A. R. (2003). Agricultural Drainage – Towards an Interdisciplinary and Integrated Approach: A Case of Controlling or Living with Floods? Banglasdesh Country Case Study Report. Washington DC: World Bank. http://web.worldbank.org/WBSITE/EXTERNAL/TOPICS/EXTARD/0,,contentMDK:20438813~pagePK:148956~piPK:216618~theSitePK:336682,00.html.

Ortiz, E., and Kellenberg, J. (2002). *Program of Payments for Ecological Services in Costa Rica.* Washington, DC: World Conservation Union.

Pagiola, S., Bishop, J. and Landell-Mills, N. (eds.). (2002). *Selling Forest Environmental Services: Market based Mechanisms for Conservation and Development.* London: Earthscan.

Parsons, G. R. and Thur, S. (in press). Valuing changes in the quality of coral reef ecosystems: a state preference study of SCUBA diving in the Bonaire National Marine Park. *Environmental and Resource Economics.*

Partidário, M. P. (2007). Scales and associated data. What is enough for SEA needs? *Environmental Impact Assessment Review,* **27**, 460–78.

Pautasso, M. (2007). Scale dependence of the correlation between human population presence and vertebrate and plant species richness. *Ecology Letters,* **10**, 16–24.

Peel, J. (2005). *The Precautionary Principle in Practice: Environmental Decision-Making and Scientific Uncertainty.* Annandale, NSW, Australia: Federation Press, p. 239.

Petterson, H. (2004). Compensation within Environmental Impact Assessment in Sweden and the United Kingdom. Unpublished Master of Science thesis. Institute of Water and Environment, Cranfield University, Silsoe.

Pimm, S. L. and Raven, P. (2000). Biodiversity – extinction by numbers. *Nature,* **403**, 843–5.

Pischke F. and Cashmore M. (2006). Decision-oriented environmental assessment: an empirical study of its theory and methods. *Environmental Impact Assessment Review,* **26**, 643–62.

Pisupati, B. and Warner, E. (2003). *Biodiversity and the Millennium Development Goals.* Gland, Switzerland: International Union for Conservation of Nature and Natural Resources (IUCN). (ISBN 955-8177-22-0.)

Pohl, C. (2005). Transdisciplinary collaboration in environmental research. *Futures,* **37**, 1159–78.

Pohl, C. and Hirsch Hadorn, G. (2006). Principles for Designing Transdisciplinary Research. Oekom Verlag, München.

Pope, J., Annandale, D. and Morrison-Saunders, A. (2004). Conceptualising sustainability assessment. *Environmental Impact Assessment Review,* **24**, 595–616.

Precautionary Principle Project. (2005). Guidelines for Applying the Precautionary Principle to Biodiversity Conservation and Natural Resource Management. Fauna and Flora International, Cambridge, UK. www.pprinciple.net.

Pritchard, D. (2005). International biodiversity-related treaties and impact assessment – how can they help each other? *Impact Assessment and Project Appraisal,* **23**, 7–16.

Pyke, C. R. (2007). The implications of global priorities for biodiversity and ecosystem services associated with protected areas. *Ecology and Society,* **12**, article 4. Available at www.ecologyandsociety.org/vol12/iss1/art4/.

Purnama, D. (2003). Reform of the EIA process in Indonesia: improving the role of public involvement. *Environmental Impact Assessment Review*, **12**, 75–88.

Putz, F. E. (1998). Halt the homogeocene. A frightening future filled with too few species. *The Palmetto*, **18**, 7–10.

Quigley, J. T. and Harper, D. J. (2006a). Compliance with Canadas' Fisheries Act: a field audit of habitat compensation projects. *Environmental Management*, **37**, 336–50.

Quigley, J. T. and Harper, D. J. (2006b). Effectiveness of fish habitat compensation in Canada in achieving no net loss. *Environmental Management*, **37**, 351–66.

Rajvanshi, A. (2002). Assessed impacts of the proposed Bodhghat Hydroelectric Project. In *Studies of EIA Practices in Developing Countries*, eds. M. McCabe and B. Sadler. Geneva: United Nations Environment Programme, pp. 281–94.

Rajvanshi, A. (2005). Quality of Biodiversity Related Information in EIA Reports for Environmental Decision-Making: The Indian Experience. Presented at the Annual Meeting of the International Association of Impact Assessment (IAIA).

Rajvanshi A. and Mathur, V. (2004a). Integrating Biodiversity into Strategic Environmental Assessment. Case Study I: Ecological evaluation of the site proposal for a nuclear power plant in Andhra Pradesh State, India. Wildlife Institute of India, Dehradun, India. Case studies compiled for the drafting of CBD guidelines on Biodiversity in SEA. www.eia.nl/ncea/pdfs/sea/casestudies/12_india_nuclear_power.pdf.

Rajvanshi A. and Mathur, V. (2004b). Integrating Biodiversity into Strategic Environmental Assessment. Case Study II: SEA of proposed Human River irrigation project, Maharashrta State, India. Wildlife Institute of India, Dehradun, India. Case studies compiled for the drafting of CBD guidelines on Biodiversity in SEA. www.eia.nl/ncea/pdfs/sea/casestudies/13_india_irrigation.pdf.

Rajvanshi, A., Mathur, V. B., Teleki, Geza C. and Mukherjee, S. K. (2001). *Road, Sensitive Habitats and Wildlife: Environmental Guidelines for India and South Asia*. Dehradun: Wildlife Institute of India and Toronto: Canadian Environmental Collaborative Ltd., p. 215. www.wii.gov.in/publications/eia/index.htm.

Rajvanshi, A., Mathur, Vinod B., Iftikhar, Usman A. (2007) *Best Practice Guidance for Biodiversity-Inclusive Impact Assessment: A Manual For Practitioners and Reviewers in South Asia*. CBBIA-IAIA Guidance Series. Published by International Association for Impact Assessment (IAIA), North Dakota, USA.

Ramsar Wetlands Convention. (2007). Guidelines for Incorporating Biodiversity-Related Issues into Environmental Impact Assessment Legislation and/or Processes and in Strategic Environmental Assessment. Ramsar Handbook for the Wise Use of Wetlands, 13. 3rd edn. Gland, Switzerland: Convention Secretariat. www.ramsar.org/lib/lib_handbooks2006_e.htm.

Rangachari, R., Sengupta, N., Iyer, Ramaswamy R., Banerji, P. and Singh, S. (2002). *Large Dams: India's Experience*. Cape Town, South Africa: Secretariat of the World Commission on Dams. www.dams.org/docs/kbase/studies/csinmain. pdf.

Raustiala, K. and Victor, D. G. (1996). Biodiversity since Rio: the future of the Convention on Biological Diversity. *Environment*, **38**, 16–20; 37–45. (Additional commentary in *Environment*, **39**(2), 3–5.)

Reeves, A. (1999). Mining – negotiating the pitfalls. *National Parks Journal*, **43**.

Regional Environmental Centre. (REC). (2002). Estonian National Development Plan for the Implementation of the Structural Funds of the European Union – A Single Programming Document 2003–2006. Strategic Environmental Assessment. (unpublished document).

Reid, H., Pisupati, B. and Baulch, H. (2004). How Biodiversity and Climate Change Interact. SciDev.Net Biodiversity Dossier Policy Brief. www.scidev.net/dossiers/index.cfm?Fuseaction=policybriefs&dossier=11.

Republic of Yemen. (2006). Directive on Environmental Impact Assessment (EIA) of Dam Projects. 2nd Draft.

Retief, F. (2007). A performance evaluation of strategic environmental assessment (SEA) processes within the South African context. Environmental Impact Assessment Review, 27, 84–100.

Reyes, V., Segura, O. and Verweij, P. (2002). Valuation of hydrological services provided by forests in Costa Rica. In Understanding and Capturing the Multiple Values of Tropical Forest, ed. P. A. Verweij. Proceedings of the International Seminar on Valuation and Innovative Financing Mechanisms in Support of Conservation and Sustainable Management of Tropical Forest.

Ricketts, T. H., Daily, G. C., Ehrlich, P. R. and Michener, C. D. (2004). Economic value of tropical forest to coffee production. Proceedings of the National Academy of Sciences of the United States of America, 101, 12579–82.

Ridder, W. de, Turnpenny, J., Nilsson, M. and von Raggamby, A. (2007). A framework for tool selection and use in integrated assessment for sustainable development. Journal of Environmental Policy and Management, 9, 423–42.

Ridker, R. G. (1967). Economic Costs of Air Pollution: Studies in Measurement. New York: Praeger.

Rio Tinto. (2004). Sustaining a Natural Balance: A Practical Guide to Integrating Biodiversity into Rio Tinto's Operational Activities. London: Rio Tinto.

Risser, P. G. (1994). Biodiversity and ecosystem function. Conservation Biology, 9, 742–6.

Robinson, J. B. and Herbert, D. (2001). Integrating climate change and sustainable development. International Journal of Global Environmental Issues, 1, 130–49.

Roe, D. (2004). The Millennium Development Goals and Conservation Managing: Nature's Wealth for Society's Health. London: International Institute for Environment and Development (IIED).

Roe, D. and Elliott, J, (2004). Poverty reduction and biodiversity conservation: rebuilding the bridges. Oryx, 38, 137–9.

Rojas, M. and Aylward, B. (2001). The Case of La Esperanza: A Small, Private Hydropower Producer and a Conservation NGO in Costa Rica. Land Water Linkages in Rural Watersheds Case Study Series, Rome: FAO.

RSPB. (1996). Good Practice Guide for Prospective Developments. Royal Society for the Protection of Birds, Bedfordshire, UK.

Runhaar, H. and Driessen P. J. (2007). What makes strategic environmental assessment successful assessment? The role of context in the contribution of SEA to decision-making. Impact Assessment and Project Appraisal, 25, 2–14.

Ryndgren, B., Kyläkorpi, L., Bodlund, B., Ellegård, A., Grusell, E. and Miliander, S. (2005). Experiences from five years of using the biotope method, a tool for quantitative biodiversity impact assessment. *Impact Assessment and Project Appraisal*, **23**, 47–54.

Sachs, J. D. and Reid, W. V. (2006). Investments toward sustainable development. *Science*, **312**, 1002.

Sadler, B. (1993). NSDS and environmental impact assessment: post Rio perspectives. *Environmental Assessment*, **1**, 29–31.

Sadler, B. (1996). International Study of the Effectiveness of Environmental Assessment. Final Report: Environmental Assessment in a Changing World: Evaluating Practice to Improve Performance. Canada, Minister of Supply Services.

Sadler, B. and Verheem, R. (1996). Strategic Environmental Assessment: Status, Challenges and Future Directions, Report 53, Ministry of Housing, Spatial Planning and the Environment, The Hague, the Netherlands.

SAIEA. (2003). Environmental Impact Assessment in Southern Africa. Southern African Institute for Environmental Assessment. Windhoek, Namibia.

SAIEA. (2003): Improving the Effectiveness of Environmental Impact Assessment and Strategic Environmental Assessment in Southern Africa. Workshop proceedings 13–26 May, Namibia. Southern African Institute for Environmental Assessment (SAIEA) and International Institute for Environment and Development (IIED).

Saksena, S., Prasad, R. and Joshi, V. (1995). Time allocation and fuel usage in Three villages of the Garhwal Himalaya, India. *Mountain Research and Development*, **15**, 57–67.

Sala, O. E., Chapin, F. S., III, Armesto, J. J., Berlow, R., Bloomfield, J., Dirzo, R., Huber-Sanwald, E., Huenneke, L. F., Jackson, R. B., Kinzig, A., Leemans, R., Lodge, D., Mooney, H. A., Oesterheld, M., Poff, N. L., Sykes, M. T., Walker, B. H., Walker, M. and Wall, D. H. (2000). Global biodiversity scenarios for the year 2100. *Science*, **287**, 1770–4.

Salafsky, N. and Wollenberg, E. (2000). Linking livelihoods and conservation: a conceptual framework and scale for assessing the integration of human needs and biodiversity. *World Development*, **28**, 1421–38.

Salzman, J. and Ruhl, J. B. (2002). Paying to Protect Watershed Services: Wetland Banking in the United States. In *Selling Forest Environmental Services: Market based Mechanisms for Conservation and Development*, eds. S. Pagiola, J. Bishop and N. Landell-Mills. London: Earthscan.

Sánchez-Azofeifa, G. A., Pfaff, A., Robalino, J. A. and Boomhower, J. P. (2007). Costa Rica's Payment for Environmental Services Program: intention, implementation, and impact. *Conservation Biology*, **21**, 1165–73.

Sanderson, E. W., Jaiteh, M., Levy, M. A., Redford, K. H., Wannebo, A. V. and Woolmer, G. (2002). The human footprint and the last of the wild. *Bioscience*, **52**, 891–904.

Sanderson, S. (2002). The future of conservation. *Foreign Affairs*, **81**, 162–73.

Sanderson, S. and Redford, K. (2003). Contested relationships between biodiversity conservation and poverty alleviation. *Oryx*, **37**, 389–90.

Sang, Don Lee. (2005). Strategic environment assessment and biological diversity conservation in the Korean high-speed railway project. *Journal of Environmental Assessment Policy and Management*, **7**, 287–98.

424 · References

Sawsan, M., Atallah-Augé, M. and El Khoury J. (2005). Biodiversity Manual. A Tool for Biodiversity Integration in EIA and SEA. Society for the protection of nature in Lebanon.

Swanson (1999). Conserving global biological diversity by encouraging alternative development paths: can development coexist with diversity? *Biodiversity and Conservation*, **8**, 29–44.

Sax, D. F., Gaines, S. D. and Brown, J. H. (2002). Species invasions exceed extinctions on islands worldwide: a comparative study of plants and birds. *American Naturalist*, **160**, 766–83.

Sayer, A. (1992). *Method in Social Science. A Realist Approach.* London: Routledge.

Scherr, S. J. (2003). Hunger, Poverty and Biodiversity in Developing Countries. A Paper for the Mexico Action Summit, Mexico City, Mexico.

Schmidt-Soltau, K. (2002). Human activities and conservation efforts in and around Korup National Park (Cameroon). The impacts of an impact assessment. *Proceedings of the 21st Annual IAIA Conference*, The Hague, the Netherlands, June 2002.

Shabman, L. and Scodari, P. (2004). Past, Present, and Future of Wetlands Credit Sales, Resources for the Future. Discussion Paper 04–48.

Sheate, W. (2003). Contributions of Sustainability Assessment to Biodiversity Scenarios in Upland Europe. Abstract for the 22nd Annual IAIA Conference, Marrakech, Morocco, June 2003.

Sheate, W. R., Rosario do Partidario, M., Byron, H., Bina, O., and Dagg S. (2008). Sustainability Assessment of Future Scenarios: Methodology and Application to Mountain Areas of Europe, Environmental Management, DOI 10.1007/s00267–007-9051-9.

Sherrington, M. (2005). Biodiversity assessment in the Oil Sands Region, northeastern Alberta, Canada. *Impact Assessment and Project Appraisal*, **23**, 71–81.

Shiva, V. (1991). *Ecology and the Politics of Survival: Conflicts over Natural Resources in India.* New Delhi: Sage Publication India Pvt. Ltd. and Tokyo, Japan: United Nations University Press, p. 365.

SIDA. (1998). Guidelines for Environmental Impact Assessments in International Development Cooperation. Department of Natural Resources and the Environment, Swedish International Development Cooperation Agency, Stockholm, Sweden.

Silver, W. L., Brown, S. and Lugo, A. E. (1994). Effects of change in biodiversity on ecosystem function in tropical forests. *Conservation Biology*, **10**, 17–24.

Singh, J. S. (2002). The biodiversity crisis: a multifaceted review. *Current Science*, **82**, 638–47.

Slootweg, R. (2003). Biodiversity in Impact Assessment: Practical Progress and an Agenda for Further Action. (unpublished report to the Secretariat of the Convention on Biological Diversity).

Slootweg, R. (2005). The biodiversity assessment framework: making biodiversity part of corporate social responsibility. *Impact Assessment and Project Appraisal (Special Issue on Biodiversity)*, **23**, 37–46.

Slootweg, R. and Kolhoff, A. (2003). A generic approach to integrate biodiversity considerations in screening and scoping for EIA. *Environmental Impact Assessment Review*, **23**, 657–81.

Slootweg, R. and Van Beukering, P. J. H. (2008). Valuation of Ecosystem Services and Strategic Environmental Assessment: Lessons from Influential Cases. Report of the Netherlands Commission for Environmental Assessment. www.eia.nl.

Slootweg, R., Hoevenaars, J. and Abdel-Dayem, S. (2007). Drainframe as a tool for integrated strategic environmental assessment: lessons from practice. *Irrigation and Drainage*, **56**, S191–S203.

Slootweg, R., Hoevenaars J. and Abdel-Dayem S. (2007). Drainframe as a tool for integrated strategic environmental assessment: lessons from practice. *Irrigation and Drainage Management*, **56**, S191–S203.

Slootweg, R., Vanclay, F. and van Schooten, M. L. F. (2001). Function evaluation as a framework for integrating social and environmental impacts. *Impact Assessment and Project Appraisal*, **19**, 19–28.

Slootweg, R., Vanclay, F. and van Schooten, M. L. F. (2003). Integrating environmental and social impact assessment. In *International Handbook of Social Impact Assessment: Conceptual and Methodological Advances*, eds. H. Becker and F. Vanclay. Cheltenham: Edward Elgar, pp. 56–74.

Slootweg, R., Kolhoff, A., Verheem R. and Höft, R. (2006). Biodiversity in EIA and SEA. Background Document to CBD Decision VIII/28: Voluntary Guidelines on Biodiversity-Inclusive Impact Assessment. Netherlands Commission for Environmental Assessment, Utrecht. (Spanish and French version published as Technical Paper No. 26 by the Secretariat of the Convention on Biological Diversity. www.biodiv.org/programmes/outreach/awareness/ts.shtml.)

Slootweg, R., van Schooten, M. L. F., Broer, W., Baron, E. and de Lange, V. (2003) (in Dutch). Biodiversity Assessment Framework: Chapter 15 Analysis of Existing Methods. Ministry of Housing, Spatial Planning and Environment, Directorate General for Environment, The Hague, the Netherlands.

Smith, A. T. P. (2002). *The Wealth of Nations*. Cambridge, MA: MIT Press and Washington, DC: World Bank.

Solbrig, O. T. (1991). The origin and function of biodiversity. *Environment*, **16**, 34–8.

Söderman, T. (2005). Treatment of biodiversity issues in Finnish environmental impact assessment. *Impact Assessment Project Appraisal*, **22**, 87–99.

Spergel, B. (2005). *Sustainable Funding for Nature Parks in the Netherlands Antilles*. Feasibility Study of a Protected Areas Trust Fund SYNOPSIS. Amsterdam: AIDEnvironment and Curaçao: Ecovision.

Spaninks, F. and Van Beukering, P. J. H. (1997). Economic valuation of mangrove ecosystems: potential and limitations. *CREED Working Paper*, **14**, 44.

Star, S. L. and Griesemer, J. R. (1989). Institutional ecology, 'translations' and boundary objects: amateurs and professionals in Berkeley's Museum of Vertebrate Zoology, 1907–39. *Social Studies of Science*, **19**: 387–420.

State of Hawaii. (2008). HB.3176, *Relating to Administrative Penalties for Damage to Stony Coral and Live Rock*. A Bill for an Act. House of Representatives, Twenty-Fourth Legislature, 2008. State Of Hawaii.

Stephens, W. and Middleton T. (2002). Why has the uptake of decision support systems been so poor? In *Crop Simulation Models. Applications in Developing Countries*, eds. R. B. Matthews and W. Stephens. Wallingford: CABI Publishing, pp. 129–47.

Stern, N. (2006). *The Stern Review: The Economics of Climate Change*. Cambridge, UK: Cambridge University Press.

Stern, N. (2007a). *Value judgments, welfare weights and discounting*. Paper B of After the *Stern Review*: Reflections and responses, 12 February 2007, Working draft of paper published on the *Stern Review* website: www.sternreview.org.uk.

Stern, N. (2007b). *Building an effective international response to climate change*. Paper C of After the *Stern Review*: Reflections and Responses. 12 February 2007. Working draft of paper published on *Stern Review* website: www.sternreview.org.uk.

Stern, N. and Taylor, C. (2007). Climate change: risk, ethics, and the *Stern Review*. *Science*, **317**, 203–4.

STINAPA (National Parks Foundation). (2008). www.stinapa.org.

Stinchcombe, K. and Gibson, R. B. (2001). Strategic environmental assessment as a means of pursuing sustainability: ten advantages and ten challenges. *Journal of Environmental Assessment Policy and Management*, **3**, 343–72.

Stolp, A. (2003). Citizen values assessment. In *International Handbook of Social Impact Assessment: Conceptual and Methodological Advances*, eds. H. Becker and F. Vanclay. Cheltenham:. Edward Elgar, pp. 231–57.

Stolp, A. (2006). Citizen values assessment. An instrument for integrating citizens' perspectives into environmental impact assessment. Thesis, Leiden University.

Strotz, R. (1968). The use of land value changes to measure the welfare benefits of land improvements. In *The Economics of Regulated Industries*, ed. J. E. Haring. Los Angeles: Occidental College.

Suratri, R. (2000). Lessons learned from protected area management partnerships in Indonesia. In *Proceedings of the 2nd Southeast Asia Regional Forum of The World Commission on Protected Areas, 2, Pakse, Lao PDR, 6–11 December 1999*, eds. A. Galt, T. Sigaty and M. Vinton. IUCN: Vientiane.

Svarpliene, A. (2002). Preserving biodiversity when building roads. *Proceedings of the 21st Annual IAIA Conference*, The Hague, the Netherlands, June 2002.

Swanson T. (2006). Conserving global biological diversity by encouraging alternative development paths: can development coexist with diversity? *Biodiversity and Conservation*, **8**, 29–44

Sweeting, A. R. and Clark, A. P. (2000). *Lightening the Lode: A Guide to Responsible Large-Scale Mining*. Washington, DC: Business and Policy Group, Conservation International.

Swartz, M. W, Brigham, C.A., Hoeksema, J. D., Lyons, K.G., Mills, M. H. and van Mantgem, P. J. (2000). Linking biodiversity to ecosystem function: implications for conservation ecology. *Oecologia*, **122**, 297–305.

Tanaka, A. (2000). EIA can be a tool to conserve natural ecosystems? – Changing Japan's ecological assessment. *Proceedings of the 19th Annual IAIA Conference, Hong Kong, June 2000*.

Tanaka A. (2001) Changing ecological assessment and mitigation in Japan. *Built Environment*, **27**: 35–41.

Tanaka, A. (2002). How ecological impact assessment guidelines should be? – Discussion on necessity of quantitative assessment for compensatory mitigation. *Proceedings of the 21st Annual IAIA Conference, The Hague, the Netherlands, June 2002.*

TCPA. (2004). Biodiversity by Design: A Guide for Sustainable Communities. Town and Country Planning Association (TCPA), 17 Carlton House Terrace London, SW1Y5AS. www.tcpa.org.uk/downloads/TCPA_biodiversity_guide_lowres.pdf.

Ten Kate, K. (2003). Biodiversity Offsets: Mileage, Methods and (Maybe) Markets. Ref: Offset KtK 27-10-03.doc. www.forest-trends. org/documents/meetings/Switzerland_2003/Shows/Kerry%20ten%20Kate_offset%20%5BRead-Only%5D.pdf.

Ten Kate, K and Laird, S. A. (2002). *The Commercial use of Biodiversity: Access to Genetic Resources and Benefit Sharing.* London: Earthscan.

Ten Kate, K., Bishop, J. and Bayon, R. (2004). *Biodiversity Offsets: Views, Experience, and the Business Case.* IUCN, Gland, Switzerland and Cambridge, UK and Insight Investment, London, UK. (ISBN: 2-8317-0854-0.)

Therivel, R. and Partidário, M. R. (eds.). (1996). *The Practice of Strategic Environmental Assessment.* London: Earthscan.

Therivel, R. and Walsh, F. (2006). The strategic environmental assessment directive in the UK: 1 year onwards. *Environmental Impact Assessment Review,* **26,** 663–75.

Thompson, R. and Starzomski, B. M. (2006). What Does Biodiversity Do? A Review for Managers and Policy Makers. Biodiversity and conservation, DOI 10.1007/s10531–005-6232–9.

Thompson, S., Treweek, J. R. and Thurling, D. J. (1997). The ecological component of environmental impact assessment: a critical review of British environmental statements. *Journal of Environmental Planning and Management,* **40,** 157–71.

Tilman, D., Knops, J., Wedin, D., Reich, P., Ritchie, M. and Siemann, E. (1997). The influence of functional diversity and composition on ecosystem processes. *Science,* **277,** 1300–02. DOI: 10.1126/science.277.5330.1300.

Timmer, V. and Juma, C. (2005). Biodiversity conservation and poverty reduction come together in the tropics – lessons learned from the Equator initiative. *Environment,* **47,** 25–44.

Tol, R. S. J. and Yohe, G. W. (2006). A review of the *Stern Review. World Economics,* **7,** 233–50.

Tol, R. S. J. and Yohe, G. W. (2007). A stern reply to the reply of the review of the *Stern Review. World Economics,* **8,** 153–9.

Tomlinson, P. (1997). From environmental statement through to implementation. *Environmental Assessment,* **5,** 39–41.

Tortajada, C. (2006). Water Transfer from the Ebro River. Case Study for the 2006 HDR.

Treweek, J. (1996). Ecology and environmental impact assessment. *Journal of Applied Ecology,* **33**(2): 191–9.

Treweek, J. (1999). *Ecological Impact Assessment.* Oxford, UK: Blackwell Science.

Treweek, J. (2001). Integrating Biodiversity with National Environmental Assessment Processes. A Review of Experiences and Methods. UNDP/UNEP Biodiversity Planning Support Programme. Working draft for comment SBSTTA 7, Montreal, November 2001.

Treweek, J. (2004). United Kingdom: Strategic Environmental Assessment of the Lower Parrett and Tone Flood Management Strategy, Somerset, England. Case study compiled for the drafting of CBD guidelines on Biodiversity in SEA. www. eia.nl/ncea/pdfs/sea/casestudies/8_uk_sea_lower_parett_and_tone.pdf.

Treweek, J. R., Thompson, S., Veitch, N. and Japp, J. (1993). Ecological assessment of proposed road developments: a review of environmental statements. *Journal of Environmental Planning and Management*, **36**(3), 295–307.

Treweek, J., Therivel, R. and Thompson, S. (2005). Principles for the use of strategic environmental assessment as a tool for promoting the conservation and sustainable use of biodiversity. *Journal of Environmental Assessment Policy and Management*, **7**, 173–99.

Treweek, J. R., Hankard, P., Roy, D. B., Arnold, H. and Thompson, S. (1998). Scope for strategic ecological assessment of trunk-road development in England with respect to potential impacts on lowland heathland, the Dartford warbler (Sylvia undata) and the sand lizard (Lacerta agilis). *Journal of Environmental Management*, **53**, 147–63.

Tietenburg, T. and Johnstone, N. (2004). Ex post evaluation of tradeable permits: methodological issues and literature review. In *Tradeable Permits: Policy Evaluation, Design and Reform*. Paris: OECD.

UMhlathuze Municipality. (2004). *Key statistics for the uMhlathuze area*. Available at www.richemp.org.za, accessed 2 December 2004.

UNECA. (2005). Review of the application of environmental impact assessment in selected African countries. United Nations Economic Commission for Africa (UNECA).

UNEP. (2001). South Asia: State of the Environment 2001. The United Nations Environment Programme.

UNEP. (2002). Environmental Impact Assessment Training Resource Manual. 2nd edn. www.unep.ch/etu/publications/EIAMan_2edition_toc.htm.

UNEP. (2006). Environmental Considerations of Human Displacement in Liberia. A guide for decision-makers and practitioners. United Nations Environment Programme, Post Conflict Branch, Geneva, Switzerland. http://postconflict. unep.ch/liberia/displacement.

UNDP. (not dated). Biodiversity and the Millennium Development Goals. www. undp.org/biodiversity/biodiversitycd/frameMDG.htm (last accessed September 2008).

UNICEF. (1997). Facts and figures 1997. www.unicef.org/factx/facright.htm.

United Nations. (2000). *United Nations Millennium Declaration*. www.un.org/ millenniumgoals/background.html.

United Nations. (2005) The Millennium Development Goals Report 2005. New York. www.unfpa.org/icpd/docs/mdgrept2005.pdf.

United Nations. (2007). *UN Millennium Development Goals*. United Nations Department of Economics and Social Affairs. (ISBN 978-92-1-101153-1.)

United Nations Economic Commission for Europe. (2003). Protocol on Strategic Environmental Assessment to the Convention on Environmental Impact Assessment in a Transboundary Context, Kiev. www.unece.org/env/eia/sea_ protocol/contents.htm.

Uprety, B. (2004). Integration of Biodiversity Aspects in Strategic Environmental Assessment of Operational Forest Management Plans in Nepal. Case study compiled for the drafting of CBD guidelines on Biodiversity in SEA. www.eia.nl/ncea/pdfs/sea/casestudies/15_nepal_water_plan.pdf.

Uprety, B. (2005). Biodiversity considerations in strategic environmental assessment: a case study of the Nepal water plan. *Journal of Environmental Assessment Policy and Management*, **7**, 247–66.

Van Beukering, P. J. H. and Cesar, H. S. J. (2004). Economic Analysis of Marine Managed Areas in the Main Hawaiian Islands. *Report for the National Oceanic and Atmospheric Administration, Coastal Ocean Program*. Washington, DC, p. 28.

Van Beukering, P. J. H. and van Heeren, A. E. (2002). *The Economic Value of the Iwokrama Forest Reserve, Guyana: A Stakeholder Perspective*. IVM Report (W02–22), Institute for Environmental Studies, Vrije Universiteit, Amsterdam, p. 82.

Van Beukering, P. J. H., Cesar, H. S. J., Dierking, J. and Atkinson, S. (2004). Recreational Survey in Selected Marine Managed Areas in the Main Hawaiian Islands. Report for the Division of Aquatic Resources (DAR) and the Department of Business, Economic Development and Tourism (DBEDT), Honolulu, p. 14.

Van Beukering, P., Brander, L., Tompkins, E. and McKenzie, E. (2007). Valuing the Environment in Small Islands – An Environmental Economics Toolkit. Joint Nature Conservation Committee (JNCC), Peterborough, p. 128. (ISBN 978 1 86107 5949.)

Van Beukering, P. J. H., Slootweg, R. and Immerzeel, D. (2008) Valuation of Ecosystem Services and Strategic Environmental Assessment. Influential Case Studies. Report of the Netherlands Commission for Environmental Assessment. www.eia.nl.

Van Bohemen, H. D. (2004). Ecological engineering and civil engineering works; a practical set of ecological engineering principles for road infrastructure and coastal management. Thesis, Delft, the Netherlands.

Van Den Tempel, R. and Brouwer, H. (in press) (in Dutch). Chapter 6: Mitigatie en compensatie; monitoring en evaluatie. [Mitigation and compensation; monitoring and evaluation]. Effectenanalyse [The analysis of effects]. In *Praktijkboek Flora-en Faunawet* [Putting the law on flora and fauna in practise]. The Hague, the Netherlands: SDU.

Van Der Wateren, T., Diederichs, N., Mander, M., Markewicz, T. and O'Connor, T. (2004). Mhlathuze Strategic Catchment Assessment, Richard Bay, South Africa. Case study compiled for the drafting of CBD guidelines on Biodiversity in SEA. *UMhlathuze Municipality*. www.eia.nl/ncea/pdfs/sea/casestudies/10_rsa_mhlathuze_strategic_catchment_assessment.pdf.

van Dijk, M. (2005). SEA of the Sigma Plan for Flood Safety and Ecological Restoration of the Scheldt River. Case study compiled for the drafting of CBD guidelines on Biodiversity in SEA. Resource Analysis, Antwerp, Belgium. www.eia.nl/ncea/pdfs/sea/casestudies/20_belgium_sigmaplan.pdf.

van Schooten, M. L. F., Vanclay, F. and Slootweg, R. (2003). Conceptualising social change processes and social impacts. In *International Handbook of Social Impact*

Assessment: Conceptual and Methodological Advances, eds. H. Becker and F. Vanclay. Cheltenham: Edward Elgar, pp. 74–107.

van Schooten, M. L. F. (2004a). The Netherlands: SEA for the National Policy Plan on Industrial and Drinking Water Supply. Case study compiled for the drafting of CBD guidelines on Biodiversity in SEA. www.eia.nl/ncea/pdfs/sea/casestudies/1nl_sea_industrial_drinking_water.pdf.

van Schooten, M. L. F. (2004b). The Netherlands: SEA on the Routing the River Meuse (Zandmaas / Maasroute). Case study compiled for the drafting of CBD guidelines on Biodiversity in SEA. www.eia.nl/ncea/pdfs/sea/casestudies/3_nl_sea_river_meuse.pdf.

van Schooten, M. L. F. (2004c). The Netherlands: SEA for the National Policy Plan on Shell Mining. Case study compiled for the drafting of CBD guidelines on Biodiversity in SEA. www.eia.nl/ncea/pdfs/sea/casestudies/2_nl_sea_shell_mining.pdf.

Van't Hof, T. (1991). Marine Parks in the Netherlands Antilles: Lessons from ten Years of Experience. *Unpublished Conference Paper. Tobago.* http://www.mina.vomil.an/Pubs/vantHof-10yrsMarParksNA.html.

Van Waarden, F. (1992). Dimensions and types of policy networks. *European Journal of Political Research*, **21**(1–2), 29–52.

van Wetten, J., Joordens, J., van Dorp, M., and Bijvoet, L. (1999). *De schaduwkant van Waddengas.* AIDEnvironment. Report assigned by Greenpeace-NL, Amsterdam.

Victorian Environment Protection Authority. (1994). Victoria's environment protection system: Innovative approaches and economic instruments. *Environmental Economics Update* (Environment Protection Authority, NSW), **4**, 1–6.

Viessman, W. (1998). Water policies for the future: an introduction. *Water Resources Update*, **111**, 4–7, Universities Council on Water Resources.

von Loesch, H. (1991). Gene wars: the double helix is a hot potato. *Ceres*, **131**, 39–44.

VROM. (1994). The Quality of Environmental Impact Statements. *EIA Series No. 47.* Ministry of Housing, Spatial Planning and Environment, The Hague, the Netherlands.

Walker, B., Holling, C. S., Carpenter, S. R. and Kinzig, A. (2004). Resilience, adaptability and transformability in social–ecological systems. *Ecology and Society*, **9**(2), 5. http://www.ecologyandsociety.org/vol9/iss2/art5/.

Wallace, K. (2007). Classification of ecosystem services: problems and solutions. *Biological Conservation*, **139**, 235–46.

Warken, J. and Buckley, R. (1998). Scientific quality of tourism environmental impact assessment. *Journal of Applied Ecology*, **35**, 1–8.

Watt A. S. (1947). Pattern and process in the plant community. *Journal of Ecology*, **35**, 1–22.

WBCSD. (2008). *Million acres of Guyanese rainforest to be saved in groundbreaking deal.* www.wbcsd.org/plugins/DocSearch/details.asp?type=DocDet&ObjectId=MjkzMTM.

WCED. (1987). Our Common Future, Report of the World Commission on Environment and Development, World Commission on Environment and

Development, 1987. Published as an Annex to General Assembly document A/42/427, Development and International Co-operation: Environment, 2 August 1987.

Wende, W. (2002). Evaluation of the effectiveness and quality of environmental impact assessment in the Federal Republic of Germany. *Impact Assessment and Project Appraisal*, **20**, 93–9.

Western Australia EPA. (2004). Environmental Offsets. Preliminary position statement No. 9.

Weston, J. (2000). EIA, decision-making theory and screening and scoping in UK practice. *Journal of Environmental Planning and Management*, **43**, 185–203.

White, G. H., Morozow, O., Allan, J. G. and Bacon, C. A. (1996). Minimisation of impact during exploration. In *Environmental Management in the Australian Minerals and Energy Industries: Principles and Practices*, ed. D. Mulligan. Sydney: UNSW Press in association with the Australian Minerals and Energy Environment Foundation, pp. 99–130.

Whittaker, R. H. (1972). Evolution and measurement of species diversity. *Taxon*, **21**, 213–51.

WHO. (2002) *WHO Traditional Medicine Strategy (2002–2005)*, World Health Organization, Geneva.

WII. (1993). Environmental Impact Assessment of HBJ Gas Pipeline Upgradation Project on Wildlife and Wildlife Habitats. *WII-EIA Technical Report 2*, Wildlife Institute of India, Dehradun, India.

WII. (1994). Impact Assessment Studies of Narmada Sagar and Omkareshwar Projects on Flora and Fauna with Attendant Human Aspects. *WII-EIA Technical Report 9*. Wildlife Institute of India, Dehradun, India.

WII. (1998). Ecological Assessment of the Proposed Mumbai – Pune Expressway. *WII – EIA Technical Report 22*. Wildlife Institute of India, Dehradun, India.

Wilcove, D.S., Rothstein, D., Dubow, Phillips, J. A. and Losos, E. (1998). Quantifying threats to imperilled species in the United States. *BioScience*, **48**, 607–15.

Wilkinson, J. and Kennedy, C. (2002). *Banks and Fees: The Status of Off-site Wetland Mitigation in the United States*. Washington, DC: Environmental Law Institute.

Williams, P. H. and Humphries, C. J. (1996). Comparing character diversity among biotas. In *Biodiversity: A biology of Numbers and Difference*, ed. K. J. Gaston. Oxford, UK: Blackwell Science Ltd., pp. 54–76.

Wilson, E. O. and Peter, F. M. (eds.) (1988). *Biodiversity*. Washington, DC: National Academy Press.

Wood, C. and Coppell, L. (1999). An evaluation of the Hong Kong environmental impact assessment system. *Impact Assessment and Project Appraisal*, **20**, 101–11.

Wood, C. M. (1995). *Environmental Impact Assessment: A Comparative Review*. Harlow, UK: Longman Higher Education.

Wood, C. M. (2003). *Environmental Impact Assessment, A Comparative Review*. 2nd edn. Harlow: Prentice Hall.

World Bank. (1993). *Costa Rica: Forest Sector Review*. Agricultural Operations Division Report 11516 CR. Washington, DC: World Bank.

World Bank. (1997). *Environmental Assessment Sourcebook Update: Biodiversity and Environmental Assessment*. Washington, DC: World Bank.

World Bank. (1999). Moldova: Poverty Assessment Technical Papers. *Report No. 19846 MD*. Washington, DC: World Bank.

World Bank. (2000). *Biodiversity and Environmental Assessment Toolkit*. Washington, DC: World Bank.

World Bank. (2001): *The World Bank Operational Manual – Operational Policy 4.04: Natural Habitats*. Washington, DC: World Bank.

World Bank. (2003). Striking a Better Balance. The World Bank Group and Extractive Industries. The Final Report of the Extractive Industries Review, Vol. I.

World Bank. (2005). Conceptual Framework and Transaction Model for a Public-Private Partnership in Irrigation in the West Delta, Egypt. Internal report.

World Bank. (2006): Environmental Impact Assessment, Regulations and Strategic Environmental Assessment Requirements. Practices and lessons learned in East and Southeast Asia. Environmental and social development. Safeguard dissemination note No. 2. Washington, DC.

WRI. (2004). *Reefs at Risk in the Caribbean*. Washington. DC: World Resources Institute (WRI).

Wunder, S. (2001). Poverty alleviation and tropical forests – what scope for synergies? *World Development*, **29**, 1817–33.

WWF. (2002). *To dig or not to dig?* Switzerland: World Wide Fund for Nature – International.

WWF. (2004). Conserving medicinal plants. *TWAS Newsletter*, **16**, 18–23.

WWF. (2006). Principles and Guidelines for Conservation Offsets, Position Paper 9. Conference of the Parties to the Convention on Biological Diversity Eighth Meeting Curitiba, 20–31 March 2006.

WWF. (2006). Analysis and Socio-Economic Assessment of the Ebro Transfers included in the Spanish National Hydrological Plan (SNHP).

WWF. (2007). *Sustainable Financing and Payments for Environmental Services*. World Wide Fund (WWF), Info Exchange, Year 3, Newsletter 19.

Yachi, S. and Loreau, M. (1999). Biodiversity and ecosystem productivity in a fluctuating environment: the insurance hypothesis. *Proceedings of the National Academy of Sciences of the United States of America*, **96**(4), 1463–68.

Yohe, G. W. and Tol, R. S. J. (2007). The *Stern Review*: implications for climate change. *Environment*, **49**, 36–42.

Young, M., Cunningham, N., Elix, J., Lambert, J., Howard, B., Grabosky, P. and McCrone, E. (1996). Reimbursing the Future: An Evaluation of Motivational, Voluntary, Price Based, Property Right and Regulatory Incentives for the Conservation of Biodiversity. Department of the Environment, Sport and Territories Biodiversity Unit; Biodiversity Series Paper, 9.

Zedan, H. (2007). Biodiversity and Changes in the Ecosystems: Consequences to Sustainable Development. Alexandria, Egypt: Conference on Environmental Economics and Sustainable Development, 25 October 2007.

Zwarts, L., Van Beukering, P. J. H., Koné, B. Wymenga, E. and Taylor, D. (2006). The economic and ecological effects of water management choices in the Upper Niger River: development of decision support methods. *Water Resources Development*, **22**, 135–56.

Zwarts, L., Van Beukering, P. J. H., Kone, B. and Wymenga, E. (eds.). (2005). *The Niger, a Lifeline: Effective Water Management in the Upper Niger Basin*. Mali: Wetlands International. (ISBN-10: 9080715069.)

Index

advocacy, 305
affected people, 99
alpha diversity, 17
alternatives, 130, 131, 175, 179, 180
Argentina Pascua Lama project, 61
aspects of biodiversity, 41, 236
  composition, 41, 43, 44, 236
  impact mechanism, 110
  key processes, 42, 50, 53, 54, 173, 207,
    209, 237
  pattern, 41, 49
  structure, 41, 45, 46, 236
Australia
  New South Wales, 61
  Pacific Highway upgrade, 269
  Striped Legless Lizards, 276
  Victorian Trust for Nature, 273
  Watheroo National Park, 276
autonomous development, 67, 180, 186,
  329

baseline, 180
Belgium, management of the Scheldt river,
  239
beneficiaries, 99, 179, 219, 272
beta diversity, 18
biodiversity
  aspects of biodiversity. See aspects of
    biodiversity
  ecosystem diversity, 19
  genetic diversity, 17
  levels of diversity, 16, 169, 170, 187,
    240
  medicinal plants, 75
  neglect in EIA, 157
  species diversity, 17
Biodiversity Action Plan, 238, 241
biodiversity offset
  'no harm' principle, 264
  approaches, 268

barriers, 284
definition, 256
market-based approaches, 271
policy context, 257
biological diversity. See biodiversity
biophysical effects, 103, 109
biophysical subsystem, 94
Bolivia, road SEA, 232
boundary concepts, 116
boundary objects, 113, 117
boundary settings, 119
Brazil, Atlantic Forest, 64
business and biodiversity, 62, 263

Cameroon
  Benue river
    Chad-Cameroon oil pipeline, 269
    Korup National Park, 342
Canada, habitat compensation, 181
capacity development, 331
carbon trading, 367, 384
Central America, free trade agreement,
  152
Chad-Cameroon oil pipeline, 269
child mortality, 74
Chile, Pascua Lama project, 61
China
  EIA practise in Hong Kong, 168
  Nu River, 61
climate change, 77, 81, 298, 328
composition. See aspects of biodiversity
connectivity, 46, 47
contingent valuation, 301, 304, 317, 396
Convention on Biological Diversity
  Article 8 (j). See indigenous knowledge
  Cartagena Protocol on Biosafety, 44
  ecosystem approach. See ecosystem
    approach
  EIA guidelines, 154
  impact assessment (article 14), 24

objectives, 20, 87, 169
SEA guidance, 210
trade impacts, 243
use of terms, 16
Costa Rica, Payments for Ecosystem
Services, 301
cost-benefit analysis, 87, 349, 370
credibility (of information), 113
cumulative effects, 39, 127, 131, 179, 186,
205, 217
currency, 283

data transfer, 358
decision making, 113, 132, 143, 150, 193,
220, 305, 320
discount rate, 363
drivers of change, 37, 173
definition, 102
direct drivers, 38, 39, 40, 103, 212, 226,
231
endogenous driver, 213
exogenous drivers, 213
indirect drivers, 41, 212, 242
trigger for SEA, 225

ecosystem approach, 25, 26, 93
ecosystem diversity. See biodiversity
ecosystem services, 34, 38, 94, 237
carrying services, 37, 96
cultural services, 37, 96
examples, 121
in scoping for EIA, 179
in screening for EIA, 172
in SEA, 212, 287
influential cases, 289
provisioning services, 37, 95
regulating services, 37, 95
supporting services, 37, 96
trigger for SEA, 224
Egypt
Alexandria wetland, 110
West Delta project, 290
EIA
alternatives. See alternatives
characteristics, 143
definition, 129
effectiveness, 132
evaluation of impacts, 130
impact analysis, 185
mitigation. See mitigation
origins, 126

reporting, 131, 190
review, 131, 192
scoping, 178, See scoping
screening, 165, See screening
state of the art, 136
energy sector, 236, 269
environmental assessment
challenges, 66
context, 147
meeting the MDGs, 77
environmental auditing, 196
environmental goods and services. See
ecosystem services
Environmental Impact Assessment. See EIA
equity, 323
Estonia
National Development Plan, 207
public involvement in local planning,
208
EU EIA directive, 133
EU Habitat Directive, 183
EU SEA directive, 127, 148, 206, 214

fisheries sector, 239
follow-up, 195
foodweb, 49
forestry sector, 236, 301
fragmentation, 47, 174
function evaluation, 88
functional groups, 49
future generations, 16, 27, 60, 98, 100, 149,
231

gamma diversity, 18
gender, 72, 73, 177
genetic diversity. See biodiversity
Guyana, Iwokrama forest carbon trading,
367

habitat, 18
higher-order effects, 105
HIV/AIDS, 75
hot spots, 34
human security, 68, 78

IAIA, 23, 160, 206
IAIA\t, 161
India
Bodhghat hydropower, 188
dams & irrigation, 237
Indira Sagar Hydropower, 269

India (*cont.*)
  Jharkhand, 73
  nuclear power facility, 236
  Western Ghats, 61, 65
indicators, 249
indigenous knowledge, 15, 25
indigenous people, 98, 177, 280
Indonesia, Kutai National Park, 273
industry sector, 155, 277
insurance hypothesis, 36
integrated assessment. *See* sustainability
  assessment
interdisciplinarity, 115
Israel, EIA practise, 168, 175

Kenya, Tana River Delta, 67
key processes. *See* aspects of biodiversity
keystone species, 50
Kiev protocol, 214
Korea (South), high-speed railway project,
  205

legal obligations, 220
legitimacy (on information), 114
Lithuania, road EIA, 155
livelihoods, 218
local knowledge, 31, 106

malaria, 75
Malaysia
  biological pollination in oil palm, 189
  dams, 73
maternal health, 74
measures of biodiversity, 17
Mexico, Monarch butterfly conservation
  fund, 273
Millennium Development Goals, 68, 69, 87,
  128, 189, 216
Millennium Ecosystem Assessment, 15, 34,
  38, 57, 66, 88, 94, 101, 214, 242, 249,
  257
mining sector, 62, 73, 162, 264, 277
mitigation, 179
  avoidance, 182, 199
  biodiversity offset. *See* biodiversity offset
  compensation, 181, 182, 199
  enhancement, 180, 199
  hierarchy, 199
  minimization, 199
  remediation, 199
monitoring, 132, 196, 249
multi-criteria analysis, 324, 347

National Biodiversity Strategy and Action
  Plan, 21, 167, 185
National Ecological Network, 241
natural capital index, 244
NBSAP. *See* national biodiversity strategy
  and action plan
Nepal, forestry plan SEA, 236
Netherlands
  management of river Meuse, 235
  National Ecological Network, 47
  National Policy on Water Supply, 232
  Oostvaardersplassen
  shell extraction, 239
  spatial planning, 209
  Wadden Sea dredging, 46
  Wadden Sea gas exploitation, 300
Netherlands Antilles, marine protected
  areas, 301
Niger, Inner Niger Delta, 342
no net loss, 23, 179, 257

offset. *See* biodiversity offset
oil and gas sector, 64, 90, 161, 264, 269,
  300, 304

Pakistan, NWFP road development, 177
Paraguay river, Hidrovia canalisation project,
  209
participation, 30, 118, 206, 208, 219, 229,
  238, 242, 329
Payments for Ecosystem Services. *See*
  Payments for Environmental Services
Payments for Environmental Services, 274,
  301, 323, 380
Philippines, Pagbilao Carbon Sink Initiative,
  371
pillars of sustainability. *See* triple bottom line
policies, plans and programmes – defintion,
  126
poverty alleviation, 26, 69
Precautionary Principle, 32, 33, 209, 239,
  300, 322, 370
primary education, 71
problemshed, 93
protected areas, 172, 221, 241, 273,
  301
protected species, 221, 236
public–private partnership, 337

range of influence, 108, 109, 229
replacement costs, 393
residual impacts, 180

resilience, 36
revealed preference, 311
risk assessment, 305

salience (of information), 113
scale, 30, 47, 48, 171, 217
scoping, 43, 52, 111, 130, 175
screening, 129
  biodiversity screening map, 172
  criteria, 169, 171, 172, 198
  decision, 171
  questions, 170
SEA
  biodiversity perspectives, 214
  biodiversity triggers, 224, 228
  characteristics, 143
  current thinking, 137
  effectiveness, 142
  general, 205
  link with planning, 139
  origins, 126
  state of the art, 145
  when SEA, 223
social effects, 104, 111
social impact assessment, 87
social impacts, 111
societal subsystem, 97
South Africa
  Durban Environmental Services
    Management Plan, 357
  principles for SEA, 211
  systematic biodiversity planning, 231
  uMhlathuze catchment assessment, 230,
    296
Spain, National Hydrological Plan, 302
spatial planning, 209, 230, 241, 296, 324
species diversity. *See* biodiversity
stakeholders, 179
  demand for ecosystem services, 98
  identification, 223
  valuing ecosystem services, 110
Stern Review, 298
Strategic Catchment Assessment, 352
Strategic Environmental Analysis, 218
Strategic Environmental Assessment. *See*
  SEA
structure. *See* aspects of biodiversity
succession, 46
sustainability assessment, 148, 150, 215, 332
sustainability impact assessment, 243
sustainable development, 60, 63, 87

sustainable financing, 306
Sweden, Stockholm urban planning, 235
systematic biodiversity planning, 231

Taiwan, Black-faced spoonbill, 155
threatened species, 173
tourism sector, 64, 67, 163
trade agreements, 243
transport sector, 155, 162, 177, 209, 232,
  241, 269
triple bottom line, 60, 87, 153

UK
  Flood management, 238
  great crested newts, 197
  local transport plan, 241
  Wareham flood management, 298
urban planning, 235
USA
  Catskill watershed restoration, 381
  Chesapeake Bay, 189
  Everglades restoration, 350
  Exxon Valdez oil spill, 304
  Hawaii *Acacia koa*, 65
  Hawaii coral reef management, 379
  Hawaii reef damage penalties, 394
  International Paper Company, 277
  Kennecott Utah mine, 277
Uzbekistan, Aral Sea wetland restoration
  project, 237, 295

valuation (of ecosystem services)
  economic valuation, 310
  identification and recognition, 306
  market-based valuation, 311
  quantification, 307
  revealed preference, 311
  societal valuation, 309
  stated preference, 311
values (of ecosystem services), 37, 100, 194
  ecological, 101
  economic, 100
  economic value, 31
  social, 100

water sector, 168, 188, 232, 235, 237, 238,
  239, 290, 298, 302, 342, 381
wetland restoration, 237, 295, 350, 371
willingness to pay, 301, 378

Yemen, EIS directive for dams, 168

Printed in the United States
By Bookmasters